LASER SCANNING: UPDATE 1

# Seventh International Meeting on Scanning Laser Ophthalmoscopy, Tomography and Microscopy

Llao Llao, Bariloche, Argentina, December 1999

---

Roberto Sampaolesi & Juan R. Sampaolesi

Department of Ophthalmology, School of Medicine, University of Buenos Aires, Argentina
Parana 1239 1er piso A y B, C.P. 1018, Capital Federal, Buenos Aires, Argentina
Phone: 54-11-4814-0266; Fax: 54-11-4814-2092/4806-4193; E-mail: imslo99arg@interar.com.ar

**Official sponsors:**
Alcon-Humphrey Instruments
Allergan
Heidelberg Engineering
Novartis Ophthalmics
Pharmacia and Upjohn
Pfortner Cornealent

**Also sponsored by:**
Fundación Argentina Oftalmológica Juan Sampaolesi
Fundación Bridas Argentina
Fundación Amalia Lacroze de Fortabat

# Laser Scanning: Update 1

First Official Publication of
the International Society of Laser Scanning: INSOLAS

Proceedings of the Seventh International Meeting
on Scanning Laser Ophthalmoscopy, Tomography and Microscopy

*Edited by*

JUAN R. SAMPAOLESI

Reprinted from *International Ophthalmology*, Volume 23, Nos. 4–6 (2001)

Kluwer Academic Publishers
DORDRECHT / BOSTON / LONDON

First South American Glaucoma Symposium, Bariloche, Argentina, March 24–28, 1966. Top: José Espildora Couso, Cipriano D'Allesandro, Enrique Malbrán, Manuel Chervin, Atilio Norbis,

Seventh International Meeting on Scanning Laser Ophthalmoscopy, Tomography and Microscopy, Bariloche, Argentina, November 30–December 3, 1999.

**A C.I.P. Catalogue record for this book is available from the Library of Congress**

ISBN 0-7923-7197-6

---

Published by Kluwer Academic Publishers,
P.O. Box 17, 3300 AA Dordrecht, The Netherlands

Sold and distributed in the North, Central and South America
by Kluwer Academic Publishers,
101 Philip Drive, Norwell, MA 02061, U.S.A.

In all other countries, sold and distributed
by Kluwer Academic Publishers
P.O. Box 322, 3300 AH Dordrecht, The Netherlands

**Cover photography:** Mr Luis Cambre

*Printed on acid-free paper*



Printed in the Netherlands

# Contents

## Optical Coherence Tomography



## Scanning Laser Doppler Flowmetry

## Indocyanine Green and Fluorescein Angiography

# Preface

The Seventh International Meeting on Scanning Laser Ophthalmoscopy, Tomography and Microscopy was held between November 30 and December 3, 1999 at the City of San Carlos de Bariloche, in the Argentine Patagonia. This scientific meeting gathered professionals from all over the world whose current research interests are confocal tomography, scanning laser Doppler flowmetry, digital angiography with indocyanine green and fluorescein, polarimetry, optical coherent tomography and many other different techniques. Moreover, according to the proposal made on the occasion of the Meeting, the International Society of Laser Scanning (INSOLAS) was formally created by voting. This is the first international society which will focus exclusively on the application of laser diagnostic technologies in ophthalmology. As originally conceived for an optimal coverage of the whole range of interests, the event was divided into a technical area, a research area, and mainly, a clinical application area for each technology. It is our wish that this book, compiling the presentations made at the Meeting and which will constitute the first publication of the INSOLAS, will be spread and distributed among all practical and clinical ophthalmologists around the world. The Executive Committee of the 7th IMSLO is particularly grateful to the following sponsors, whose financial support has made this event possible: Alcon-Humphrey Instruments, Heidelberg Engineering, Allergan, Novartis Ophthalmics, Pharmacia & Upjohn and Pfortner Cornealent.

Our thanks also to the following entities: Fundación Argentina Oftalmológica Juan Sampaolesi, Fundación Bridas Argentina and Fundación Amalia Lacroze de Fortabat.

Finally, we would also like to thank Miss Stella Maris Salvatori and Mr. Federico Korendij for the time and efforts devoted to the organization of the event.

ROBERTO SAMPAOLESI
Chairman

*International Ophthalmology* **23:** 183–189, 2001.
*J.R. Sampaolesi (ed.), Laser Scanning: Update 1*, 1–7.



# Basic principles of laser Doppler flowmetry and application to the ocular circulation *

Charles E. Riva
*Institute of Research in Ophthalmology, Sion Faculty of Medicine, University of Lausanne, Switzerland*

*Key words:* Doppler effect, optical mixing spectroscopy, flowmetry, ocular circulation, blood flow regulation

## Introduction

The study of the hemodynamics in the human eye requires the measurement of blood flow in the various tissues of this organ. Ideally, the measurement technique should be reproducible, accurate, sensitive enough to be able to reveal early pathological alterations, and clinically applicable. Furthermore, its spatial resolution should permit flow measurement in the microcirculation without interference from the larger vessels and its temporal response should be fast enough to allow the investigation of the process of flow regulation evoked by various physiological stimulations.

Among the various techniques which have been applied in humans to investigate ocular microvascular hemodynamics, only laser Doppler velocimetry/flowmetry and the blue field simulation techniques are truly non-invasive since they do not require the injection of a dye or the insertion of an electrode into the eye. This paper will be devoted to a review of the basic principles underlying laser Doppler flowmetry.

## The Doppler effect

This effect is the basis of LDF. The Austrian physicist Christian Doppler first described it in 1842 in an article entitled *On the Colored Light of Double Stars and Some Other Heavenly Bodies*. The Doppler effect describes the frequency shift that a wave undergoes when emitted from an object which is moving away or towards an observer. It manifests itself, for example, in the increase in the pitch of the siren of an ambulance, when this vehicle moves towards an observer.

The application of the Doppler effect to optics goes back to the end of last century, with the measurement of the velocity of galaxies. Today, with the advent of the laser, an instrument that provides optical waves of extreme purity, i.e. of practically single frequency (the frequency is number of waves/sec), it is possible to detect Doppler shifts with a very high resolution and over a range of velocities from $\mu$m/sec to many km/sec. The first report on the application of the Doppler effect to blood flow goes back to 1972 (Riva et al., 1972). These authors succeeded in measuring the velocity of red blood cells (RBCs) in a rabbit retinal arteriole. Blood velocity in the human ONH microcirculation was first described in 1982 (Riva, Grunwald and Sinclair). The technique was then extended to the measurement of blood flow in the vascular bed of the cat ONH (Riva et al., 1989a, 1992). Michelson and Schmauss (1995) first reported LDF measurements of blood flow in the microcirculation of the human retina by means of a scanning laser ophthalmoscope.

## The magnitude of the Doppler shift from RBCs in retinal vessels

Consider a RBC moving in the direction **V**, with velocity |**V**| (Figure 1). A laser beam of frequency $f_i$ (number of electromagnetic vibrations/sec) is incident on this RBC. This beams makes an angle $\alpha_i$ with **V**. The RBC scatters the laser light in various directions. In the scattering direction defined by the angle $\alpha_s$ with **V**, the frequency of the light differs from $f_i$ by an amount $\Delta f = n\,\mathbf{V}\,(\cos\alpha_s - \cos\alpha_i)/\lambda_i$. $\Delta f$ is the Doppler shift. $\lambda_i$ is the wavelength of the laser within

* Supported in part by the Swiss National Fund for Scientific Research No. 32-43157.

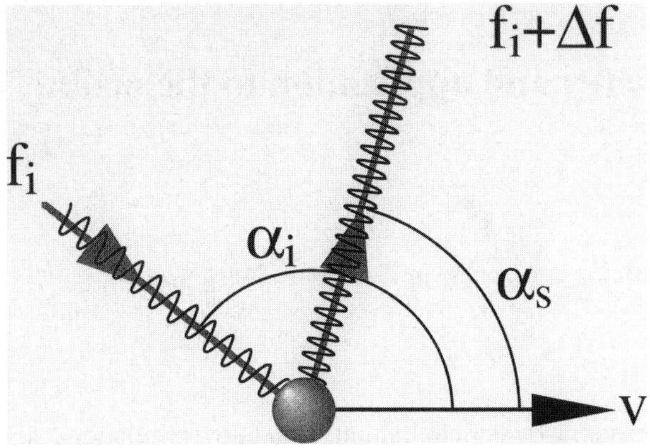


*Figure 1.* The Doppler effect. $f_i$ is the frequency of the light on a particle. The direction of this light is defined by the angle $\alpha_i$ with the direction, V, of the particle velocity. $f_i + \delta f_i$ is the frequency of the scattered light in the direction defined by the angle $\alpha_s$.

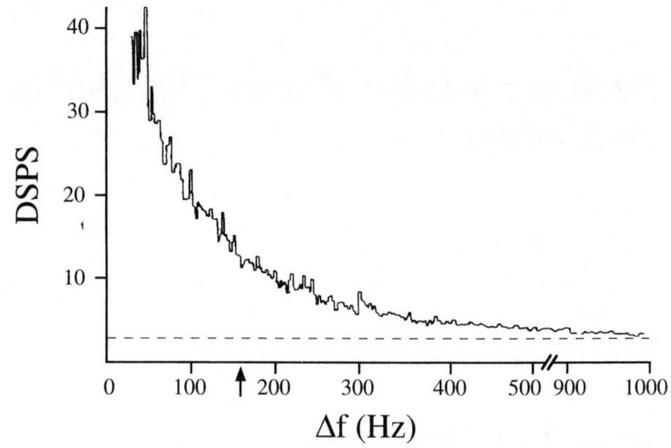


*Figure 2.* The spectrum (DSPS) of the light scattered from red blood cells (RBCs) moving in a capillary bed. Black arrow is the frequency shift corresponding to the first moment of the DSPS. This frequency is proportional to the mean velocity of the RBCs.

the blood. $\lambda_i = c/nf_i$. c = velocity of light in vacuo. n = index of refraction of the light in the blood.

Let us assume $|\mathbf{V}| = 5$ cm/sec, which is typical for the mean velocity of RBCs moving at the center of a retinal artery. We assume that the angles $\alpha_i$ and $\alpha_s$ are equal to 90° and 80°, respectively values typical for the scattering geometry pertaining to measurements in vessels close to the optic disc in the human eye. With $\lambda_i = 632.8$ nm (helium-neon laser) and $n_{blood} = 1.33$, we find $\Delta f = 13.500 \times n_{blood} = 17.955$ Hz (1 Hz = 1 cycle/sec). Although such a shift is $\ll f_i$ fi ($\approx 5 \times 10^{14}$ Hz), its detection can be achieved using optical mixing spectroscopy (Benedek, 1967).

### The spectrum of Doppler shifts at the output of the photodetector in the case of the movement of RBCs in the microvascular bed of a tissue

When a laser beam illuminates RBCs moving in a net of capillaries with various velocities and directions, the light scattered by the RBCs and reaching the detector consists of a summation of waves with various Doppler shifts ($\Delta f_i$). The Doppler Shift Power Spectrum (DSPS) is the plot of the power of the Doppler shifted waves versus the Doppler shift frequency. In general, the DSPS is a function that decreases monotonically with frequency (Figure 2). The following generally applies:

1. Most of the light emerging from the tissue is not Doppler shifted since it has been scattered by static structural components. Only 1–5% of the detected light has been scattered by the moving RBCs.
2. The non-Doppler shifted light acts as a reference signal (frequency $f_r$) which is optically mixed at the photodetector surface with the light scattered by the RBCs moving with velocity $v_i$ (frequency $f_r + \Delta f_i$). This process is called heterodyne mixing. Since the reference light is much more powerful than the light scattered by the RBCs, this mixing produces a current at the output of the detector that consists of a summation (spectrum) of waves oscillating at the frequencies $(f_r + \Delta f_i) - f_r = \Delta f_i$. Thus, the photodetector current contains precisely the Doppler shifts that one wants to measure. The DSPS is extracted from the signal by Fourier analysis of the photocurrent. The process of optical mixing is schematically represented in Figure 3.
3. The light scattered by the RBCs contains Doppler shift frequencies which can be positive or negative, depending on the direction of movement of the RBCs with regard to the incident and scattered light directions. As the heterodyne mixing used so far in LDF provides only the absolute value of the Doppler shifts, the DSPS extends only in the positive frequency domain (Figure 2).

### The hemodynamical parameters which are derived from the DSPS

Analysis of the DSPS allows the determination of the following hemodynamical parameters: **Vel** (units of Hz), which is proportional to the mean velocity of the RBCs within the volume sampled by the laser light; **Vol**, which is proportional to the number of RBCs in this volume; and **F** = constant × **Vel** × **Vol**. Both

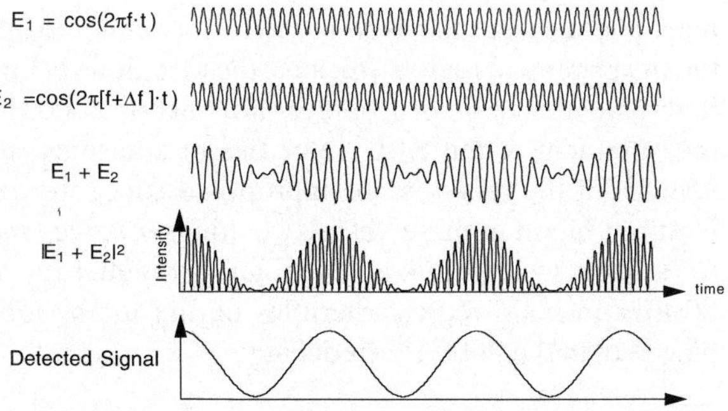


*Figure 3.* Schematic representation of light mixing technique to determine the frequency of the Doppler shifted scattered light. $E_1$ represents the amplitude of the wave scattered by the non-moving tissue (reference beam), the frequency of which ($f_i$) is the same as that of the light incident on the RBCs. $E_2$ is the amplitude of the light scattered by the RBCs which is Doppler shifted by an amount $\Delta f$ due to the movement of the particle. These 2 waves are made to interfere on the surface of a square-law photodetector to obtain the wave with amplitude $E_1 + E_2$. The output of the detector is proportional to the intensity $|E_1 + E_2|^2$. However, the photodetector, due to its limited time response, can transmit only the envelope of the fluctuations. The signal detected fluctuates at the frequency $\Delta F$, the difference between the frequency of the reference beam and that of the scattered light.

**Vol** and **F** are expressed in arbitrary units. This is due to the fact that the laser light incident on the tissue is reflected, scattered and absorbed in various amounts. These last two processes influence its penetration into the tissue. This penetration may differ from one region to the other. Thus, spatial or temporal variations in tissue structure and vascularization will affect the measurement of the hemodynamical parameters. Furthermore, meaningful comparison of the hemodynamical parameters obtained from different tissues presupposes that these tissues have similar optical properties.

The determination of these parameters is based on the theory of Bonner and Nossal (1990), which constitutes the basis of LDF. This theory requires that the light scattered by the RBCs is collected at a distance **r** from the site of incidence of the probing laser beam. In the most extended application of LDF, that of skin blood flow, **r** = 1–2 mm.

The measured flux is generally called blood flow. However, what is measured is the flux of the RBCs. In experiments where the hematocrit is maintained constant, both RBC flux and blood flow will remain proportional to each other. Should the hematocrit change during an experiment, then the changes in the proportionality coefficient should be taken into account in relating blood flow to RBC flux.

## LDF measurement modes

Two modes of LDF measurements have been developed: a) the continuous mode allowing on-line, continuous recording of the Doppler signal when the laser is focused on a discrete area of the optic disc, the subfoveal region of the choroid or iris vascular bed; b) the scanning mode which provides a two-dimensional image of the RBC flux in the capillaries of the optic nerve head (ONH) and retina.

### *A. Continuous recording of the LDF signal in the ONH and subfoveal choroid*

In the first report on the application of the LDF to the ONH (Riva et al., 1992), an optical system adapted to a standard fundus camera (Riva et al., 1989b) delivers a laser beam at a discrete site of the optic disc. Another optical system focuses the light scattered by the RBCs and the tissue structures at the aperture of an optical fiber placed in the retinal plane of the camera. This fiber guides the light to a photodetector. An area of the fundus (30° in diameter) is illuminated in redfree light allowing the observation and positioning of the laser beam at the disc, away from visible blood vessels. The effective diameter at the disc from which the scattered light is collected is approximately 150 $\mu$m.

Sebag et al. (1985, 1986) used a similar setup to determine the velocity of RBCs in the ONH. They combined this measurement with a determination of relative blood volume by means of fundus reflectometry in order to obtain an estimation of the change in blood flow induced by experimental optic nerve atrophy and neurogenic optic atrophy.

Recently, a confocal laser Doppler flowmetry module mounted on a scanning laser ophthalmoscope (SLO, TopSS, Laser Diagnostic Technologies, San Diego, CA) was described (Petrig et al., 1998). Figure 4 shows a schematic diagram of the device. Measurements of the hemodynamical parameters in the ONH, choriocapillaris and retinal vessels obtained with this device were presented.

With both the fundus camera and the SLO based LDF systems, a NeXT computer system with dedicated software was used for the LDF analysis (Petrig and Riva, 1996). This software allows averaging of the Doppler signal in phase with the heart cycle so that a precise measurement of RBC flux during the systolic and diastolic phases can be obtained. Furthermore, an algorithm automatically removes the signal during blinks. Recordings of **Vel**, **Vol** and **F** obtained from the

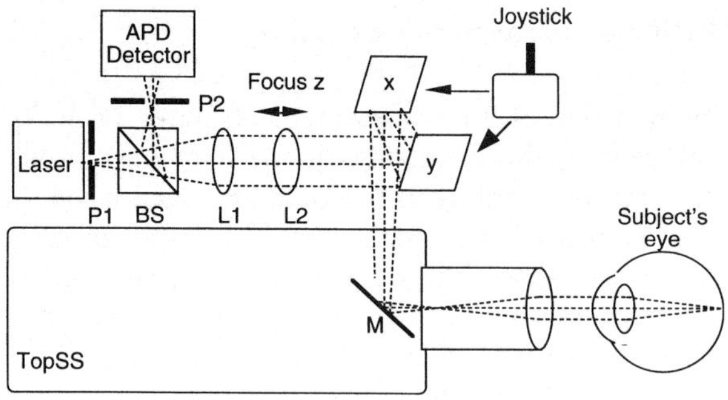


*Figure 4.* Schematic diagram of the LDF confocal module inserted into the TopSS (Petrig and Riva, 1998). Cold mirror M couples the module's light path into that of the TopSS.

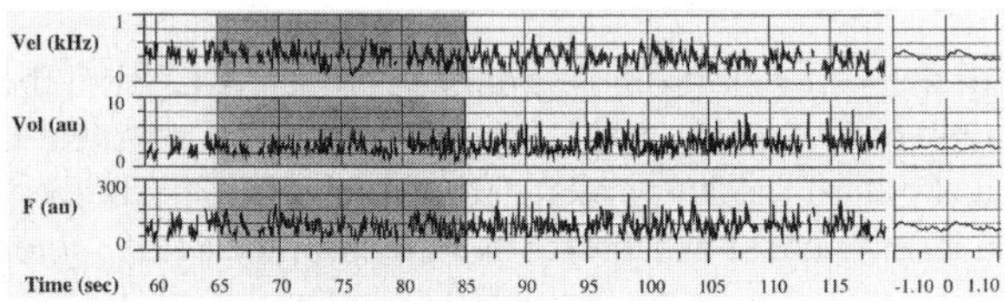


*Figure 5.* **Vel**, **Vol** and **F** obtained from theoptic nerve head of a normal volunteer, in an area deprived of large vessels. The changes in these hemodynamical parameters during 2 heart cycles are shown at the right. They were derived from a 20 sec recording segment (shaded area) using the pulse pressure wave from the ear to average all data with the same phase of the cardiac cycle and repeating this operation for all phases.

ONH of a normal volunteer are shown in Figure 5. The change in these hemodynamical parameters during 2 heart cycles is shown at the right. They were derived from a 20 sec recording segment (shaded area). One can observe the pulsatile nature of the ONH blood flow in this subject. This LDF analysis software allows the measurements of Doppler shift frequencies up to 40 kHz.

### *B. Scanning Laser Doppler Flowmetry (SLDF)*

The SLDF combines the techniques of LDF and scanning laser ophthalmoscopy (Michelson and Schmauss, 1995; Zinser, 1999). It is based on the principle of confocal microscopy and has a nominal depth resolution of approximately 300 $\mu$m. The Heidelberg Retinal Flowmeter (HRF) performs quick measurements at 16,384 different locations in a two-dimensional grid and provides an image of the retinal perfusion. The perfusion map (see Zinser, 1999 for a color picture of such a map) can be used to provide a qualitative visualization of the network of perfused vessels and capillaries. It is also possible to define a measurement region of variable size (a 100 $\times$ 100 $\mu$m is commonly used), place it interactively anywhere in the perfusion map, and measure the average perfusion values inside the measurement region. Because the technique is limited to the detection of Doppler shifts below 2000 Hz, measurement of the RBC velocities in arterioles and venules of the retina is not appropriate, since the velocity of blood in these vessels (> mm/sec) gives rise to frequency shifts that are significantly higher (up to 50 kHz in main retinal arterioles during the systolic phase) than the 2000 Hz frequency.

## LDF sample volume

A central question in the application of LDF to tissue blood flow is the depth of the volume of tissue sampled by the laser light. Depending upon this depth, different vascular beds may contribute to the signal. This is, for example, the case with the ONH which is supplied by two main sources of blood: the superficial layers by the central retinal artery and the deeper layers by the posterior ciliary arteries. These two vascular bed may have different regulatory capabilities in response to various physiological stimuli. Furthermore, the deep layers of the ONH appear to be particularly susceptible to ischemic disorders, including glaucoma.

Theoretical analysis suggests that the average depth $\langle z \rangle$ probed by emerging photons before they exit a tissue and are detected is equal to 0.4 $s^{1/2}$ $m^{-1/4}$ (Bonner and Nossal, 1990). m is the dimensionless absorption coefficient per unit scattering length within the tissue and s = **r**/L. L is the mean free path of a photon within the tissue, i.e. the average path a photon travels before being scattered by a RBC. **r** is the separation between the incident laser beam and the site of detection of the scattered light. Increasing r results in an increase of the sampled depth. For tissues such as the skin and using a laser at 800 nm and probe separation of 500 $\mu$m, the analysis predicts a depth of measurement of 0.5–1 mm. For brain tissue, a depth greater than 1 mm is suggested by the experiments of Skarphedinsson et al. (1988). With a probe separation of 150 $\mu$m, application of the above equation gives a depth of approximately 260 $\mu$m in the skin and around. 520 $\mu$m in the brain (grey matter). These calculations are not valid when **r** approaches zero.

Investigations on a model system suggests that when the light collecting aperture coincides with the tissue volume illuminated by the probing laser, layers of the ONH tissue as deep as 300 $\mu$m contribute to the LDF signal (Koelle et al., 1993). In man, however, the depth of tissue sampling in the ONH remains to be experimentally confirmed. Measurements in humans have shown that, when the intraocular pressure (IOP) is rapidly increased to systolic ophthalmic artery pressure in a normal volunteers, **F** decreases nearly to zero (Riva et al., 1997a). This observation strongly suggests that the LDF technique detects predominantly the motion of RBCs within the intraocular region of the ONH and not beyond it, where blood flow is not affected by the IOP (Geijer and Bill, 1979).

In order to obtain a better understanding of which part of the ONH circulation is measured by LDF, studies were conducted in rhesus monkey eyes (Petrig et al., 1999). By means of the fundus camera LDF system, the flow parameters were measured at baseline and then after experimental occlusion of the central retinal artery, posterior ciliary arteries and combined occlusion of the central retinal artery and posterior ciliary arteries. The findings of this study suggest that LDF is predominantly sensitive to blood flow changes in the superficial layers of the ONH and less to those layers of the prelaminar and deeper regions of the ONH, and their relative proportions are still unknown. The weaker signal from the deep layers cannot be separated from the dominant signal from the superficial layers to exclusively study the circulation in the deep layers.

## Linearity of LDF

Valid measurements of the changes in blood flow are obtained if **F** varies linearly with the actual RBC flux or blood flow. To test this for the ONH, **F** was measured while the mean ocular perfusion pressure ($PP_m$ = mean ophthalmic artery pressure minus IOP), i.e. the pressure that drives the blood through the eye, was rapidly decreased in cats by means of a lethal injection of pentobarbital. It was assumed that during this rapid decrease, the ocular resistance would remain constant, so that the change in blood flow would be only due to the change in $PP_m$. The measurements demonstrated a linear decrease in **F** as a function of $PP_m$ ( Riva et al., 1992).

Linearity between actual flow and **F** has been documented in various tissues, such as the skin, skeletal muscle, cerebral cortex, nerves and others (Shepherd and Öberg, 1990). This property is expected from theoretical arguments in so far as the fraction of light that is Doppler shifted by RBCs is small compared to the total scattered light that is detected. Laser Doppler signals recorded from the ONH, a tissue where capillaries occupy only approximately 2.5% of the volume (Quigley et al., 1982) show that this condition is fulfilled, since the power of the fluctuating photocurrent, the one due primarily to shifted light, was consistently found to represent not more than 10% of the total photocurrent.

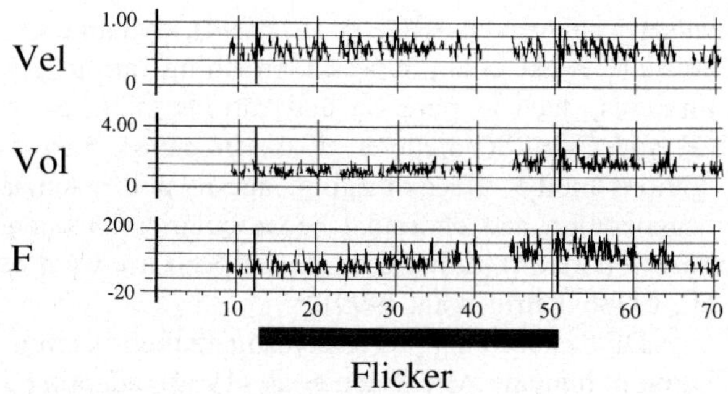


*Figure 6.* Effect of diffuse luminance flicker on **Vel**, **Vol** and **F** measured from a small area (diameter at the disk ≈150 $\mu$m in diameter of the ONH in a normal volunteer.

## Applications of LDF

The high spatial and temporal resolution of the LDF technique, particularly in the continuous mode, makes this technique most suitable for the investigation of the process of regulation of blood flow in response to various physiological stimuli.

### *a. ONH*

The application of LDF to the physiology and pathology of the human ONH circulation includes the investigation of the response of blood flow to a) decreases in $PP_m$ induce by increases in IOP (Pillunat et al., 1997; Riva et al., 1997a); b) increases in $PP_m$, produced by increases in systemic blood pressure with isometric exercises (Movaffaghy et al., 1998), c) hyperoxia, carbogen and mixtures of $O_2$ and $CO_2$ (Harris et al., 1996; Lietz et al., 1998), and increased neuronal activity (Riva et al., 1996), as illustrated in Figure 6.

### *b. Choroid and iris*

LDF measurement of choroidal blood flow in the foveal region of the human fundus is more recent. Non-confocal (Riva et al., 1994) and confocal (Geiser et al., 1999) techniques have been used. Studies have been performed in humans on the effect of increases and decreases of $PP_m$ (Riva et al., 1997b, 1997c), valsalva maneuvers (Riva et al., 1994), dynamic exercise (Lovasik et al., in press), breathing various gas mixtures, such as pure $O_2$ and various mixtures of $O_2$ and $CO_2$ (Schmetterer et al., in press). Investigations of the effect of aging, age-related macular degeneration and choroidal neovascularization have been reported (Grunwald et al., 1998a; Grunwald et al., 1998b; Petrig et al., 1999).

LDF has been applied to investigate the iris circulation in humans. Michelson et al. (1994), adapting a laser Doppler system designed for the measurement of skin blood flow, has investigated the effect of acute physical exercise on iris blood flow. Chamot et al. (1999) have used a slit-lamp based LDF system to determine the flux of RBCs through the iris microcirculatory network and the effect of increased IOP on this flux (Chamot et al., 2000).

In conclusion, LDF is a powerful technique to investigate noninvasively changes in blood flow in the ONH, retina, subfoveal choroid and iris. The technique has a high sensitivity and its temporal response allows to demonstrate changes in flux during the cardiac cycle and in response to changes in various physiological parameters, such as perfusion pressure, breathing conditions and neuronal activity. These capabilities open new avenues in the understanding of the regulation of blood flow in the various tissues of the eye.

## Acknowledgements

The author thanks Pascale Evêquoz and Martial Geiser for their help in preparing this review.

## References

1. Benedek GB. Optical mixing spectroscopy with applications to problems in physics, chemistry, biology, and engineering. In the jubilee volume in honor of Alfred Kastler, Polarization, Matter and Radiation, Paris, 1969, Presses Universitaires de France, p. 49.
2. Bonner RF, Nossal R: Principles of laser-Doppler flowmetry. In: Shepherd AP, Öberg PA (eds) Developments in Cardiovascular Medicine: Laser-Doppler Blood Flowmetry . Vol. 107. Boston: Kluwer Academic Publishers, 1990; 73–92. Chamot S, Movaffaghy AM, Petrig BL, Rica CE. Blood flow in the human iris measured by laser Doppler flowmetry. Microvasc Res 1999; 57: 153–161.
3. Chamot S, Movaffaghy M, Petrig BL, Riva CE. Iris blood flow response to acute decreases in ocular perfusion pressure: a laser doppler flowmetry study in humans. Exp Eye Res. In press.
4. Geijer C, Bill A. Effects of raised intraocular pressure on retinal, prelaminar, laminar and retrolaminar optic nerve blood flow in monkeys. Invest Ophthalmol Vis Sci. 1979; 118: 1030–1036.
5. Geiser MH, Diermann U, Riva CE. Compact laser Doppler choroidal flowmeter. J Biomed Optics 1999; 4: 459–464.
6. Grunwald JE, Hariprasad SM, DuPont J. Effect of aging on foveolar choroidal circulation. Arch Ophthalmol 1998a; 116: 150–154.
7. Grunwald JE, Hariprasad SM, DuPont J, et al. Foveolar choroidal blood flow in age-related macular degeneration. Invest Ophthalmol Vis Sci 1998b; 39: 385–390.
8. Harris A, Anderson DR, Pillunat L, et al. laser Doppler flowmetry measurement of changes in human optic nerve head blood flow in response to blood gas perturbations. J Glaucoma 1996; 5: 258–265.
9. Koelle J, Riva CE, Petrig BL, Cranstoun SD. Depth of sampling in the optic nerve head using laser Doppler flowmetry. Lasers Med Sci 1993; 8: 49–54.
10. Lietz, Hendrikson P, Flammer J, Orgül S, Haefliger IO. Effect of carbogen, oxygen and intraocular pressure on Heidelberg retina flowmeter parameter "Flow" measured at the papilla. Ophtalmologica 1998; 212: 149–152.
11. Lovasik JV, Kergoat H, Riva CE, Petrig BL. Choroidal blood flow in man after increased systemic perfusion pressure induced by dynamic exercise. Invest Ophthalmol Vis Sci. In press.
12. Michelson G, Groh M, Grundler A. Regulation of ocular blood flow during increases of arterial blood pressure. Br J Ophthalmol 1994; 78: 461–465.
13. Michelson G and Schmauss B. Two dimensional mapping of the perfusion of the retina and optic nerve head. Br J Ophthalmol 1995; 79: 1126–1132.
14. Movaffaghy A, Chamot SR, Petrig BL, Riva CE. Blood Flow in the Human Optic Nerve Head during isometric exercise. Exp Eye Res 1998; 67: 561–568.
15. Petrig BL, Riva CE. Optic nerve head laser Doppler flowmetry: Principles and computer analysis. In Ocular Blood Flow. Glaucoma-Meeting 1995, Kaiser HJ, Flammer J, Hendrickson P (eds) (Karger Basel), 1996, pp. 120–127.
16. Petrig BL, Riva CE, Lorenz B, Movaffaghy A, Harbarth UP, Dreher AW. Confocal laser Doppler system for measurement of blood velocity in retinal vessels and flow in the optic nerve through the undilated pupil. Lasers and Light 1998; 8: 137–142.
17. Petrig BL, Riva CE, Hayreh SS. Laser Doppler flowmetry and optic nerve head blood flow. Am J Ophthalmol 1999; 127: 43–425.
18. Petrig BL, Chamot S, Riva CE, Sickenberg M, Ballini JP, Van den Bergh H. Subfoveal blood flow in choroidal neovascularization compared to normal fovea as measured by Topss-based laser Doppler flowmeter. Invest Ophthalmol Vis Sci 1999; 40: S561.
19. Pillunat EL, Anderson DR., Knighton RW, Joos KM, Feuer WJ. Autoregulation of human optic nerve head circulation in response to increased intraocular pressure. Exp Eye Res 1997; 64: 737–744.
20. Quigley HA, Hohmann RM, Addicks EM. Quantitative study of optic nerve head capillaries in experimental disk pallor. A. J Ophthalmol 1982; 93: 689–699.
21. Riva CE, Ross B, Benedek GB. Laser Doppler measurements of blood flow in capillary tubes and retinal arteries. Invest Ophthalmol Vis Sci 1972; 11: 936–944.

22. Riva CE, Grunwald JE, Sinclair SH. Laser Doppler measurement of relative blood velocity in the human optic nerve head. Invest Ophthalmol Vis Sci 1982; 22: 241–248.
23. Riva CE, Shonat RD, Petrig BL, Pournaras CJ, Barnes GB.Noninvasive measurement of the optic nerve head circulation. Ocular Blood Flow in Glaucoma. Lambrou GN, Greve EL, eds, Kugler and Ghedini Publications, Berkley, 1989a; 129–134. Riva CE., Petrig BL, Grunwald JE. Retinal blood flow. In Laser-Doppler Blood Flowmetry,
24. Shepherd AP and Öberg PA (eds), Volume 107 of Developments in Cardiovascular Medicine, Kluwer Academic Publishers, Boston, Chapter 19, pp. 349–383, 1989b.
25. Riva CE, Harino S, Petrig BL, Shonat RD. Laser Doppler flowmetry in the optic nerve. Exp Eye Res 1992; 55: 499–506.
26. Riva CE, Cranstoun SD, Grunwald JE, Petrig BL. Choroidal blood flow in the foveal region of the human ocular fundus. Invest Ophthalmol Vis Sci 1994; 35: 4273–4281.
27. Riva CE, Mendel MJ, Petrig BL. Flicker-induced optic nerve blood flow change. In: Kaiser HJ, Flammer J, Hendrickson P (eds) Ocular Blood Flow. Glaucoma-Meeting 1995. Basel: Karger. 1996: 128–137.
28. Riva CE, Hero M, Titze P, Petrig BL. Autoregulation of human optic nerve head blood blow in response to acute changes in ocular perfusion pressure. Graefe's Arch Clin Exp Ophtalmol 1997a; 235: 618–626.
29. Riva CE, Titze P, Hero M, Petrig B.L. Effect of acute decreases of perfusion pressure on choroidal blood flow in humans. Invest Ophthalmol Vis Sci 1997b; 38: 1752–1760.
30. Riva CE, Titz, P, Hero M, Petrig BL, Movaffaghy A. Choroidal blood flow during isometric exercises. Invest Ophthalmol Vis Sci 1997c; 38: 2338–2343.
31. Sebag J, Feke GT, Delori FC, Weiter JJ. Anterior optic nerve blood flow in experimental atrophy. Invest Ophthalmol Vis Sci 1985; 26: 1415-1422.
32. Sebag J, Delori, FC, Feke GT, Goger DG, Fitch K, Tagawa H, Deupree D, Weiter JJ, McMeel JW. Anterior optic nerve blood flow decreases in clinical neurogenic Optic atrophy. Opthalmology 1986; 93: 858–865.
33. Shepherd AP, Öberg PA, Laser-Doppler Blood Flowmetry. Kluwer Academic Publishers, 1990, Boston.
34. Skarphedinsson JO, Harding HS, Thore UP. Repeated measurements of cerebral blood flow in rats. Comparison between the hydrogen clearance method and laser Doppler flowmetry. Acta Physiol Scan 1988; 134: 133–142.
35. Zinser, G. Scanning laser Doppler flowmetry. Current Concepts on Ocular Blood Flow in Glaucoma. Pillunat LE, Harris A, Anderson DR, Greve EL (eds). Kugler Publications, The Hague, The Netherlands, 1999, pp. 197–204.

*International Ophthalmology* **23:** 191–197, 2001.
*J.R. Sampaolesi (ed.), Laser Scanning: Update 1*, 9–15.



# Optical Coherence Tomography (OCT): principles of operation, technology, indications in vitreoretinal imaging and interpretation of results

P.E. Stanga & A.C. Bird
*The Institute of Ophthalmology, University College London and Moorfields Eye Hospital, Medical Retina Dept., London, UK*

Optical coherence tomography (OCT) is a relatively new non-contact, non-invasive technique for *in vivo* examination of the human retina. It is based on the Michaelson interferometer. It is analogous to B-scan ultrasound except that it measures optical rather than acoustic reflectivity. It has a depth resolution of approx. 10 $\mu$m compared to 150 $\mu$m of B-scan ultrasound. High-frequency ultrasound has a resolution of approximately 20 $\mu$m, but due to the limited penetration of 4 mm it can only be applied to the anterior segment. Neither computed tomography nor magnetic resonance imaging achieve a similar resolution to that of OCT.

The commercial OCT uses a superluminescent diode (NOT a laser) with a central bandwidth of 850 nm. Cross-sectional tomographs of the logarithm of reflectivity in the retina are displayed in real time as two-dimensional false-color images. Each 2-dimensional OCT scan is comprised of 100 A-scans spanning a length of 1–10 mm (adjustable by the user) and which can be acquired in 2.5 seconds.

Relative high reflectivity (RHR) layers correspond to areas of horizontal retinal elements such as the nerve fiber layer at the retinal surface or deeper plexiform layers and a single layer of retinal pigment epithelium (RPE) and choroid. Relative low reflectivity (RLR) layers correspond to the nuclear layers and a single layer of photoreceptor inner and outer segments. Warm colors (red to white) represent areas of RHR, while cold colors (blue to black) represent areas of RLR. A good correlation has been found between retinal morphology and macular OCT imaging. Resolution of retinal images is dependent not only on the resolving power of the instrument, but also on the contrast in relative reflectivity of adjacent structures. The OCT system cannot discriminate between the images of adjacent tissues that possess matching relative reflectivity, i.e., the RPE and choroid, and the photoreceptor outer segments and photoreceptor nuclei. RLR of photoreceptor inner and outer segments may be due to the vertical macrostructure and the regularly aligned microstructural elements.

OCT can demonstrate the presence of edema where it is not seen on biomicroscopy or angiographically. A very important feature of the OCT system is that it provides information on the retinal structures. The location of fluid accumulation in relation to the different retinal layers may be determined and the response to treatment without the need to perform invasive studies such as fluorescein angiography may be objectively monitored. At the same time it may be possible to explain why some patients respond to treatment while others do not. OCT may become a valuable tool in determining the minimum maintenance dose of a certain drug in the treatment of CMO, and may demonstrate retinal changes that explain the recovery in some patients without angiographically demonstrable improvement and lack of recovery in others.

OCT can aid in identifying, monitoring and quantitatively assessing various posterior segment conditions including macular edema, age-related macular degeneration, full and partial-thickness macular hole, epiretinal membrane, intaretinal exudate, idiopathic central serous chorioretinopathy, RPE detachment, detachment of the neurosensory retina, macular lesions associated with optic nerve head pits or glaucoma.

The OCT system seems to be a promising tool for the study of intraretinal fluid accumulation.

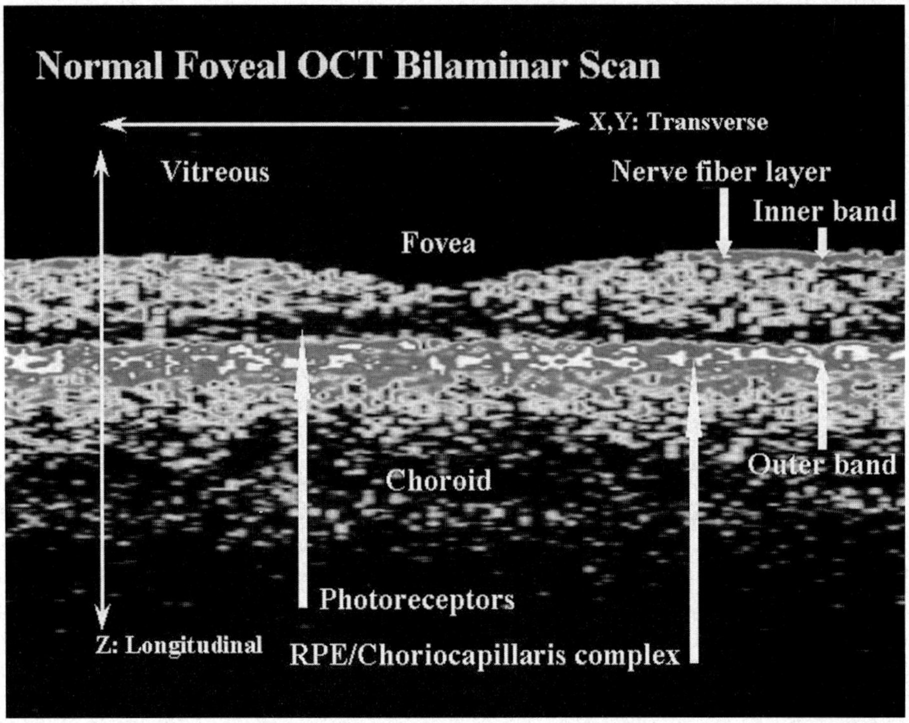


*Figure 1.*

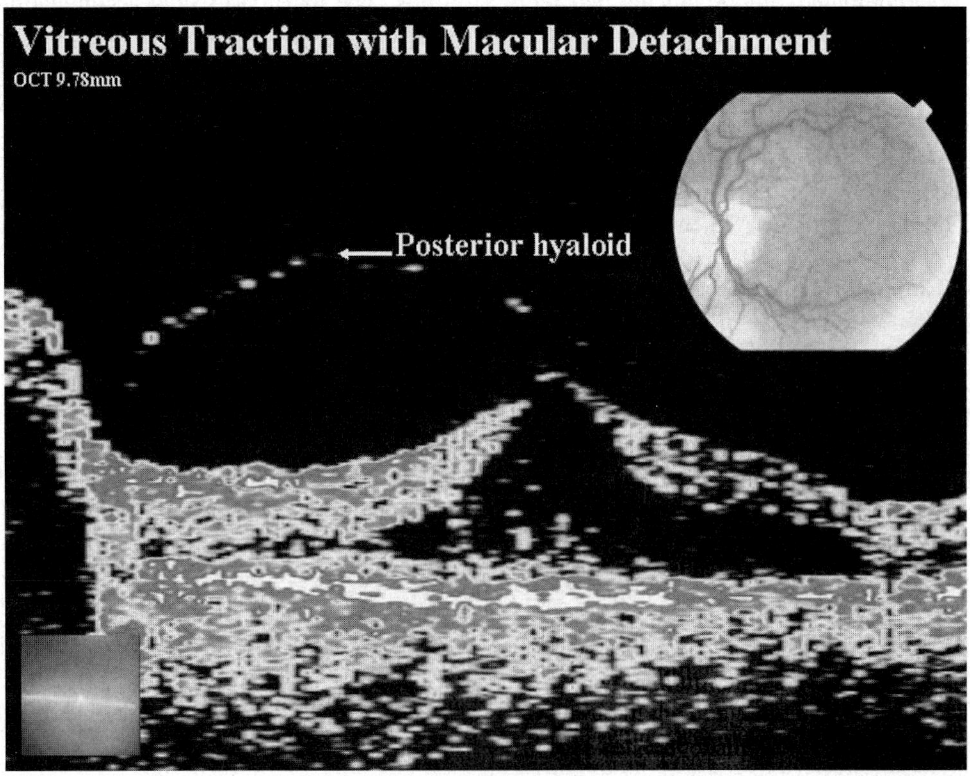


*Figure 2.*

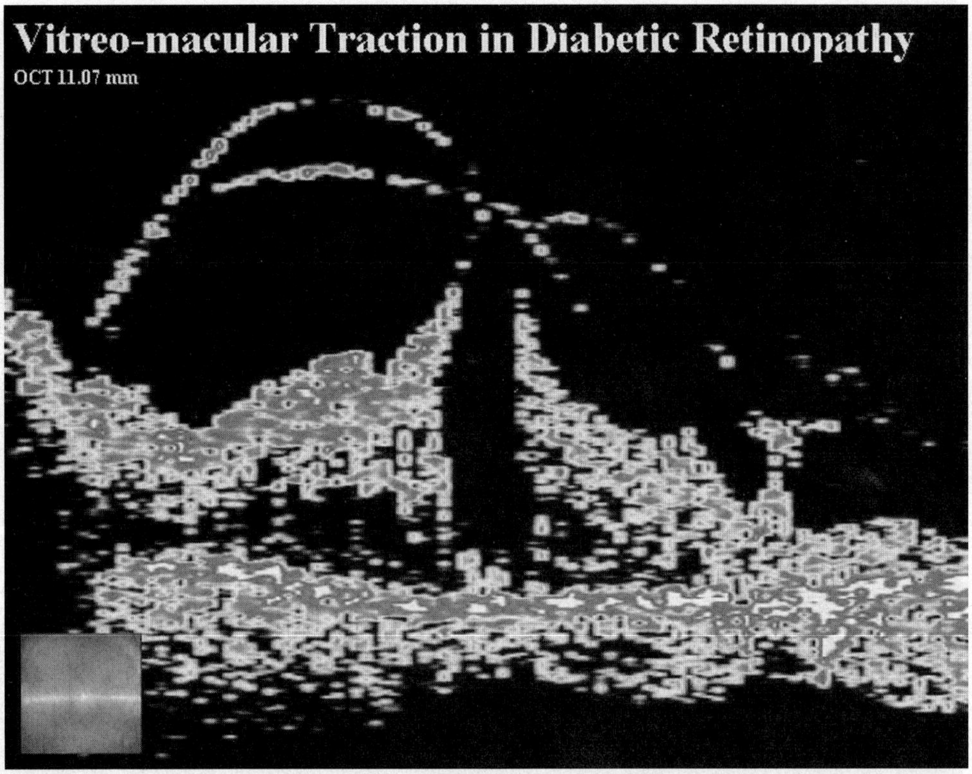


*Figure 3.*

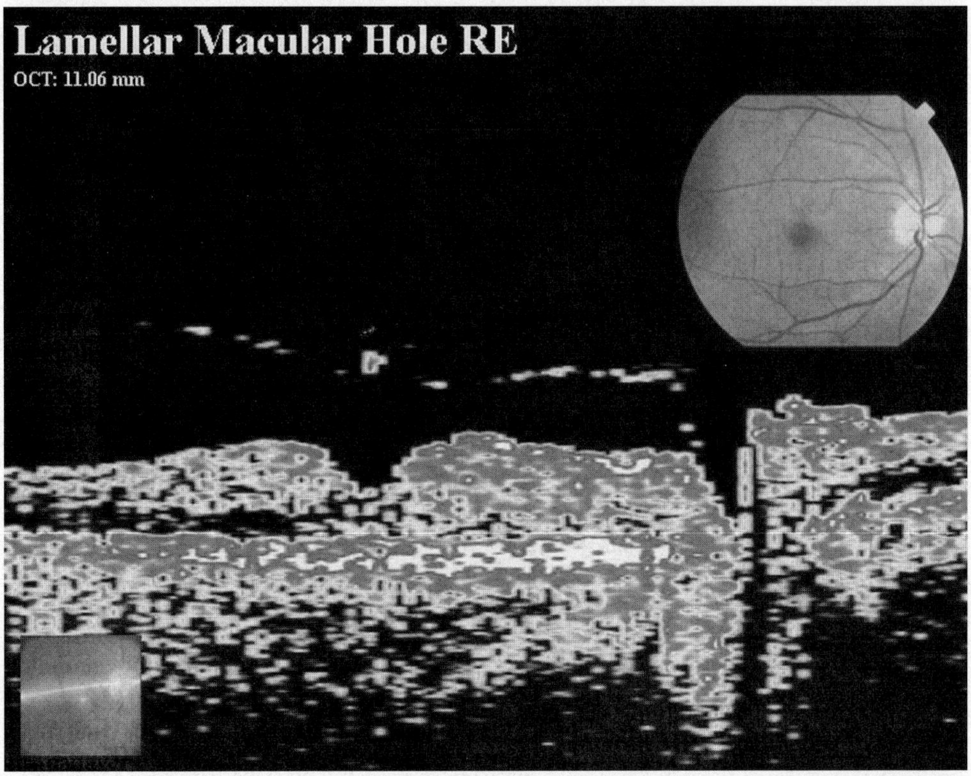


*Figure 4.*

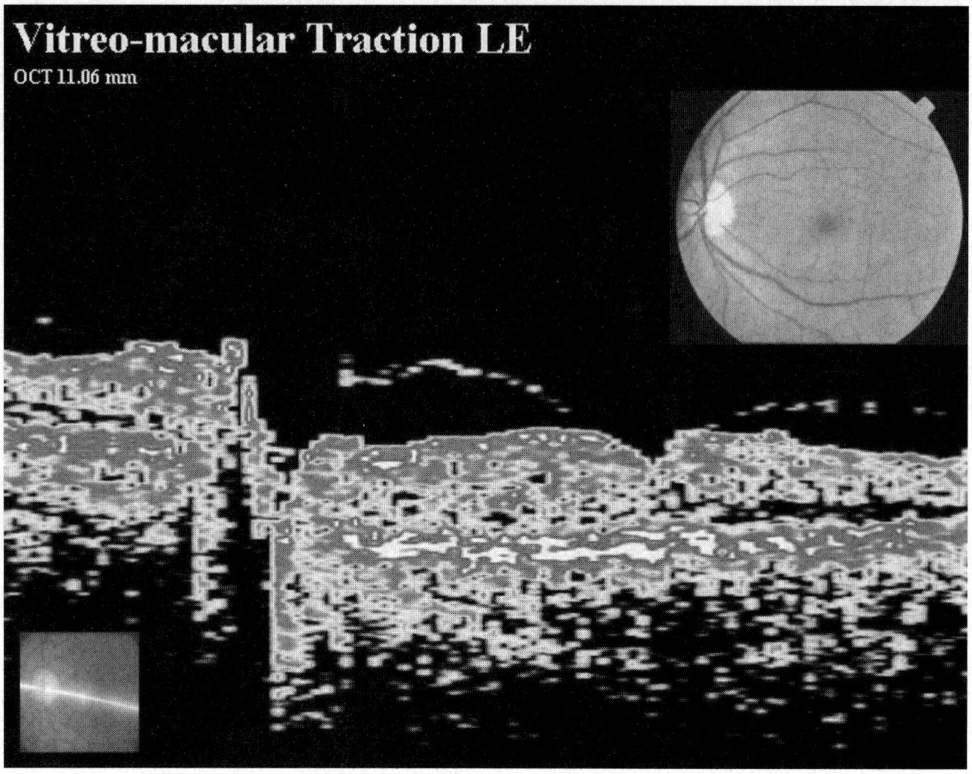


*Figure 5.*

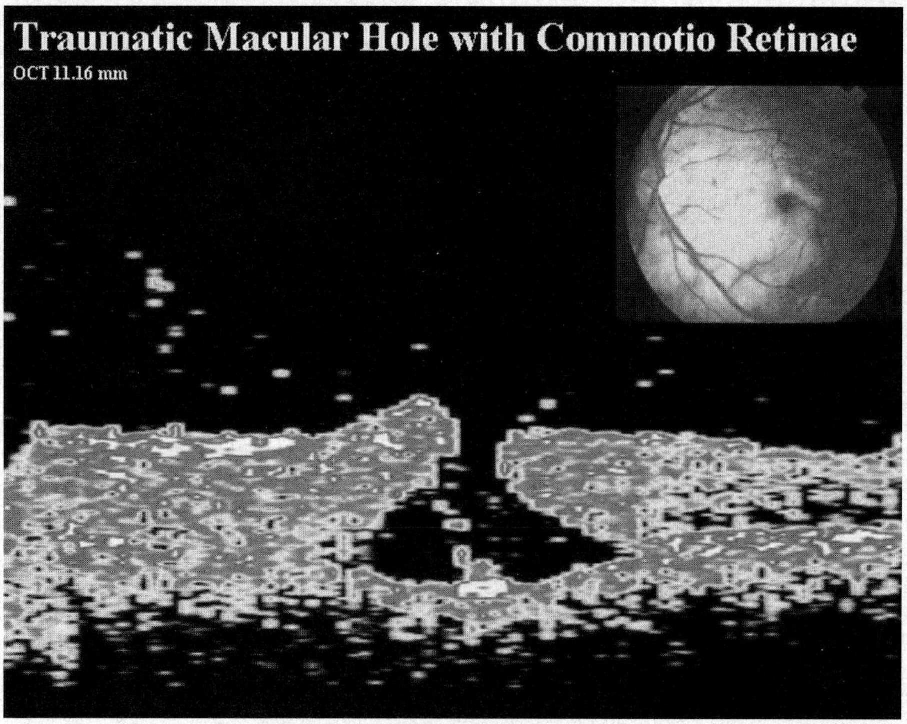


*Figure 6.*

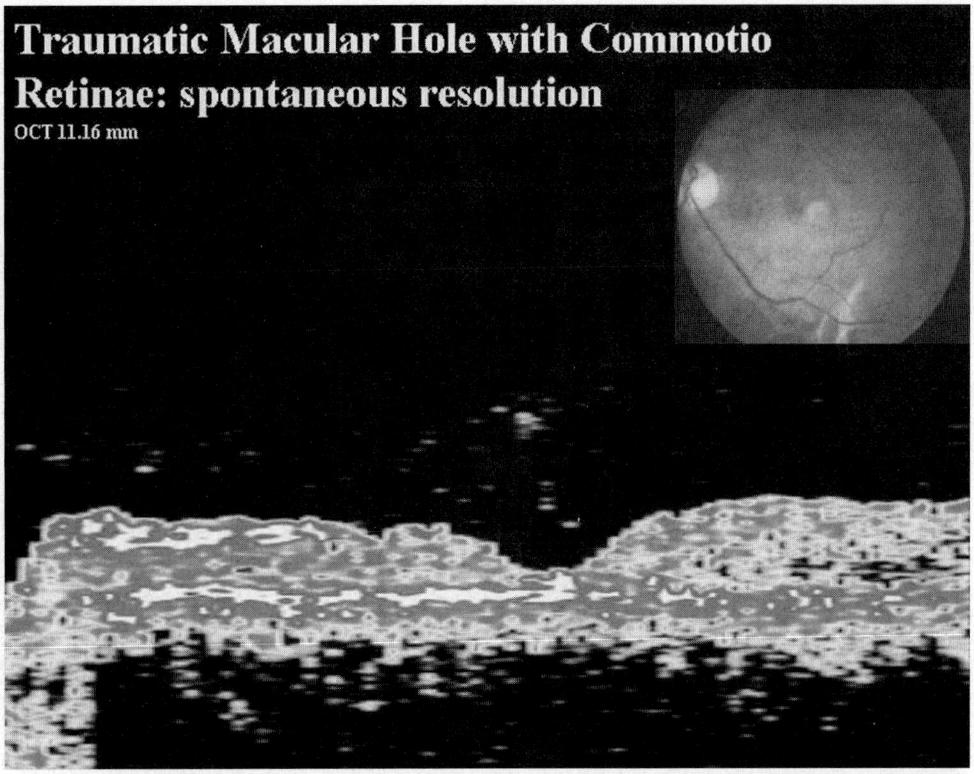


*Figure 7.*

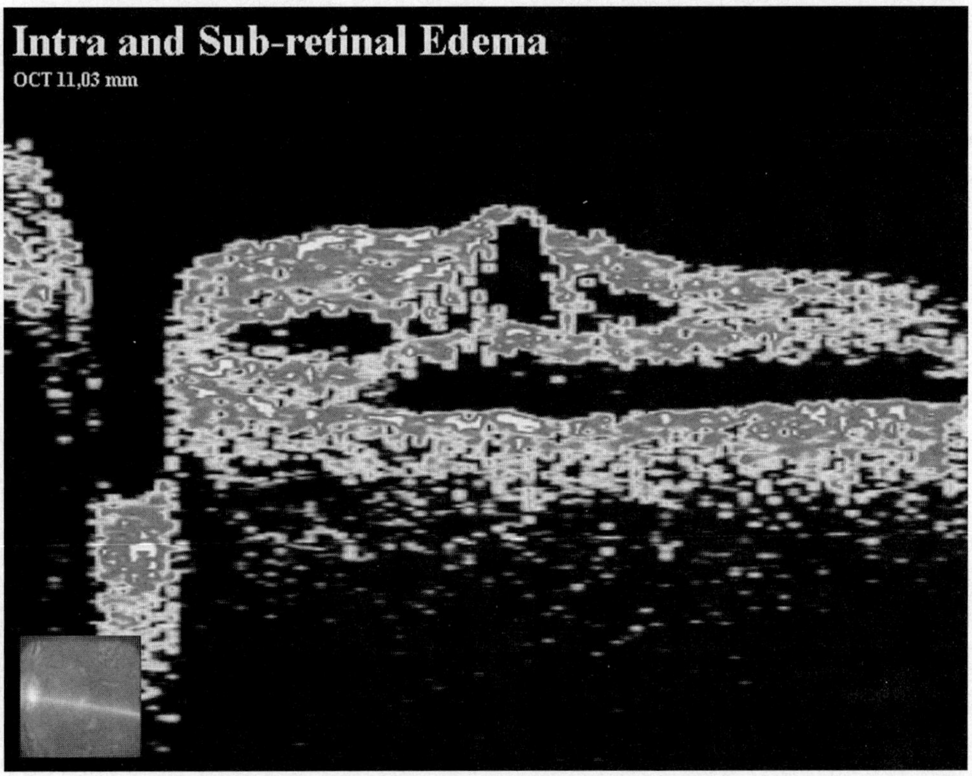


*Figure 8.*

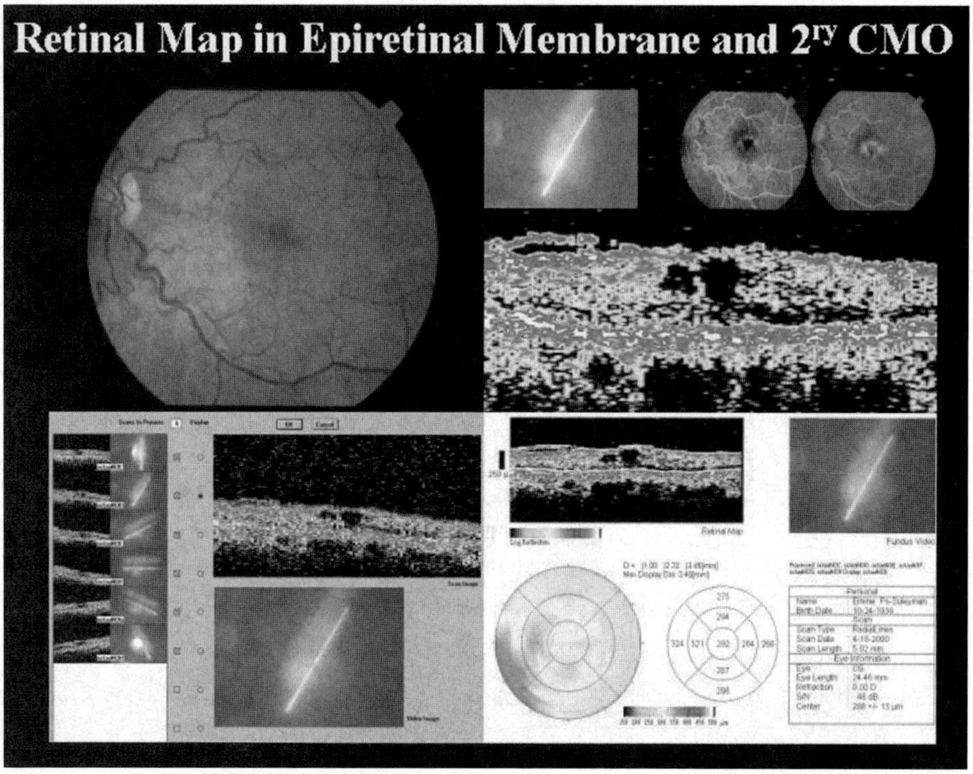


*Figure 9.*

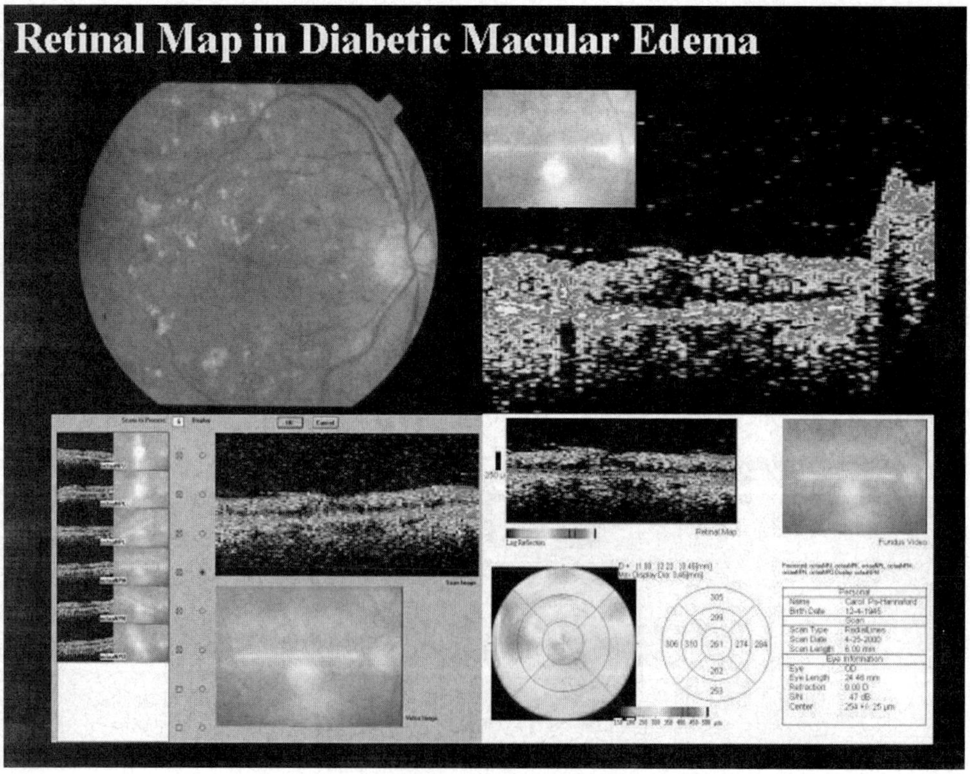


*Figure 10.*

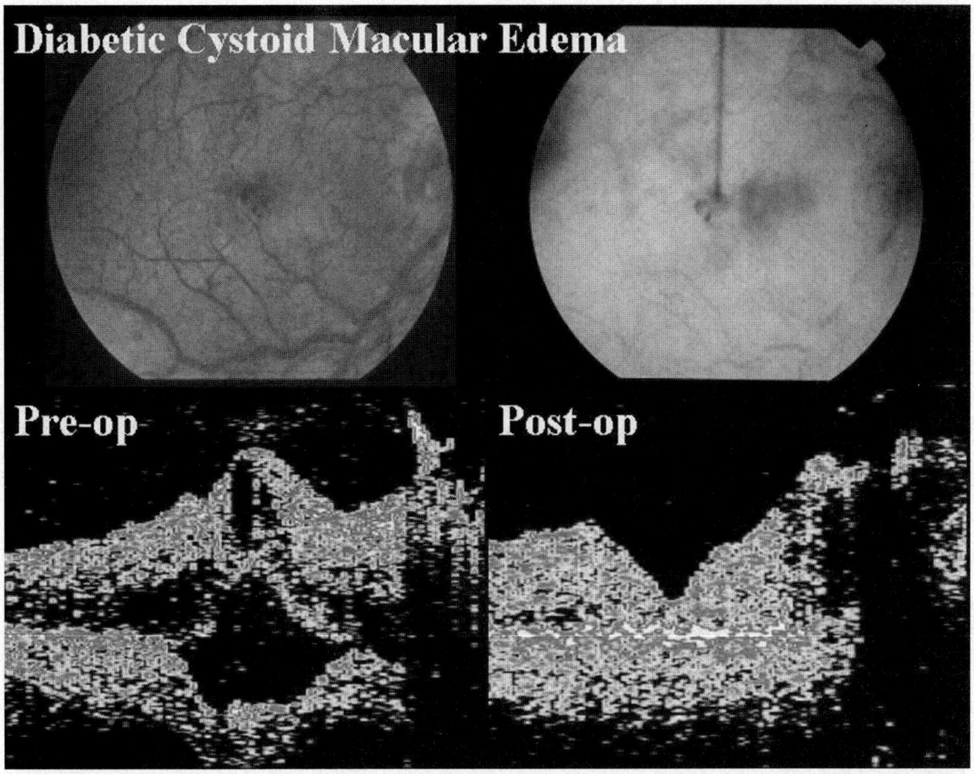


*Figure 11.*

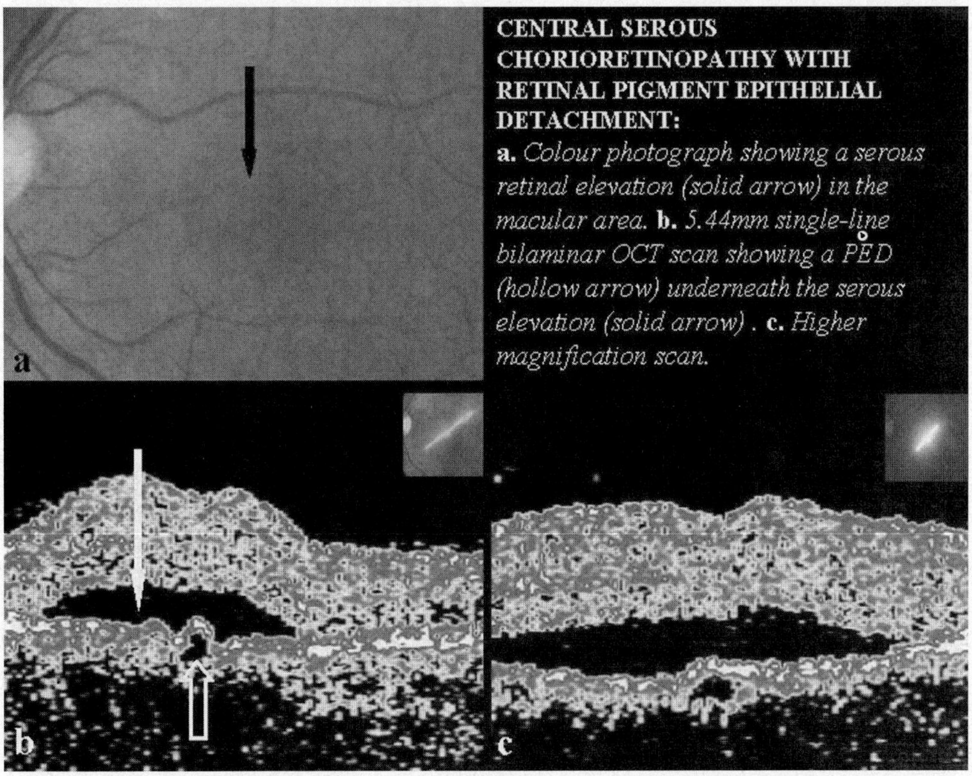


*Figure 12.*

International Ophthalmology **23:** 199–206, 2001.
J.R. Sampaolesi (ed.), Laser Scanning: Update 1, 17–24.



# Confocal microscopy of the human cornea *in vivo*

Barry R. Masters & Matthias Böhnke
*Universitäts-Augenklinik, Inselspital, University of Bern, 3010 Bern, Switzerland*

*Key words:* three-dimensional *in vivo* confocal microscopy, human cornea

## Abstract

*In vivo*, scanning-slit, confocal microscopy offers improved resolution and has resulted in new discoveries of corneal pathology at the cellular level. The ability to provide high resolution, real-time images of the full thickness of the living human cornea gives the clinician and the researcher an important new tool.

## Introduction

The fundamental principles of noninvasive instrumentation to investigate the eye are developed in several books [1–3]. There are classic works which provide the reader with an excellent introduction to the biomicroscopy of the eye and the clinical use of the slit lamp [4–7]. Background information on confocal microscopy may be found in several sources [8–10]. A recent reviews the historical development of optical techniques used for biomicroscopy of the eye, and develops the historical and theoretical foundations of confocal microscopy [11]. This paper critically discusses the following topics: the practical techniques of clinical confocal microscopy, the clinical findings of normal, subclinical and pathological cases, the post surgical cornea, and corneal preservation.

## Biomicroscopy of the living eye from the slit lamp to the confocal microscope

There is a direct and interesting lineage from the confocal microscope developed by Goldmann, to the development of the specular microscope by Maurice, Koester and others, to the various types of clinical confocal microscopes [11]. These instruments represent partial solutions to the problem of how to image thin optical sections from a 500 micron thick, transparent, moving object-the human cornea *in vivo*.

The optical principle of the confocal microscope can be shown in a schematic diagram (Figure 1). This diagram shows a confocal microscope with two microscope objectives; however for clinical *in vivo* use, a single objective lens is used to both illuminate and detect the light scattered and reflected from the thick object (cornea). A point source of light is focused to a point within the cornea by lens L1. The focused light illuminated a small volume at P1. A second lens, L2 collects the light from the same small volume P1 and focuses the collected light from P1 at the slit S2, placed in front of the light detector. Both apertures S1, and S2 are co-focused ("confocal") on the same point in the focal plane that simultaneously is being illuminated and detected. The confocal microscope discriminates against light that is not in the focal plane. The out-of-focus light collected by lens L2 cannot enter the aperture at S2; thereby limiting the amount of out-of-focus light that is detected. This provides the physical basis for the strong discrimination against out-of-focus light in a confocal microscope. A confocal microscope (two conjugate slits or apertures) provides increased lateral and depth resolution [8–10]. A confocal microscope provides two enhancements compared to a standard light microscope: enhanced lateral resolution and enhanced axial resolution. The latter property is the basis of its capability to optically section a thick, highly scattering specimen such as the cornea.

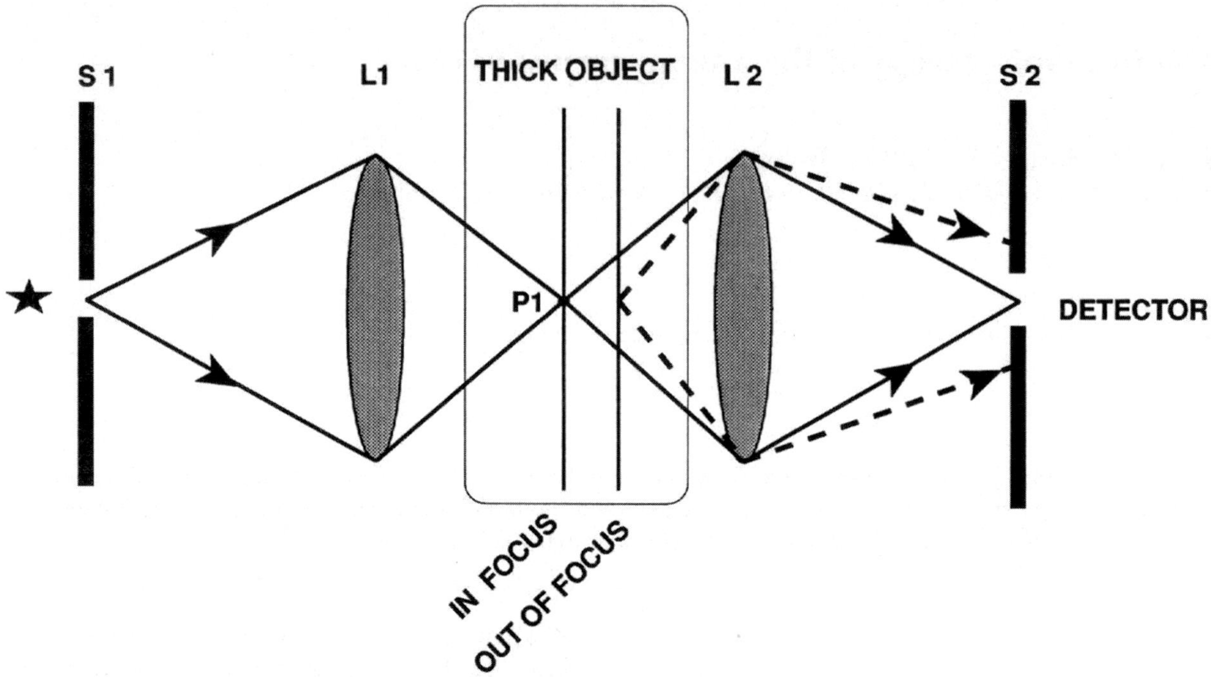


*Figure 1.* The optical principle of a confocal microscope. S1 and S2 are confocal apertures. L1 and L2 are focusing lens for illumination and detection respectively. P1 is the focal volume that is illuminated with the point source of light from S1 and focused at P1 with lens L1. P1 is imaged by lens L2 to form an image at the aperture S2. The light from focal volume P1 can enter the aperture S2. The out of focus light (dotted lines) only partially enters the aperture S2; thus, the depth discrimination.

## Development of the scanning-slit confocal microscope

A book in which the history of confocal microscopy is documented as a series reprinted milestone papers and patents from 1884 [Paul Nipkow patent] to the present was recently published [12]. This work shows the history of the various types of confocal microscope and the papers and patents that describe the development of the scanning-slit confocal microscope [13].

An alternative to point scanning uses a slit of illumination that is scanned over the back focal plane of the microscope objective [13]. This system has that advantage that many points are scanned in parallel, markedly decreasing the scanning time. Another important advantage is the superior light throughput compared with that in the point-scanning Nipkow disk systems. The disadvantage is that the microscope is truly confocal only in the axis perpendicular to the slit height.

Thaer developed a scanning slit confocal microscope based on an oscillating double sided mirror (Figure 2). This design has the advantage of variable slit widths, a high numerical aperture objective, and a halogen lamp as the light source [13]. Scanning and descanning of the cornea is accomplished with an oscillating two sided mirror. This design follows from the original microscope constructed over twenty years ago by Svishchev in Moscow [14, 15].

The image of a slit is scanned over the back focal plane of the microscope objective. The slit width can be varied in order to optimize the balance of optical-section thickness and image brightness. The instrument is based on the double-sided mirror which is used for scanning and descanning. This confocal microscope used a halogen lamp for illuminating the slit. The detector is a video camera that acquires images at video rates. This confocal microscope can image basal epithelial cells and the adjacent wing cells in the living human cornea due to its high light throughput. The design consists of two adjustable slits placed in conjugate planes of the confocal microscope. Both scanning of the illumination slit over the back focal plane of the microscope objective and descanning of the reflected light from the object is accomplished with an oscillating two-sided mirror.

For corneal imaging, a 40×, a 63× or a 100× high numerical aperture, water immersion microscope ob-

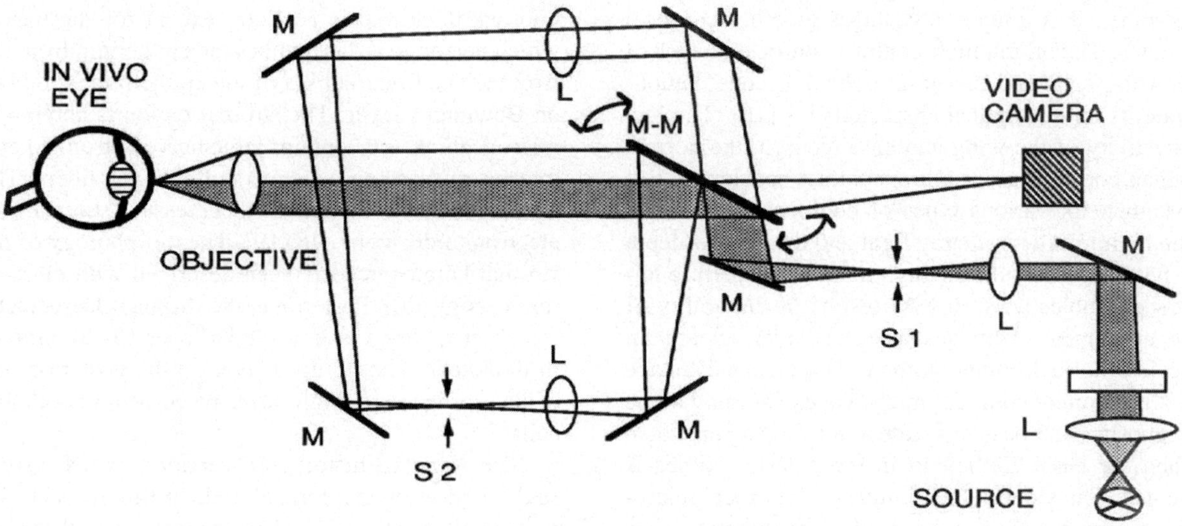


*Figure 2.* Schematic diagram illustrating the optical system of a clinical real-time scanning-slit confocal microscope. The light source is a halogen lamp. S1 and S2 are confocal slits, and the L's are lenses. Objective is a water immersion microscope objective. One half of the objective is used for the illumination of the cornea; the other half of the objective is used to collect the light form the cornea. The illumination light path is shown by black dots, and the collection light path is shown in white. The M's are fixed front-surface mirrors; M-M is an oscillating double-sided mirror that is used for both scanning and descanning. The descanned image is detected by an intensified video camera.

jective (Zeiss, Germany) can be used. The resolution of the different objectives and the thickness of the optical section to some extent are influenced by the light levels used and the geometry and the reflectivity of the structures studied.

The following design parameters were incorporated into the video-rate, scanning slit confocal microscope.

- Nonapplanating, high numerical aperture, water immersion microscope objectives, Zeiss 40×, Zeiss 63×, Zeiss 100× microscope objectives are used.
- A methylcellulose gel to optically couple the tip of the microscope objective to the cornea. There is no applanation or direct physical contact, which deforms the cornea, and introduces artificial folds and ridges in the cornea, between the objective and the surface of the cornea.
- One half of the numerical aperture is used for illumination, and one half of the numerical aperture is used for collection of the reflected and fluorescence light.
- Optical sectioning in the plane of the cornea is obtained with two sets of conjugate slits. The slit heights are variable and adjustable.
- An oscillating, two sided mirror (bilateral scanning) is used for scanning the image of the slit over the back focal plane of the microscope objective, and for descanning the reflected and back scattered light collected by the microscope objective from the focal plane in the specimen.
- The light source is a 12 volt halogen lamp.
- The scanning is synchronized with the read-out of an interline CCD camera in order that the full vertical resolution of the intensified CCD camera could be utilized.

What are the advantages of using a scanning slit confocal microscope such as is described and demonstrated in this paper? Slit scanning confocal microscopes have a much higher light throughput than confocal microscopes based on Nipkow disk. This has two consequences. First, the illumination incident on the patient's eye can be much less. This allows for a much longer duration of the use of the confocal microscope on the patient's eye without the severe patient discomfort and high light intensity that is necessary with the use of the confocal microscope based on the Nipkow disk. Second, it is possible to image the low reflecting layer of wing cells that are immediately adjacent to the basal epithelial cells in the normal human cornea. This layer of wing cells has been imaged, at video-rates, as single video frames without the need for any analog or digital image processing using the video-rate scanning slit confocal microscope. No other video-rate confocal microscope has been able to image these wing cells in the normal, *in vivo* human cornea. This new confocal instrument has unique advantages of other confocal systems. The bright, high-contrast confocal images of the wing cells and the basal epithelial cells demonstrate its unique optical characteristics [16]. The low reflectivity of the wing and basal cells in the normal human cornea present a low-contrast benchmark test specimen for various types of confocal microscopy. The high rejection of stray light and the narrow depth of field coupled with the high numerical aperture microscope objective (1.0 NA) results in the ability of the instrument to image these cell layers clearly in the live normal human cornea. The clear advantage of slit scanning confocal microscopes for ophthalmic diagnostics and basic eye research is best appreciated when the basal epithelium in the anterior cornea is imaged. The video-rate scanning slit confocal microscope provides high-contrast, high resolution images of both the wing cell and basal epithelial cells in the normal *in vivo* human eye.

In 1990, Masters suggested that an internal lens system would permit focusing at different depths within the cornea; however in order to keep light losses to a minimum, and therefore maximize the light sensitivity of the z-scanning instrument it was decided to use a z-scanning microscope objective [8]. This capability of a z-scanning confocal has been implemented into the clinical video-rate, slit-scanning confocal microscope. In addition to microscopic pictures with a lateral movement of the scanning slit, a confocal measurement of the tissue reflectivity can be performed by an automated z-scan through all layers of the cornea. For this procedure, the lateral scan is switched off. Reflectivity as recorded by a photomultiplier can be used to measure corneal haze, which may be of special interest in keratorefractive procedures.

## Confocal images of the structures of the human cornea

The human cornea consists of the following layers: superficial epithelial cells, wing cells, basal epithelial cells, Bowman's layer, stroma, Descemet's membrane, and endothelial cells, from the tear film on the anterior side of the cornea to the posterior side of the cornea adjacent to the aqueous humor [17]. The unique optical properties of the cornea are consistent with its morphology [16]. The normal cornea is avascular. At the anterior side is the corneal epithelium (about 50 microns thick at the central corneal region), under the epithelium is Bowman's layer. It is 10–16 microns thick and is acellular except for the nerves which perforate it. It separates the epithelium from the stroma. The basal lamina of the epithelium is located on Bowman's layer. The stromal region is about 450 microns thick and contains large nerves, stromal keratocytes, and orthogonal layers of collagen fibers. The architecture of human corneal nerves was studied with electron microscopy [18, 19]. The morphology of the stromal keratocytes has been described with electron microscopy [20]. Posterior to the stroma is Descemet's membrane, which is an acellular layer 15–20 microns in thickness. The limiting layer on the posterior side of the cornea is a single layer of corneal endothelial cells.

The standard histological sections, which give a sagittal view of the corneal and, in this respect, are quite similar to the clinical biomicroscopy with the slit lamp. As the corneal layers and pathological findings extend laterally, the clinician is used to move the slit lamp laterally as well, change the width and illumination angle of the slit and thus collect three-dimensional information about changes in tissue reflectivity and their location. In contrast, the confocal microscope collects optical sections parallel to the corneal surface; all layers of the cornea can be imaged by changing the z-position of the objective. Consequently, the confocal flying slit microscope is an instrument which (i) allows an immediate tangential view of the selected corneal layers under study and (ii) extends the diagnostic principle of biomicroscopy into the microscopic range.

An optical section of an *in vivo* human cornea shows the intermediate cells of the anterior epithelium (Figure 3). An oblique optical section of the human cornea *in vivo* showing basal epithelial cells and three nerves anterior to Bowman's layer (Figure 4). An interesting optical section of the cornea shows a bifurcating cornea nerve in the anterior stroma (Figure 5). An confocal image of the *in vivo* human cornea at the level of the midstroma shows several keratocyte nuclei (Figure 6). In the lower left portion of the image (arrowhead) is a cell with several proceses. This may be a inflamatory cell in the storma. The endothelial mosaic of the central cornea can be observed with the *in vivo* confocal microscope (Figure 7).

A more complete atlas of confocal microscopy of the normal cornea, in addition to typical images of corneal disease, corneal alterations induced by surgery, contact lens wear, and aging is available [11]. In this extensive review the normal cornea is shown in two sets of images; whole mount, fixed, stained light microscopy and *in vivo* scanning slit confocal micro-

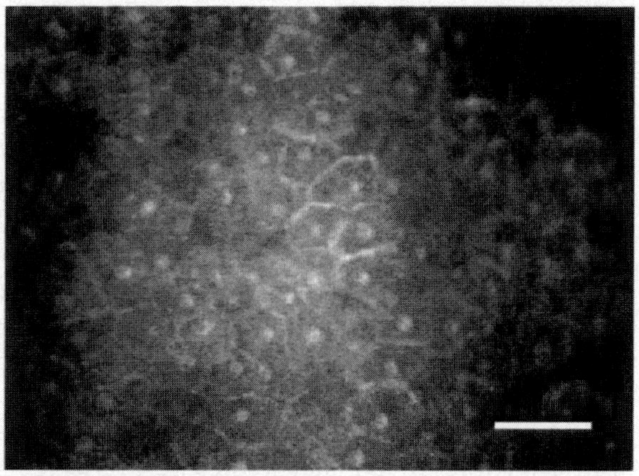



*Figure 3.* (Sv37) A confocal optical section of the human cornea *in vivo* showing the intermediate cells with their bright nuclei. The cell borders are highly reflective. The scale bar is 50 microns.

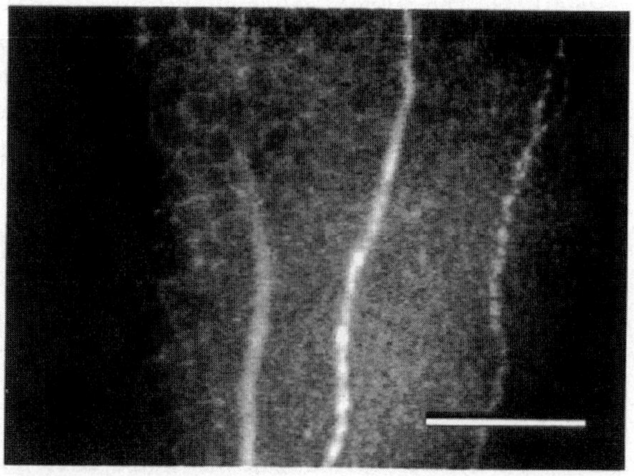



*Figure 4.* (BRM123) A slightly oblique confocal section of an *in vivo* human cornea showing the polygonal cell borders of the basal epithelial cells on the left region of the image. The three bright linear structures are nerves situated between Bowman's layer and the basal epithelial cells. The scale bar is 50 microns.

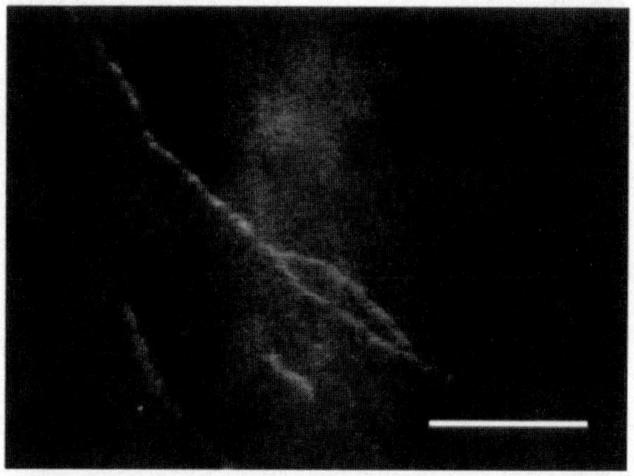



*Figure 5.* (BRM176) A confocal optical section of an in vivo human cornea showing a nerve in the anterior stroma. The nerve bifurcates and then the two sections recombine into a single nerve. The scale bar is 50 microns.

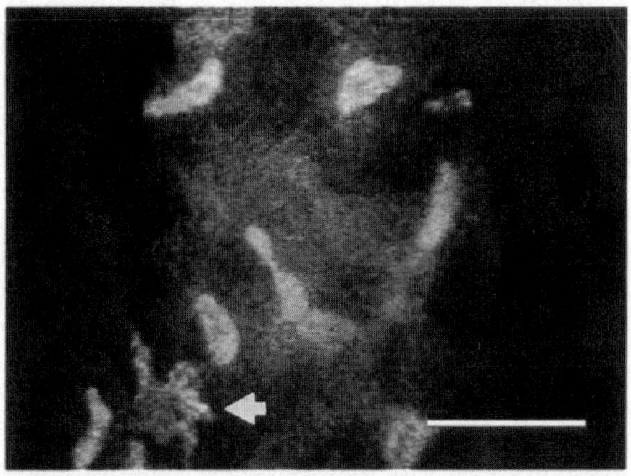



*Figure 6.* (BRM072) A confocal image of the in vivo human cornea in the mid periphery of the stroma. The highly reflective nuclei of several stromal keratocytes are shown with various shapes. In the lower left region (arrowhead) there is a cell with several processes. The scale bar is 50 microns.

scopy. This is another example of the importance of correlative microscopy for the definitive identification of ocular structures as observed in biomicroscopy.

The images obtained with the scanning-slit *in vivo* confocal microscope are dependent on at least four considerations: (1) the experience and skill of the observer, (2) the type of confocal microscope used, (3) the magnification and numerical aperture of the microscope objective, and (4) the type of image averaging and image processing and enhancement employed. Finally, it has been our experience that even with a newly installed commercial confocal microscope the image quality of a confocal microscope is highly dependent on the optical alignment of the components.

## Investigation of the cornea

### *Corneal alterations due to long term contact lens wear*

This investigation shows how the use of a real-time, scanning-slit confocal microscope with a high numer-

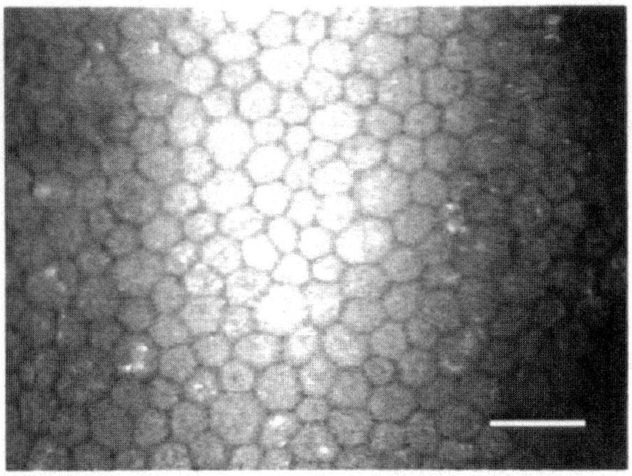

M10

*Figure 7.* (M10) A confocal image of the *in vivo* human cornea showing the endothelial mosaic of the central corneal The highly reflective dots are pigment particles. The scale bar is 50 microns.

ical aperture water immersion objective and single frame review led to the discovery of a new corneal degeneration [21]. The absence of frame averaging or other digital image processing was critical to the observation, since frame averaging would preclude the detection of the microdots that led to the discovery. A new type of chronic stromal aberrations has been observed in subjects with long-term contact lens wear. A real-time, scanning slit confocal microscope with a 50×/1.0 NA water immersion objective was used to observe the cornea. The corneal optical sections from the superficial epithelium to the endothelium were recorded in real time without any further image processing and were reviewed frame by frame. This study confirmed the presence of epithelial microcystic changes and alterations of endothelial cell morphology which was previously described by others. The new and important finding of this study was the presence of highly reflective panstromal microdot deposits in the corneal stroma. The dots were highly reflective and had a size of 0.3 to 0.6 microns with a round-to-polygonal shape. The density and the size of the microdot deposits scales with the duration of contact lens wear. In patients wearing soft contact lenses for longer than 6 years, the microdots were observed all cased investigated. The microdots may be lipofuscin or some other high reflective matter. This stromal microdot degeneration may be the early state of a significant corneal disease, which eventually may affect large numbers of patients after decades of contact lens wear. A quantitative analysis of the density distribution of the microdots has been reported [22].

*Corneal alterations following photorefractive keratectomy (PRK)*

A frequent and very legitimate question is what new clinical observations and discoveries have been directly linked to the use of the clinical confocal microscope. The work of Bshnke et al. provides an interesting illustration of how confocal microscopy led to the observation of persistent stromal changes after myopic photorefractive keratectomy (PRK) in zero haze corneas [23]. In photo- refractive keratectomy treated patients and contact lens wearers, the basal epithelial cells sporadically showed enhanced reflectivity. However, rods and needles were observed in all photorefractive keratectomy treated patients, irrespective of previous contact lens wear. In contact lens wearing controls, there were highly reflective granules which were scattered throughout the thickness of the stroma; however, rods and needles were never observed [21]. The authors conclude that after 8 to 43 months post photorefractive keratectomy there are abnormal reflective bodies that persist beyond the time that acute wound healing would be expected to be complete. The clinical significance of these findings in the context of visual acuity and long term status of the cornea is unknown.

A slit-scanning *in vivo* confocal microscope was used to assess human corneal morphological characteristics after photorefractive keratectomy [24]. Each layer of the cornea was studied. The minimum follow-up time was 12 months. Fine linear structures were observed in the anterior stroma and in the midstroma, and a thin hyperreflective scar was noted after one month post PRK. These structures were more marked at 4 months, but were still present up to 26 months. The extension of these structures to the midstroma indicated that permanent corneal changes caused by PRK affect deeper stromal layers than the immediate subepithelial region. Anterior stromal keratocyte density increased significantly 1 and 4 months after PRK. The midstromal and posterior keratocyte densities and endothelial cell densities were not affected. The significance of this investigation is that long-term alterations of the cornea in the midstroma could be observed over time with an *in vivo* confocal microscope.

*Alterations of the human cornea during examination with an applanating confocal microscope*

The use of confocal microscope for the examination of the human eye *in vivo* is not without alterations of the eye. The study by Auran et al. illustrates the flattening- induced effects of an applanating microscope objective. They reported corneal bands and ridges throughout the cornea following the contact with an applanating microscope objective [25].

In addition to the previously discussed mechanical flattening with the use of an applanating microscope objective there are several other sources of morphological and physiological alteration with the use of confocal microscopy. First the use of anesthetic drops which contain preservatives effects the cell junctions in the corneal epithelium. Second, the index matching gel used between the tip of the microscope objective and the tear film of the cornea may dehydrate the surface cell layers of the cornea. Third, if microscope objective is not sterilized between patient examinations, there is a risk of bacterial and viral transmission. Finally, repeated daily examinations may lead to a low grade cellular reaction to the combined insults of anesthetic drops and the index matching gel. We mention these possibilities as a matter of caution.

One way to mitigate these alterations of the cornea is to use a noncontact confocal microscope [26]. For example, a long working distance air microscope objective could be designed for *in vivo* observation of the cornea. The use of a noncontact microscope objective has many potential benefits for clinical observation of patients. There is not need for an index matching fluid; thus there is no physical contact with the ocular surface. There is no need for the use of anesthetic drops in the patients eye. There is minimal chance of bacterial and viral transmission from patient to patient. For these reasons we suggest and promote the development and use of noncontact confocal microscopy for the observation of the living eye.

### *Three-dimensional confocal microscopy*

Three-dimensional confocal microscopy provides a unique opportunity to visualize the *in vivo* cornea, ocular lens and optic nerve in three-dimensions [27–35]. It is the increases axial resolution of the confocal microscope that permits the sequential acquisition of thin optical sections. The stack of optical sections can be converted to a three-dimensional volume in a computer. It is then possible to observe the volume from any arbitrary angle and direction. It is the three-dimensional visualization of living cells and tissues that resulted in the ever increasing popularity of confocal microscopy.

One problem that occurs in the acquisition of a series of optical sections across the full thickness of the human cornea *in vivo* is the lack of image registration between sets of successive images. This is the result of ocular motions due to rotations and the more significant problem of the motion of the globe along the optical axis due to the cardiac pulse. One approach to this problem is to stabilize the globe with a low-vacuum suction cup system for stabilizing the eye in front of the microscope objective during the z-scan through the full thickness of the cornea [36]. The authors had developed software for correction of background illumination and small axial movements, and also calculated the keratocyte density.

## Acknowledgements

The authors thank Dr. Andreas A. Thaer for his collaboration in the development of the clinical confocal microscope, and Hans Kuenzli and his staff for their excellent work with the photographic prints.

## References

1. Masters BR. Noninvasive Diagnostic Techniques in Ophthalmology. Academic Press, New York, 1990.
2. Rosenblum WM, Benjamin WJ. Selected papers of special significance to optometry. SPIE Press, Bellingham, 1992.
3. Straub W, Kroll P, Küchle HJ. Augenärzliche Untersuchungsmethoden. Ferdinand Enke Verlag, Stuttgart, 1995.
4. Berliner ML. Biomicroscopy of the Eye. Hafner Publishing Company, New York, 1966.
5. Vogt, A. Lehrbuch und Atlas der Spaltlampenmikroskopie des lebenden Auges, Volume 1. Julius Springer, Berlin, 1930.
6. Vogt A. Textbook and Atlas of Slit Lamp Microscopy of the Living Eye. J. P. Wayenborgh, Bonn, 1981.
7. Martonyi CL, Bahn CF, Meyer RF. Clinical Slit Lamp Biomicroscopy and Photo Slit Lamp Biomicrography. Time One Ink, Ltd, Ann Arbor, 1985.
8. Wilson T: Confocal Microscopy. Academic Press, London, 1990.
9. Corle TR, Kino GS. Confocal Scanning Optical Microscopes and Related Imaging Systems. Academic Press, San Diego, 1996.
10. Webb RH. Confocal optical microscopy. Rep. Prog. Phys. 1996; 59: 427–471.
11. Böhnke, M, Masters, BR. Confocal microscopy of the cornea. Progress in Retinal and Eye Research 1999; 18: 553–628.
12. Masters BR. Selected Papers on Confocal Microscopy. The SPIE Press, Bellingham, 1996.
13. Masters BR, Thaer AA. Real-time scanning slit confocal microscopy of the *in vivo* human cornea. Appl. Optics. 1994; 33: 695–701.
14. Svishchev GM. Microscope for the study of transparent light-scattering objects in incident light. Optics and Spectroscopy 1969; 26: 171–172.
15. Svishchev GM. Image contrast in a microscope with synchronous object scanning by slit field diagrams. Optics and Spectroscopy 1971; 30: 188–191.

16. McCally RL, Farrell RA. Light scattering from cornea and corneal transparency, in BR Masters (ed.), Noninvasive Diagnostic Techniques in Ophthalmology. Springer-Verlag, New York, 1990; 1989–210.
17. Maurice, DM. The cornea and sclera, in H Davson (ed.), The Eye, 3rd edn. Academic Press, Orlando, 1984; 1–158.
18. Müller LJ, Pels L, Vrensen GFJM. Ultrastructural organization of human corneal nerves. Invest Ophthalmol Vis Sci. 1996; 37: 476–488.
19. Müller LJ, Vrensen GFJM, Pels L, Cardozo BN, Willekens B. Architecture of human corneal nerves. Invest Ophthalmol Vis Sci. 1997; 38: 985–994.
20. Müller LJ, Pels L, Vrensen GFJM. Novel aspects of the ultrastructural organization of human corneal keratocytes. Invest Ophthalmol Vis Sci. 1995; 36: 2557–2567.
21. Böhnke M, Masters BR. Long-term contact lens wear induced a corneal degeneration with microdot deposits in the corneal stroma. Ophthalmology 1997; 104: 1887–1896.
22. Cadez R, Frueh B, Böhnke M. Quantifizierung intrastromaler Mikroablagerungen bei Langzeitträgern von Kontaktlinsen. Klin Mbl Augenhlk 1998; 212: 257–258.
23. Böhnke M, Thaer A, Shipper I. Confocal microscopy reveals persisting stromal changes after myopic photo refractive keratectomy in zero haze cornea. Br J Ophthalmol 1998; 82: 1393-1400.
24. Frueh BE, Cadez R, Böhnke M. *In vivo* confocal microscopy after photorefractive keratectomy in humans. Arch Ophthalmol 1998; 116: 1425–1431.
25. Auran JD, Koester CJ, Rapaport R, Fkorakis GJ. Wide field scanning slit *in vivo* confocal microscopy of flattening induced corneal bands and ridges. Scanning 1994; 16: 182–186.
26. Massig JH, Preissler M, Wegener AR, Gaida G. Real-time confocal laser scan microscope for examination and diagnosis of the eye *in vivo*. Appl. Opt. 1994; 33: 690–694.
27. Masters BR, Böhnke M. Video-rate, scanning slit, Confocal microscopy of the living human cornea *in vivo*: Three-dimensional confocal microscopy of the eye, in PM Conn (ed.), Methods in Enzymology, Confocal Microscopy. Academic Press, New York, 1999; 307: 536–563.
28. Masters BR, Paddock SW. Three-dimensional reconstruction of the rabbit cornea by confocal scanning optical microscopy and volume rendering. Applied Optics 1990; 29: 3816–3822.
29. Masters BR, Farmer MA. Three-dimensional confocal microscopy and visualization of the *in situ* cornea, Computerized Medical Imaging and Graphics 1993; 17: 211–219.
30. Fitzke FW, Masters BR. Three-dimensional visualization of confocal sections of *in vivo* human fundus and optic nerve. Curr. Eye Res. 1993; 12: 1015–1018.
31. Masters BR, Sasaki K, Sakamoto Y, Kojima M, Emori Y, Senft S, Foster M. Three dimensional volume visualization of the *in vivo* human ocular lens showing localization of the cataract. Ophthalmic Res 1996; 28(2): 120–126.
32. Masters BR. Optical tomography of the *in vivo* human lens: Three-dimensional visualization of cataracts. J. Biomedical Optics 1997; 1(3): 289–295.
33. Masters BR, Senft SL. Transformation of a set of slices rotated on the common axis to a set of z-slices: Application to three-dimensional visualization of the *in vivo* human lens. Computerized Medical Imaging and Graphics 1997; 21(3): 145–151.
34. Masters BR. Three-dimensional confocal microscopy of the living *in situ* rabbit cornea. Optics Express 1998; 3(9): 351–355. URL: <http://www.opticsexpress.org>
35. Masters BR. Three-dimensional microscopic tomographic imaging of the cataract in a human lens *in vivo*. Optics Express 1998; 3(9): 332–338. URL: <http://www.opticsexpress.org>
36. Stave J, Slowik C, Somodi S, Hahnel C, Grummer G, Guthoff R. Keratinocyte density of the cornea *in vivo*. Automated measurement with a modified confocal microscopy MICROPHTHAL. Klin Monatsbl Augenheilkd 1998; 213: 38–44.

*International Ophthalmology* **23:** 207–208, 2001.
*J.R. Sampaolesi (ed.), Laser Scanning: Update 1*, 25–26.




# Today's clinical application of scanning laser technologies

William Eric Sponsel
*Director of Clinical Research, University of Texas Health Science Center, San Antonio, TX, USA*

## Introduction

Scanning laser technology has been embraced primarily by clinical researchers. Their work has been critical to the development and clinical adoption of scanning laser methodology. The ultimate goal of all this work is to improve patient care, which will require widespread deployment of scanning lasers in eye clinics of all kinds. Although progress has been impressive, the rate of real integration of scanning laser methods into the clinical setting needs to be increased.

There are various reasons clinical integration of scanning laser technology has not kept up with its otherwise encouraging rate of development in other areas. Below I will outline some specific goals, and grade our progress in each area, according to my own experience and bias.

## Grading our progress

Goal 1: To produce technologically astounding, robust, and beautiful clinical tools Grade: A.

This is an area of continuing strength. The manufacturers have tended to produce ingenious and technically reliable hardware. After-purchase service has been generally very good, and great people seem to be employed by each company to maintain an ongoing interface with clinicians and researchers. Continual technologic advances have characterized progress among a majority of those engaged in scanning laser marketing. Machines are becoming progressively more portable, and despite constant improvements, the price of instruments has continually fallen.

Goal 2: To gradually supplant dark-room based photographic services in major clinical centers. Grade: D-minus.

There seem to be three key problems here:

- Perception: The resolution of scanning laser micrographs, tomographs, and angiograms is judged by the average clinician on the basis of poor-quality bubble-jet printouts. This adversely affects
- Economics: Repeat clinical referrals are obtained by sending reports with beautiful pictures, and the funds to purchase a quality printer later may be elusive in the absence of a clear referral-based revenue stream (a vicious cycle).
- Tradition: Real eye doctors look at discs, fundi, and angiograms bending over an x-ray box. Abstract numeric and graphical displays on their own may impress some younger clinicians, but will not appeal to most traditionalists unfamiliar with their interpretation. This latter point is all the more apt when the subspecialist to whom a patient has been referred is also unclear about the relevance of a given numeric index. High quality images will tend to reassure our colleagues that we are at least employing data of high precision in our analyses, an that we must therefore be on the right track.

Let us consider each of these in greater detail:

*Perception:* Widespread SLO adoption will only occur after major training centers actually use scanning lasers for routine daily clinical tasks. Quality hardcopy visuals should support and supplement all SLO data analyses, even those intended for research purposes only. Real digital stereo pairs, either standard photographs or scanning laser emulations thereof, should always be printed for reference alongside interpretive or graphic SLO displays. These do not need to be large if the printer resolution is adequate. Stereo interpretation must be facilitated by presenting all such reference images as side-by-side offset pairs. Simultaneous or sequential stereo pairs are, after all, the longstanding clinical standard, and we cannot hope to make progress educating residents about the potential of the various new scanning laser analyses without reference to traditional stereo pairs. Once our trainees see these new capabilities for themselves they will realize the full clinical potential of these remarkable instruments, and will insist on having one when they open their own practice. It is therefore ultimately self-defeating for manufacturers to supply instruments without a high-resolution printer and slide-maker. I strongly believe this is the single factor which has most seriously retarded the adoption of scanning laser technology in ophthalmic and optometric practices worldwide.

*Economics:* Facing reality has not been the forte of clinicians in recent years. If we acknowledge the economics of clinical practice and employ appropriate reverse psychology [1] I believe SLO use will increase dramatically. The idea that any new methodology, regardless of its sophistication or added diagnostic utility, can be reimbursed at a level significantly higher than an established traditional method (over a sustained period) is untenable. This may have been the trend over the past 20 years, but those days are over. Scanning laser ophthalmoscopy, despite its multitude advantages, must pay for itself at or below current stereo fundus photography reimbursement levels and it can! Darkroom facilities are very expensive, and if proper costing of clinician time and administrative effort are considered, evaluating and collating slides is enormously time-consuming and inefficient. One really doesn1t need to expound the multitude of new capabilities the SLO can provide to make a strong case for institutional purchase. Companies and physicians alike would be wiser to highlight how these machines can accomplish traditional tasks more cost-effectively, and enjoy their many spinoffs without scaring administrators with the exotic potential of these amazing devices.

What else is required?

- Means to progressively integrate old and new photographic records alongside new digital records (slide scanner + software).
- Ability to input quality anterior segment imaging (i.e., co-development of SLO software with high-resolution stereo slitlamp and external photo capabilities).
- Confidence (i.e., reliability, longevity, service record, positive word of mouth).
- Goal 3: To provide clinically-relevant numeric/statistical analyses and graphic output. Grade: B-plus.

This is the realm of SLO development with the most widespread investment of effort. Creative approaches to data analysis and presentation have characterized this component of SLO) evolution. Beautiful graphics assist the practitioner in all phases of data interpretation. Nevertheless, there is still room for improvement:

- Our algorithms have overlooked a fundamental principle for assessing paired organs, one which all clinicians intuitively (subconsciously) apply when studying the fundi; bilateral symmetry.

*International Ophthalmology* **23:** 209–210, 2001.
*J.R. Sampaolesi (ed.), Laser Scanning: Update 1*, 27–28.




# Heidelberg Retina Tomograph measurements before and after non penetrating surgery

Y. Lachkar, Ph. Demailly* & O. Barrault
**Author for correspondence: Fondation Hôpital Saint Joseph, Institut Glaucome, Paris, France*

The purpose of this study is to evaluate quantitative changes in the optic disk parameters using the Heidelberg Retina Tomograph (HRT) after intraocular pressure reduction with non penetrating deep sclerectomy (NPDS).

## Material and methods

Sixty-six eyes of 44 patients with primary open angle glaucoma were analysed. The mean age was around 60 years in both groups with no statistical difference. The patients were divided into 2 groups, according to the percentage reduction of the IOP after surgery:

- Group 1: 36 eyes with IOP reduction > 35%.
- Group 2: 30 eyes with IOP reduction < 35%.

Parameters analysed with the HRT were:

- C/D area ratio.
- RNFL.
- Cup shape.

We chose these 3 parameters because they seemed to be the best to use for the follow-up; in particular, mean cup depth seemed difficult to analyse because of the changes of the reference plane.

HRT was performed before 1–3 months and 3 months after surgery.

Before surgery, there are the values of these parameters:

- in both groups
- in groups 1 and 2.

There was no difference statistically significant between the 2 groups in spite of worse values in the group 2.

## Results

In group 1, there was no difference statistically significant before and after surgery. Nevertheless, there was a mild improvement of C/D area ratio after surgery.

The results of group 2 showed no difference in all parameters.

Concurrently, the visual field parameters at octopus programme G2 did not show any significant difference before and after surgery in both groups, in all parameters.

## Discussion

Changes in the C/D ratio after control of IOP are well known in congenital glaucoma. In the same way, lowering of IOP can result, in certain cases, an improvement of the visual field.

After trabeculectomy, several studies suggested that a lowering of IOP could result in an improvement of Optic Nerve Head. Armaly in 1974, at international glaucoma symposium in Albi [1], suggested, with stereo photographies, that an improvement of ONH could be observed. Spaeth [5] in 1976 also suggested this improvement of the ONH. With new device, Paranhos and Caprioli (Arvo 1999) [3–4] and Hitchings et al. [2] have demonstrated an improvement of HRT parameters: C/D area ratio, cup volume, mean cup depth, before and after trabeculectomy. Kitazawa et al. [7] in 1999, with scanning laser polarimetry have also showed an improvement of the thickness of NFL. Coleman [6], in a very recent study, has also analysed changes in optic disc after trabeculectomy. He found differences 2 weeks after surgery, but not at 8 months. At this date, we found similar results. So improvement of optic disc topography 2 weeks measurements a few days after the operation.

## Conclusion

Improvement in optic nerve head appearance seems to be related to the IOP reduction whatever the filtration technique used.

An adequate IOP reduction obtained by NPDS with fewer complications could stop the visual field defects progression and improve, in certain cases, certain parameters of ONH.

Is it an argument to propose an earlier surgery?

## References

1. Armaly, M. The optic disc in open angle glaucoma. International Glaucoma Symposium, Albi, 1974: 153–170. Edited by Diffusion Générale de Librairie, Marseille, France.
2. Kotecha, A, Siriwardena, D, Garway-Heath, DF, Hitchings, R, Khaw, P, Fitzke, FW. Segmental optic disc and field changes following trabeculectomy. Arvo 1999 (5173).
3. Lima, M, Paranhos Jr, A, Salim, S, Osorio, P, Caprioli, J, Shields, MB. Heidelberg Retina Tomograph measurements and intraocular pressure before and after trabeculectomy – regression analysis. Arvo 1999 (395-B 355).
4. Lima, M, Paranhos Jr, A, Salim, S, Osorio, P, Caprioli, J, Shields, MB. Quantitative analysis of disc measurements by Heidelberg Retina Tomograph before and after trabeculectomy. Arvo 1999 (3474-B 332).
5. Spaeth, G, Fernandes, E, Hitchings, R. The pathogenesis transient of permanent improvement in the appearance of the optic disc following glaucoma surgery, in EI Greve (ed.), Glaucoma Symposium Diagnosis and Therapy, 1979. Dr W. Junk, The Hague, 1980: 111–126 (Doc Ophthalmol Proc Ser, 22).
6. Topousiz, F, Peng, F, Kotas-Neumann, R, Garcia, R, Sanguinet, J, Fei Yu, Coleman, A. Longitudinal changes in optic disc topography of adult patients after trabeculectomy. Ophthalmology 1999; 106: 1147–1151.
7. Yamada, N, Tomita, G, Kitazawa, Y. Changes in the nerve fiber layer thickness after trabeculectomy. Arvo 1999 (396-B 356).

*International Ophthalmology* **23:** 211–217, 2001.
*J.R. Sampaolesi (ed.), Laser Scanning: Update 1*, 29–35.



# Variability of topographic measurements after trabeculectomy in primary angle closure glaucoma with the laser tomographic scanner

Walter Meerhoff & Guillermo Meerhoff
*Canelones 1228, Montevideo, Uruguay*

## Aim

To demonstrate optic disc change after trabeculectomy in a case of acute angle closure glaucoma.

## Methods

A 50 year old patient who consulted 20 days after the onset of an acute glaucoma was studied. With an intraocular pressure of 47 mmhg and without a cornea edema, a scanning laser tomograph prior to trabeculectomy was carried out. Thirteen days later, with an intraocular pressure of 10 mmhg, a tomographic follow up was obtained.

## Results

Significant changes in all the stereometric parameters were found. Reversibility of the movement of the blood vessels with respect to the optic disc is remarkable.

Two situations stand out: (1) the particularity of a patient with an intraocular pressure above 40 mmhg and with 20 days of evolution who was able to be studied with a tomography; (2) the reversibility of the changes of the optic disc due to the mechanical effect of the intraocular pressure.

## Conclusion

The marked transformation that the intraocular pressure effects on an optic disk without previous glaucomatous damage is evident.

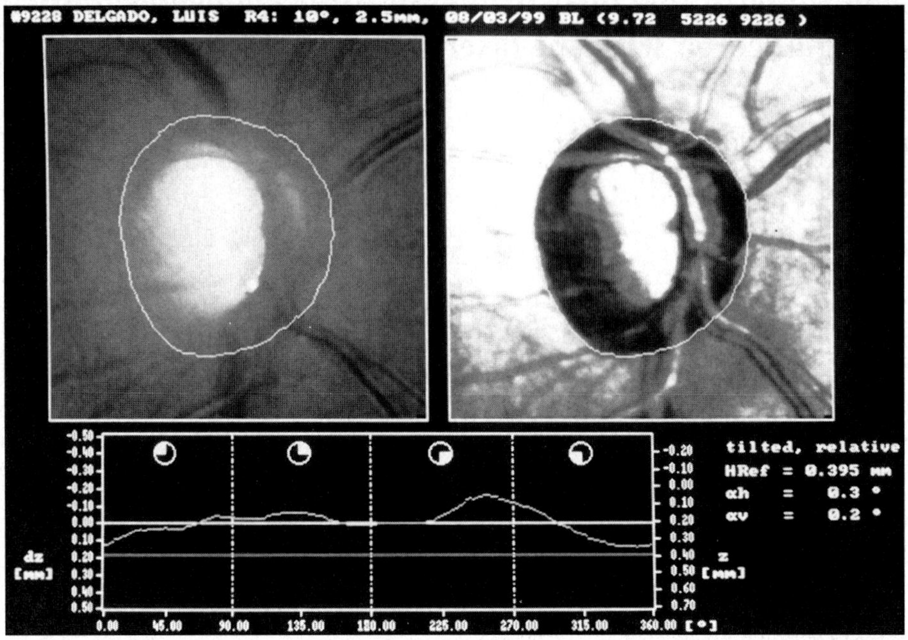


*Figure 1.* Preop.

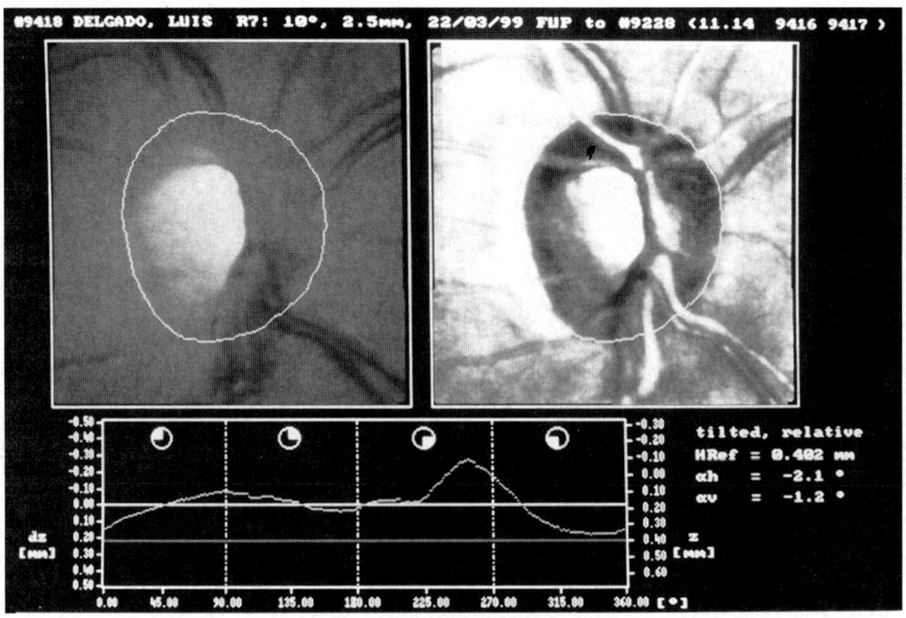


*Figure 2.* Postop.

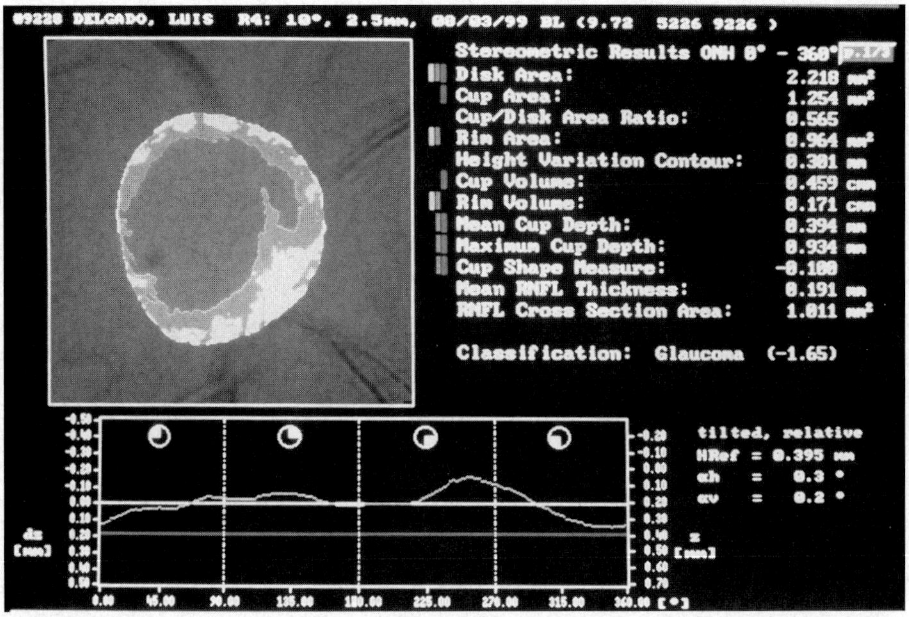


*Figure 3.* Preop.

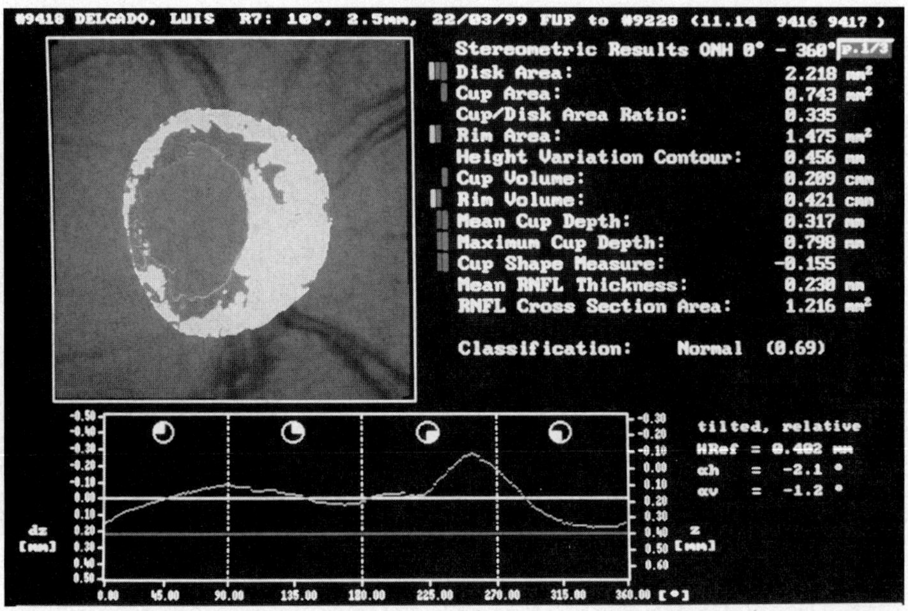


*Figure 4.* Postop.

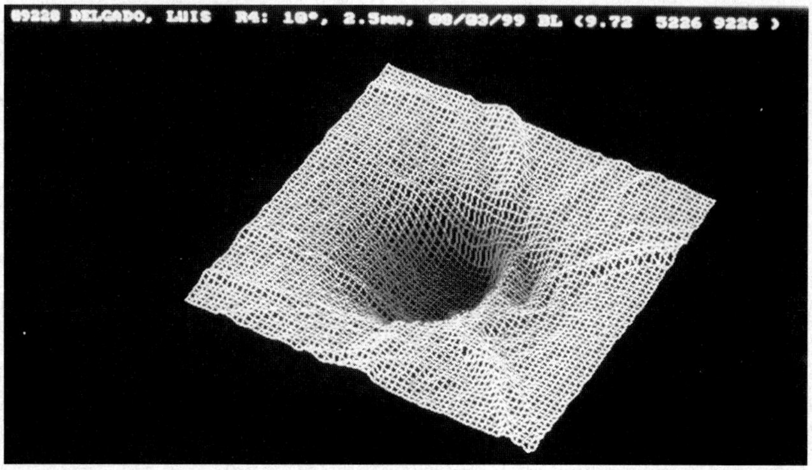


*Figure 5.* Preop.

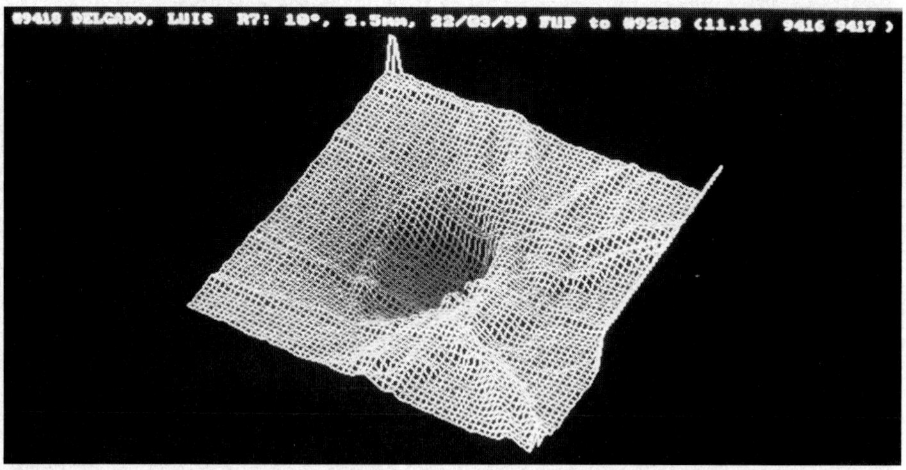


*Figure 6.* Postop.

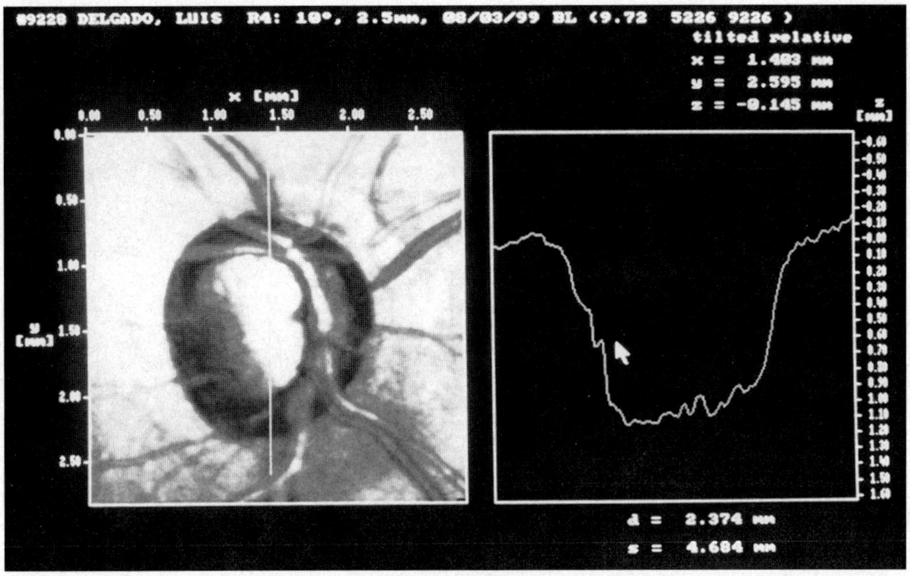


*Figure 7.* Preop.

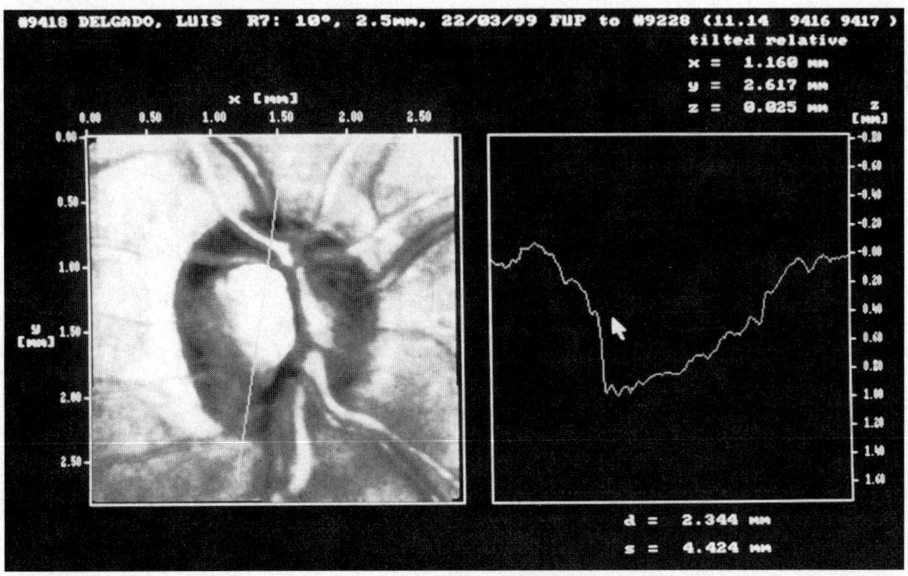


*Figure 8.* Postop.

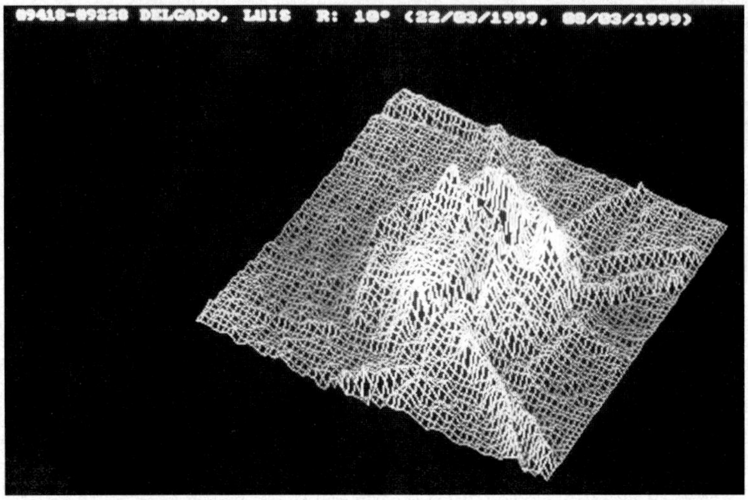


*Figure 9.*

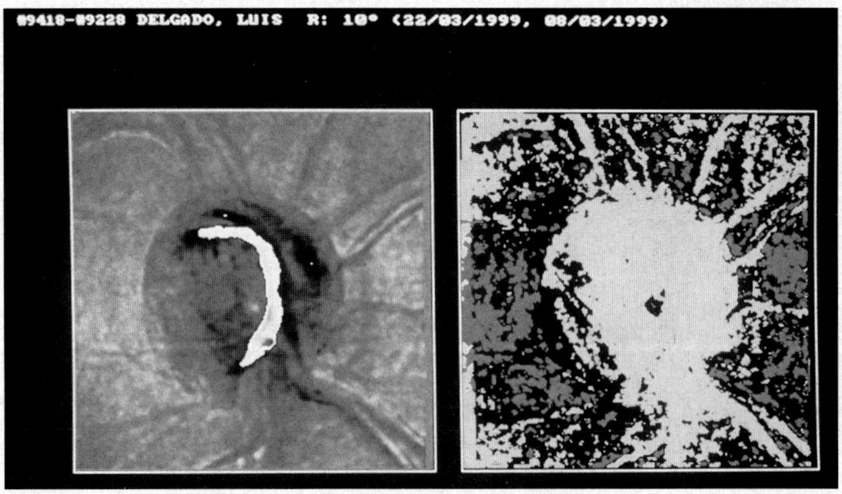


*Figure 10.*

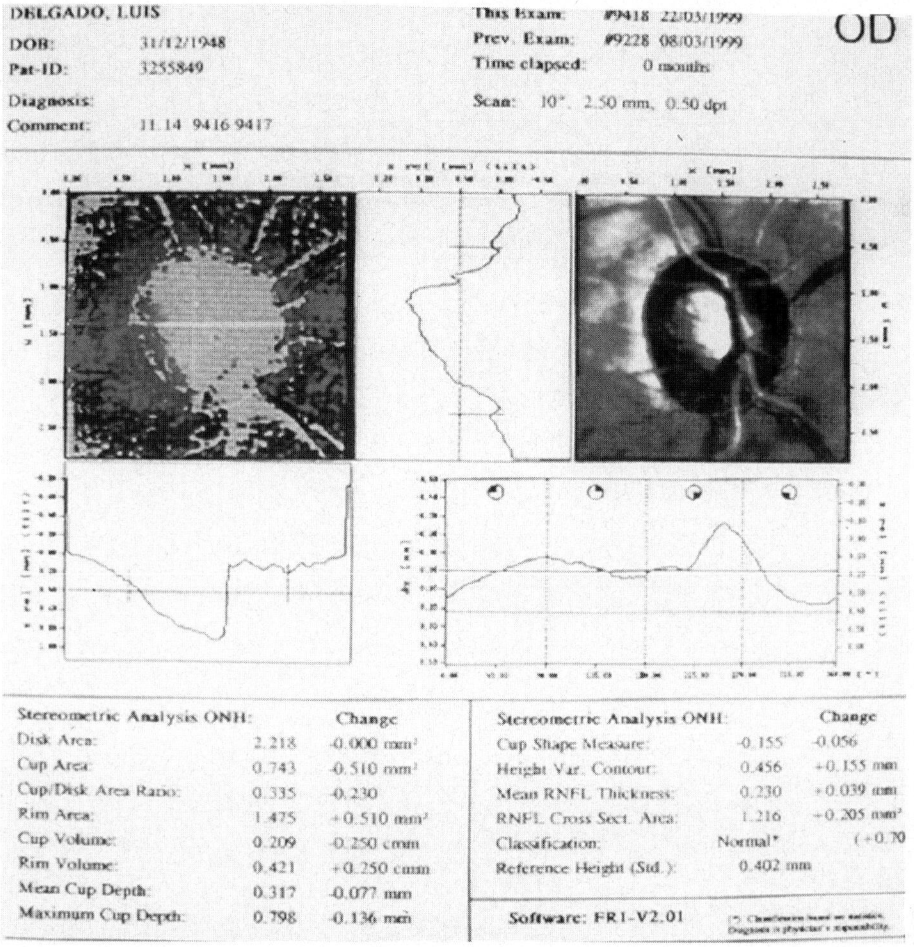


*Figure 11.*

| Stereometric Analysis ONH: | | Change | Stereometric Analysis ONH: | | Change |
|---|---|---|---|---|---|
| Disk Area: | 2.218 | -0.000 mm² | Cup Shape Measure: | -0.155 | -0.056 |
| Cup Area: | 0.743 | -0.510 mm² | Height Var. Contour: | 0.456 | +0.155 mm |
| Cup/Disk Area Ratio: | 0.335 | -0.230 | Mean RNFL Thickness: | 0.230 | +0.039 mm |
| Rim Area: | 1.475 | +0.510 mm² | RNFL Cross Sect. Area: | 1.216 | +0.205 mm² |
| Cup Volume: | 0.209 | -0.250 cmm | Classification: | Normal* | (+0.70) |
| Rim Volume: | 0.421 | +0.250 cmm | Reference Height (Std.): | 0.402 mm | |
| Mean Cup Depth: | 0.317 | -0.077 mm | | | |
| Maximum Cup Depth: | 0.798 | -0.136 mm | **Software: FR1-V2.01** | (*) Classification based on satistics. Diagnosis is physician's responsibility. | |

*Figure 12.*

International Ophthalmology **23:** 219–227, 2001.
J.R. Sampaolesi (ed.), Laser Scanning: Update 1, 37–45.




# Reliability in the use of the Heidelberg Retina Tomograph

Guillermo Meerhoff & Walter Meerhoff
*Canelones 1228, Montevideo, Uruguay*


## Aim

How to achieve the minimal variation in the stereometric values among different mean topographies in normal patients.

## Methods

The variation in a day and in a variable stretch of time up to 29 months was studied in 36 normal subjects.

The mean topographies used were obtained with the least standard deviation possible.

Among the mean topographies compared, those which had the least height variation in the relative reference plane with respect to the mean height of the peripapillary retinal surface were chosen.

## Results

As the ability to detect a change in the topography of the optic nerve heads depends to a large extent on the test-retest variability of the measurements, we found that using these parameters the reliability improved substantially. The variation found using these parameters in a day or in the follow up was minimal.

## Conclusion

Using this procedure a significant increase in the reliability of the method is achieved.


The Heidelberg Retina Tomograph (HRT) is a confocal laser scanning microscope which is used to acquire and analyse three-dimensional images of the posterior segment of the eye. The instrument enables a quantitative description of the fundus topography and time related changes to it. The most important routine clinical application of the Heidelberg Retina Tomograph is to describe the morphology of the papilla and to follow-up glaucomatous changes [2].

The objective of this study is to achieve the best reproducibility of a three dimensional assessment of the optic nerve head with the HRT in normal patients.

## Materials and methods

The HRT uses a diode laser beam with a wavelength of 670 nm. A three-dimensional image is generated as a series of 32 consecutive and equidistant two-dimensional optical section images, with each of the 32 two-dimensional images of 256–256 picture elements. From this series, the integrated computer calculates the height position of each pixel, which covers a rectangle of 10-10 of the area of interest. The width of each pixel is approximately 11 $\mu$m.

## Parameters

The reproducibility of the following parameters using the software version 2.01 was calculated:

- cup area (mm$^2$): area below the reference plane,
- rim area (mm$^2$): area above the reference plane,
- cup volume (mm$^3$): volume below the reference plane,
- rim volume (mm$^3$): volume above the reference plane,
- mean cup depth (mm): mean depth inside the contour (relative to the curved surface).

The mean standard deviation of three and two series for the height position of each pixel was calculated.

## Patients

Thirty-six normal eyes were studied. No history of IOP elevation (above 21 mmHg), no family glaucomatous history. Normal visual field. All subjects were phakic, had ocular clear media, with a pupil size that average 3mm, and with a refractive error below 1 diopter. Optic disc appearance was not a restriction criterion for entry into the study.

To analyse the reproducibility of optic disc measurements the authors made eight recordings of each eye for the population studied in a day: 26 eyes; and three to four recordings in the follow up examination for the evolution population: 10 eyes. So the time interval between the examinations was very variable. 26 eyes were scanned 8 times a day with time intervals ranging from minutes to hours, and other 10 eyes, the evolution population had a follow up of up to 29 months.

A good quality mean topographical image was defined as one in which the optic disc morphology was clearly visible, with an overall standard deviation (SD) of less than 16 $\mu$m. The aim was to obtain mean topographies with a SD below 10 $\mu$m.

The height of the reference plane is always indicated relative to the mean of the peripapillary retinal surface. The latter is specified as the mean height of a so-called reference ring. Among the mean topographies compared, those which had the least relative height variation in the reference plane were chosen.

Thus three mean topographies of each eye studied in a day were obtained . The first from three series (M1), the second (M2) from three different series, and the third one (M3) with only two series different from the ones used in the first mean topography. The disc margin was then defined interactively by one observer in the first mean topography, and then transported automatically by the computer of the tomographic scanner onto the other two mean topographies.

For the evolution population the mean topographies were obtained with only two series. The disc margin of the base line mean topography was exported to the follow up series. The two series that presented the least variation in the relative height of the reference plane were selected to obtain the follow up mean topography (M2).

Mean, standard deviation, and coefficient of variation of the various topographic parameters were computed for each disc in the study using the mean topographies selected. The coefficient of variation is

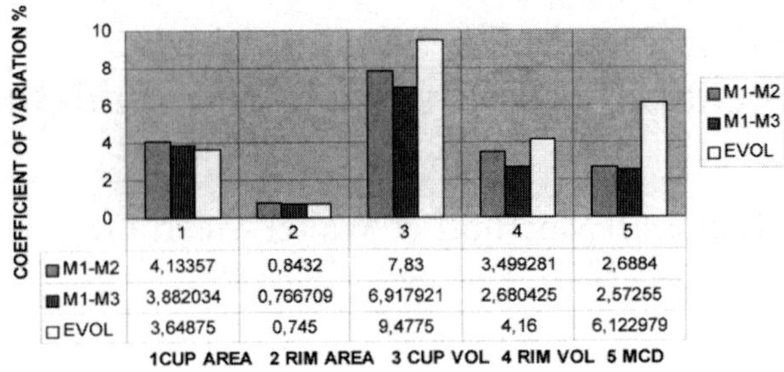


*Figure 1.* Coefficient of variation %.

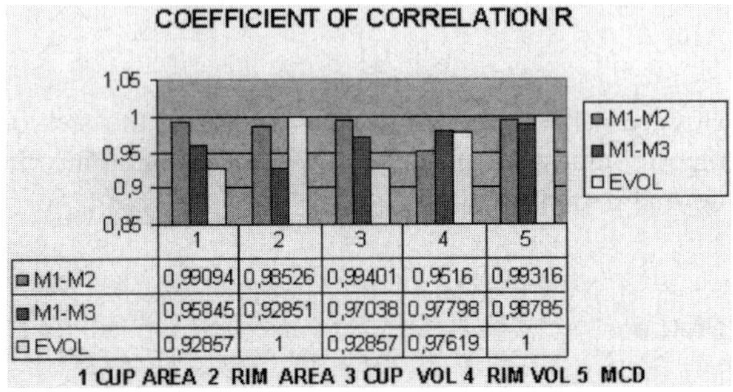


*Figure 2.* The Spearman range correlation coefficient.

given as a standard deviation divided by the mean and reported in percent.

## Statistics and results

Through the D'Agostino test it was observed that the population does not have a normal distribution. Therefore non parametric tests were used.

The mean ± standard deviation for age was 32.17 ± 13.54 years.

The mean standard deviation for the height position of each picture element for M1, M2, M3, was 12.98 ± 2.44, 11.53 ± 1.57, 8.6 ± 1.97, and for the evolution population it was M1 9.57 ± 3.1, M2 10.3 ± 3.16 respectively.

As a descriptive measurement the coefficient of variation (for the cup volume, rim volume, rim area, cup area, and mean cup depth) is given as standard deviation divided by the mean and reported in percent (Figure 1).

Studying separately the samples, the descriptive measurements of average, standard deviation, coefficient of variation, range and median were obtained (Table 1). With the Wilcoxon signed addition range the existence of significant differences among the medians of the samples (between M1–M2 and M1–M3)

*Table 1.* Reproducibility of topographic measurements

| | Cup area | | | Rim area | | |
|---|---|---|---|---|---|---|
| | M1 | M2 | M3 | M1 | M2 | M3 |
| Average | 0.42863158 | 0.43757895 | 0.45152941 | 1.68784211 | 1.67910526 | 0.59411008 |
| SD | 0.31920547 | 0.32632283 | 0.34013437 | 0.20954614 | 0.21670817 | 0.22328237 |
| CF | 0.74470825 | 0.74574619 | 0.75329394 | 0.17466527 | 0.1290617 | 0.13138348 |
| Range | 0.099/1.298 | 0.105/1.317 | 0.089/1.318 | 1.284/2.126 | 1.266/2.141 | 1.271/2.135 |
| Median | 0.272 | 0.283 | 0.285 | 1.689 | 1.673 | 1.684 |
| | Cup Vol. | | | Rim Vol. | | |
| | M1 | M2 | M3 | M1 | M2 | M3 |
| Average | 0.11594737 | 0.12221053 | 0.127 | 0.38394737 | 0.38789474 | 0.38494118 |
| SD | 0.15126962 | 0.16017996 | 0.16554984 | 0.06322049 | 0.0707051 | 0.07652979 |
| CF | 1.30464037 | 1.31068874 | 1.30354206 | 0.16465926 | 0.18227908 | 0.19880905 |
| Range | 0.003/0.604 | 0.004/0.645 | 0.003/0.639 | 0.25/0.514 | 0.236/0.501 | 0.246/0.526 |
| Median | 0.042 | 0.044 | 0.044 | 0.38 | 0.379 | 0.376 |
| | MCD | | | | | |
| | M1 | M2 | M3 | | | |
| Average | 0.18347368 | 0.187 | 0.18647059 | | | |
| SD | 0.10271501 | 0.10392198 | 0.21065081 | | | |
| CF | 0.55983514 | 0.55573251 | 0.58647086 | | | |
| Range | 0.091/0.504 | 0.088/0.505 | 0.09/0.505 | | | |
| Median | 0.145 | 0.143 | 0.141 | | | |

SD = standard deviation.
CV = coefficient of variation.

*Table 2.* The Spearman range correlation coefficients

| | M1–M2 | P | M2–M3 | P |
|---|---|---|---|---|
| Cup area | 29.46507944 | 4.93286E-16 | 11.21659844 | 2.8063E-09 |
| Rim area | 19.77468453 | 3.60117E-13 | 8.297053306 | 2.214E-07 |
| Cup Vol. | 98.215797 | 7.34358E-25 | 26.52332995 | 2.8442E-15 |
| Rim Vol. | 11.53377388 | 1.84044E-09 | 12.65862595 | 4.4244E-10 |
| MCD | 34.61843786 | 3.32399E-17 | 41.79390362 | 1.4007E-18 |

were studied. There were no statistically significant differences ($p < 0.01$).

The Spearman range correlation coefficients were calculated for the study of the way the HRT qualifies M1–M2, M1–M3, and for the evolution population, concluding that the HRT significantly qualifies as equal the samples M1–M2, M1–M3 and the evolution population ($p < 0.01$) (Table 2). The evolution population has a non significant disparity. According to the Spearman range correlation coefficient we obtained values closer to 1 with M1–M2 than with M1–M3 but without significant differences between them (Figure 2).

## Discussion

A mean topographic image was obtained using three or two different series. The standard deviation of the mean height (average of the three or two acquired series) is given for each pixel. This represents the test-retest variability of the three or two images of each point of the whole image (not only of the optic disc).

Lusky and coworkers [4] obtained three to five series to study the reproducibility. They found that the mean standard deviation was 30.1 $\mu$m in healthy eyes. Kruse and associates 3 shown a mean coefficient of variation for the optic cup volume of 9.5%. Weinreb et al. [7] used five normal eyes. For each one five images over five visits were recorded; the mean standard

**Heidelberg Retina Tomograph**
**Initial Exam Report**

HEIDELBERG ENGINEERING

**MEERHOFF, ANDRES** — **Exam:** #12995 27/09/1999 — **OD**

**DOB:** 07/08/1973
**Pat-ID:** 3323157

**Diagnosis:** — **Scan:** 10°, 2.00 mm, -1.25 dpt

**Comment:** 11.16 2419 12880 12881

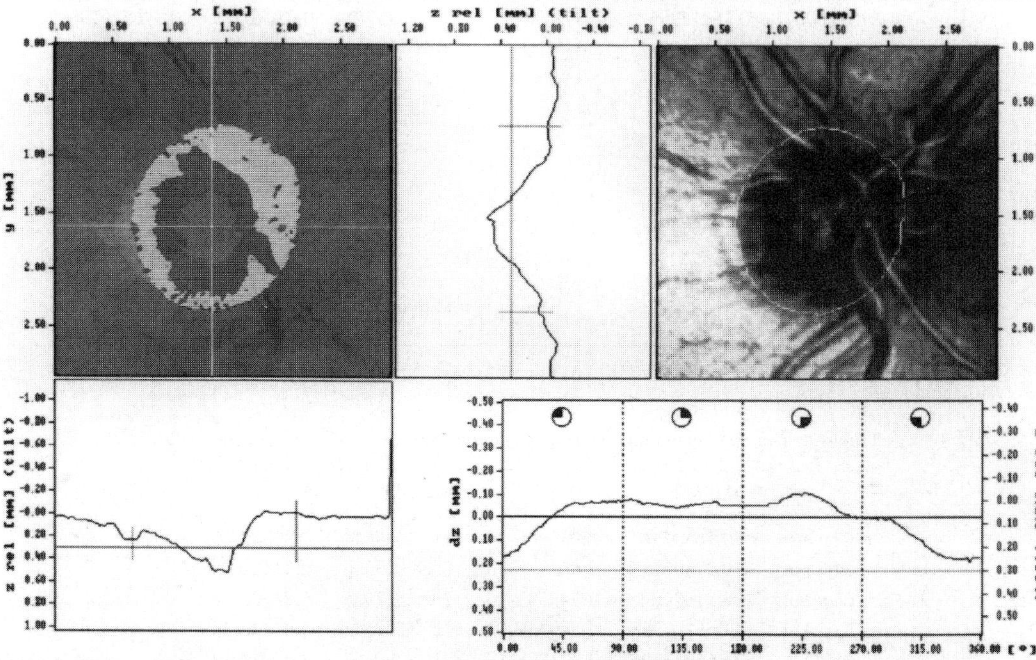


| **Stereometric Analysis ONH:** | |
|---|---|
| Disk Area: | 1.899 mm² |
| Cup Area: | 0.246 mm² |
| Cup/Disk Area Ratio: | 0.129 |
| Rim Area: | 1.653 mm² |
| Cup Volume: | 0.026 cmm |
| Rim Volume: | 0.367 cmm |
| Mean Cup Depth: | 0.144 mm |
| Maximum Cup Depth: | 0.450 mm |

| **Stereometric Analysis ONH:** | | |
|---|---|---|
| Cup Shape Measure: | | -0.229 |
| Height Variation Contour: | | 0.298 mm |
| Mean RNFL Thickness: | | 0.233 mm |
| RNFL Cross Section Area: | | 1.141 mm² |
| Classification: | Normal* | (+1.59) |
| Reference Height (Std.): | | 0.302 mm |

**Software: IR1-V2.01**

(*) Classification based on statistics. Diagnosis is physician's responsibility.

**Signature/Date:**

CETAO MONTEVIDEO URUGUAY

*Figure 3A.*

**Heidelberg Retina Tomograph**
**Follow-Up Exam Report**

HEIDELBERG ENGINEERING

**MEERHOFF, ANDRES**
**DOB:** 07/08/1973
**Pat-ID:** 3323157
**Diagnosis:**
**Comment:** 9.60 12882 12883 12884

**This Exam:** #12996 27/09/1999
**Prev. Exam:** #12995 27/09/1999
**Time elapsed:** 0 months
**Scan:** 10°, 2.00 mm, -1.25 dpt

OD

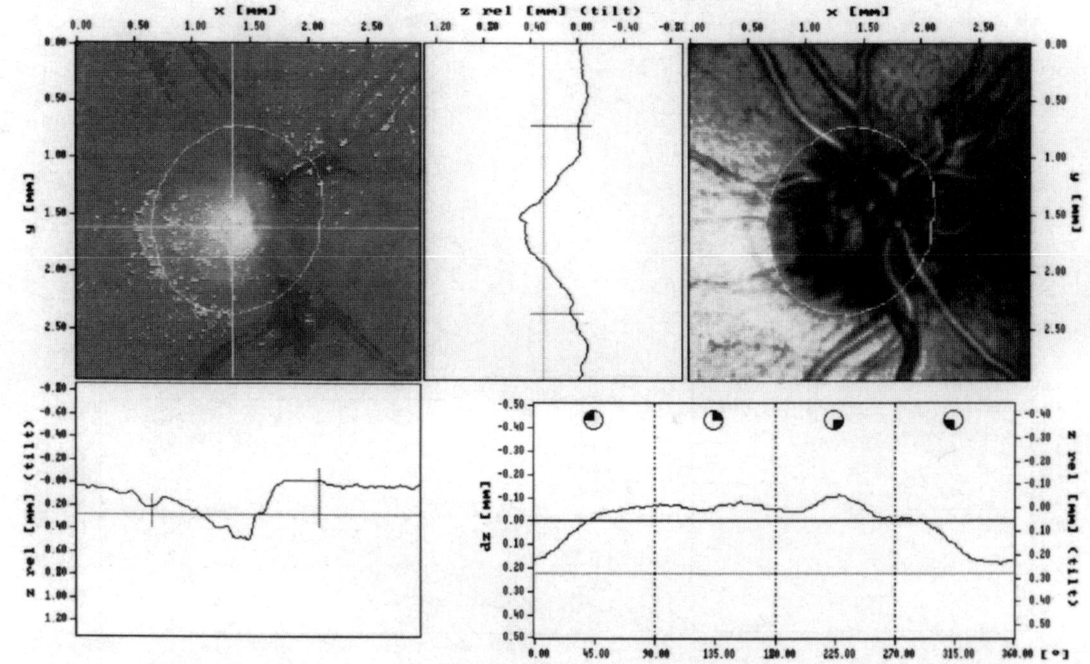

| **Stereometric Analysis ONH:** | | **Change** |
|---|---|---|
| Disk Area: | 1.899 | +0.000 mm² |
| Cup Area: | 0.250 | +0.004 mm² |
| Cup/Disk Area Ratio: | 0.132 | +0.002 |
| Rim Area: | 1.649 | -0.004 mm² |
| Cup Volume: | 0.026 | +0.001 cmm |
| Rim Volume: | 0.358 | -0.010 cmm |
| Mean Cup Depth: | 0.143 | -0.002 mm |
| Maximum Cup Depth: | 0.450 | -0.001 mm |

| **Stereometric Analysis ONH:** | | **Change** |
|---|---|---|
| Cup Shape Measure: | -0.234 | -0.005 |
| Height Var. Contour: | 0.301 | +0.003 mm |
| Mean RNFL Thickness: | 0.228 | -0.005 mm |
| RNFL Cross Sect. Area: | 1.116 | -0.024 mm² |
| Classification: | Normal* | (+1.52) |
| Reference Height (Std.): | 0.281 mm | |

**Software: FR1-V2.01**

(*) Classification based on statistics. Diagnosis is physician's responsibility.

**Signature/Date:**

CETAO MONTEVIDEO URUGUAY

*Figure 3B.*

## Heidelberg Retina Tomograph
## Follow-Up Exam Report

HEIDELBERG ENGINEERING

**MEERHOFF, ANDRES**

**DOB:** 07/08/1973
**Pat-ID:** 3323157
**Diagnosis:**
**Comment:** 9.91 12883 12886 12887

**This Exam:** #12998 27/09/1999
**Prev. Exam:** #12995 27/09/1999
**Time elapsed:** 0 months
**Scan:** 10°, 2.00 mm, -1.25 dpt

OD

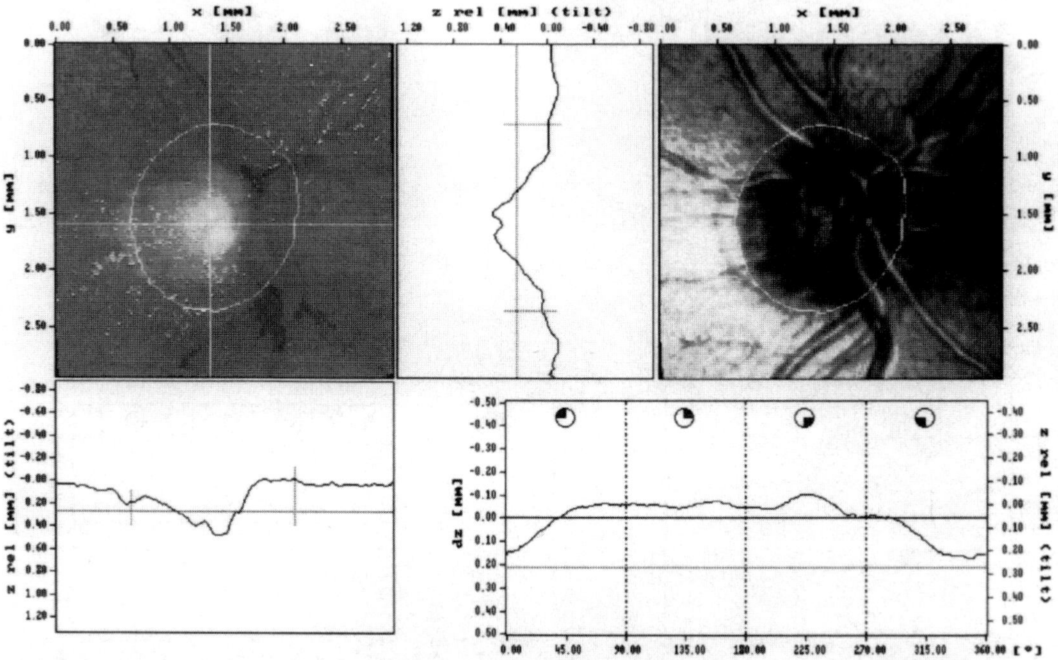


| Stereometric Analysis ONH: | | Change |
|---|---|---|
| Disk Area: | 1.899 | +0.000 mm² |
| Cup Area: | 0.275 | +0.029 mm² |
| Cup/Disk Area Ratio: | 0.145 | +0.015 |
| Rim Area: | 1.624 | -0.029 mm² |
| Cup Volume: | 0.029 | +0.003 cmm |
| Rim Volume: | 0.332 | -0.035 cmm |
| Mean Cup Depth: | 0.138 | -0.006 mm |
| Maximum Cup Depth: | 0.444 | -0.006 mm |

| Stereometric Analysis ONH: | | Change |
|---|---|---|
| Cup Shape Measure: | -0.245 | -0.017 |
| Height Var. Contour: | 0.279 | -0.019 mm |
| Mean RNFL Thickness: | 0.214 | -0.019 mm |
| RNFL Cross Sect. Area: | 1.048 | -0.092 mm² |
| Classification: | Normal* | (+1.56) |
| Reference Height (Std.): | 0.270 mm | |

**Software: FR1-V2.01**

(*) Classification based on statistics. Diagnosis is physician's responsibility.

**Signature/Date:** ______________________

CETAO MONTEVIDEO URUGUAY

*Figure 3C.*

# Heidelberg Retina Tomograph
# Follow-Up Exam Report

HEIDELBERG ENGINEERING

**MEERHOFF, ANDRES**

**DOB:** 07/08/1973
**Pat-ID:** 3323157
**Diagnosis:**
**Comment:** 7.04 12882 12884

**This Exam:** #13001 27/09/1999
**Prev. Exam:** #12995 27/09/1999
**Time elapsed:** 0 months
**Scan:** 10°, 2.00 mm, -1.25 dpt

OD

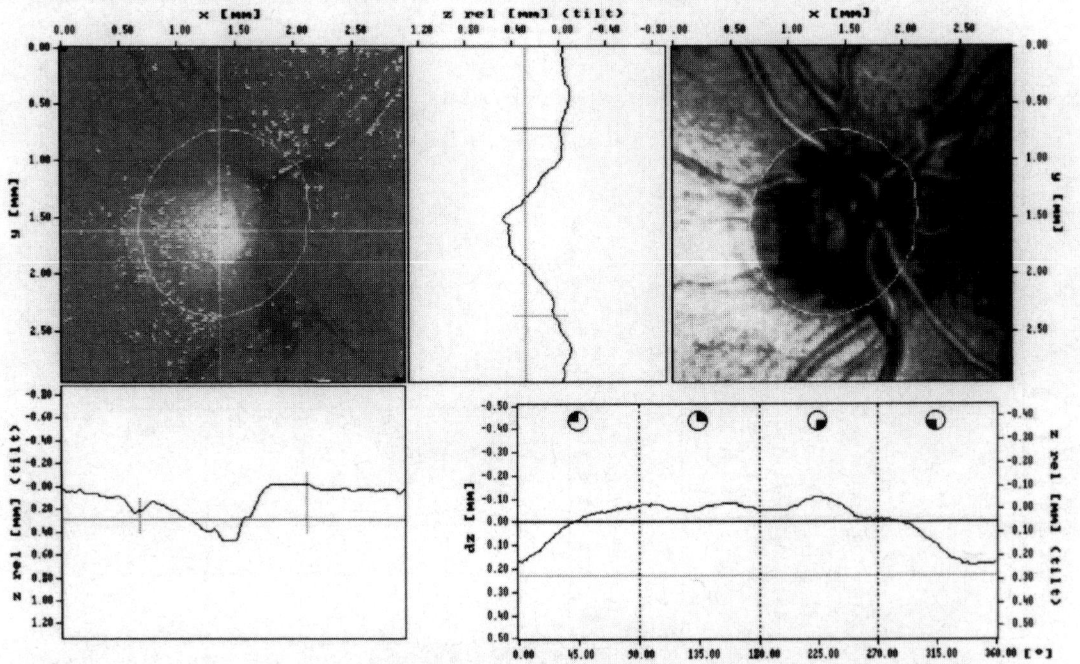


| Stereometric Analysis ONH: | | Change |
|---|---|---|
| Disk Area: | 1.899 | -0.000 mm² |
| Cup Area: | 0.236 | -0.010 mm² |
| Cup/Disk Area Ratio: | 0.124 | -0.005 |
| Rim Area: | 1.663 | +0.010 mm² |
| Cup Volume: | 0.023 | -0.002 cmm |
| Rim Volume: | 0.364 | -0.003 cmm |
| Mean Cup Depth: | 0.141 | -0.004 mm |
| Maximum Cup Depth: | 0.434 | -0.016 mm |

| Stereometric Analysis ONH: | | Change |
|---|---|---|
| Cup Shape Measure: | -0.222 | +0.006 |
| Height Var. Contour: | 0.290 | -0.008 mm |
| Mean RNFL Thickness: | 0.232 | -0.001 mm |
| RNFL Cross Sect. Area: | 1.134 | -0.006 mm² |
| Classification: | Normal* | (+1.53) |
| Reference Height (Std.): | 0.284 mm | |

**Software: FR1-V2.01**

(*) Classification based on statistics. Diagnosis is physician's responsibility.

**Signature/Date:** ______________________

CETAO MONTEVIDEO URUGUAY

*Figure 3D.*

## Heidelberg Retina Tomograph
## Follow-Up Exam Report

HEIDELBERG ENGINEERING

**MEERHOFF, ANDRES**

**DOB:** 07/08/1973

**Pat-ID:** 3323157

**Diagnosis:**

**Comment:** 11.11 12885 12886 12887

**This Exam:** #12997 27/09/1999

**Prev. Exam:** #12995 27/09/1999

**Time elapsed:** 0 months

**Scan:** 10°, 2.00 mm, -1.25 dpt

OD

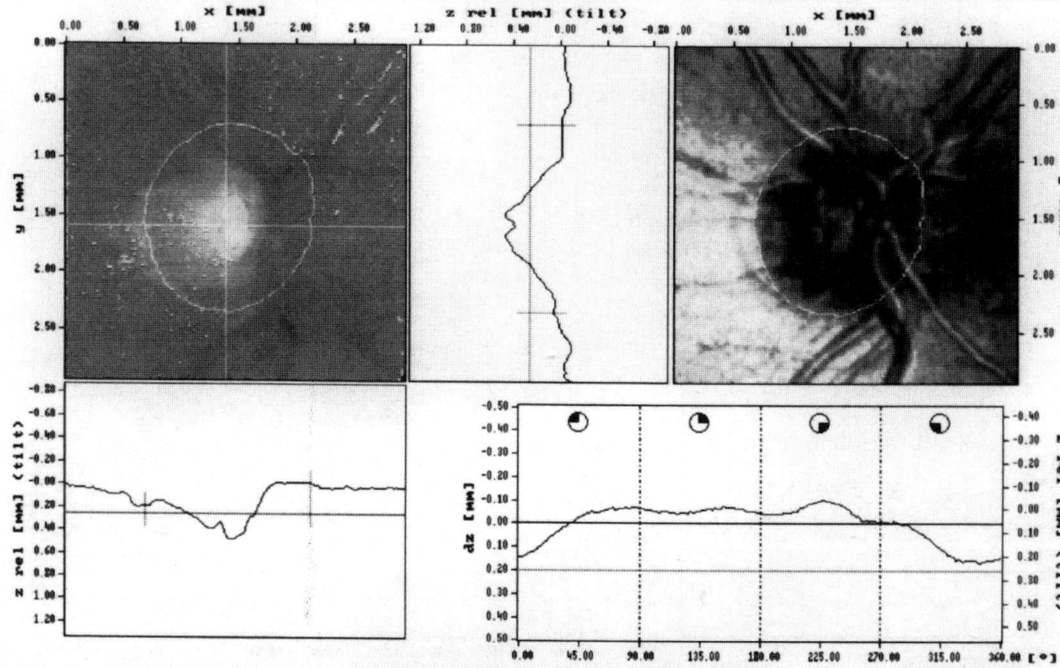


| Stereometric Analysis ONH: | | Change |
|---|---|---|
| Disk Area: | 1.899 | +0.000 mm² |
| Cup Area: | 0.292 | +0.047 mm² |
| Cup/Disk Area Ratio: | 0.154 | +0.025 |
| Rim Area: | 1.606 | -0.047 mm² |
| Cup Volume: | 0.030 | +0.004 cmm |
| Rim Volume: | 0.318 | -0.049 cmm |
| Mean Cup Depth: | 0.138 | -0.007 mm |
| Maximum Cup Depth: | 0.435 | -0.015 mm |

| Stereometric Analysis ONH: | | Change |
|---|---|---|
| Cup Shape Measure: | -0.237 | -0.009 |
| Height Var. Contour: | 0.274 | -0.025 mm |
| Mean RNFL Thickness: | 0.206 | -0.027 mm |
| RNFL Cross Sect. Area: | 1.009 | -0.132 mm² |
| Classification: | Normal* | (+1.34) |
| Reference Height (Std.): | 0.265 mm | |

**Software: FR1-V2.01**

(*) Classification based on statistics. Diagnosis is physician's responsibility.

**Signature/Date:**

CETAO MONTEVIDEO URUGUAY

*Figure 3E.*

deviation was 22.5 $\mu$m. Chauhan et al. [1] recorded three images per eye in 30 normal eyes with a mean standard deviation of 25.94 $\mu$m; the lowest variability was found in the retinal areas (flat slopes), and the highest was found along the cup border and at blood vessels. Orgul et al. [5] found that the most important source of test-retest variability was the variability in the horizontal shift of the direction of the imagining. Rohrschneider et al. [6] studied 13 normal eyes; the mean standard deviation was $22 \pm 6$ $\mu$m with a coefficient of variation for rim area of 4.8, for cup area 3.4, for cup volume 4.6, and 3.3 for mean cup depth.

This study is intended to evaluate the variation of different parameters analysing the coefficient of variation in a day or during several months.

To re-examine an eye, special care is taken to do so in an identical way to the first time. That is why before re-examining an eye, one of the series that was part of the first mean topography is studied beforehand. To do this, the first mean topography has to consist of very well defined images, and therefore the standard deviation for each pixel should be kept as low as possible ($<16$ $\mu$m).

It was noted that the difference of the rim vol. and the cup vol. in different series was related to the variation of the relative height of the reference plane (with respect to the peripapilar retina). Consequently, the minimal variation in each eye is sought.

To study the reproducibility of a mean topography obtained with only 2 series the difference of the relative height of the reference plane has to be minimal. In this study the disc margin drawn in the first mean topography was exported and then imported to the remaining series of the same eye. The 2 series that presented the least variation were selected. With these 2 series the third mean topography was obtained for the population studied in a day and the second mean topography for the evolution population. Figures 3A to E show a case studied in a day.

The reproducibility obtained with the second mean topography was the same as the one obtained with the third mean topography (for the eyes studied in a day), but we recommend using a mean topography obtained with three series (Figure 2) [7].

The evolution population consisted of only ten eyes, statistically a low number, but in a descriptive way we can say that the results are similar to those of the eyes studied in a day.

These measures are considered extremely practical and reliable in clinical application.

## References

1. Chauhan BC, LeBlanc RP, McCormick TA, et al. Test-retest variability of topographic measurements with confocal scanning laser tomography in patient with glaucoma and control subjects. Am J Ophthalmol 1994; 118: 9–15.
2. Heidelberg Retina Tomograph, operation manual. Software Version 2.01-0, January 1997.
3. Kruse FE, Burk RO, Völcker HE, Zinser G, Harbarth U. Reproducibility of topographic measurements of the optic nerve head with laser tomographic scanning. Ophthalmology 1989; 96: 1320–1324.
4. Lusky M, Bosem ME, Weinreb RN: Reproducibility of optic nerve head topography measurements in eyes with undilated pupils. J Glaucoma 1993; 2: 104–109.
5. Orgul S, Cioffi GA, Bacon DR, et al. Sources of variability of topometric data with a scanning laser ophthalmoscope. Arch Ophthalmol 1996; 114: 161–164.
6. Rohrschneider K, Burk ROW, Kruse FE, et al. Reproducibility of the optic nerve head topography with a new laser tomograph scanning device. Ophthalmology 1994; 101: 1044–1049.
7. Weinreb RN, Lusky M, Bartsh DU, et al. Effect of repetitive imagining on topographic measurements of the optic nerve head. Arch Ophthalmol 1993; 111: 636–638.

International Ophthalmology **23:** 229–232, 2001.
J.R. Sampaolesi (ed.), Laser Scanning: Update 1, 47–50.




# Normal pressure glaucoma. Open angle glaucoma

Jorge Vasco
*Universidad de Antioquia, Medellín, Colombia*

## Introduction

One hundred and fifty years ago Argyl-Robertson [1] postulated the elevation of the eye pressure as the main cause of optic nerve damage in glaucoma patients. Since then the therapeutical approach has been directed towards lowering the intra-ocular pressure, (IOP).

New evidence however, has demonstrated that other factors play an important role in the pathogenesis of this disease. Recent research has pointed out that abnormalities of the anterior portion of the optic nerve and its passage through the lamina cribrosa and the posterior scleral rim and canal, constitute a significant factor in its development.

Bioengineering, biological chemistry and histopathological examinations of the optic nerve in glaucomatous eyes have demonstrated that the distensible colagenous tissue in this area becomes hard, compact and highly resistant [2–6]. The elastic fibers that surround the lamina cribrosa and create its pores, loose their straightness, look wavy and become curved, reducing the area through which the axons pass [7].

This generates an effect of compression, which affects the blood flow patterns at the central retinal vessels and probably the axonal flow. Studies on the microcirculation at the level of the laminar and prelaminar zones, have showed that these areas are prone to ischemia [8–11]. The following study was done in an attempt to decrease the external compression effects on the structures mentioned above.

## Subjects and methods

Twenty one eyes of 15 patients with normal pressure glaucoma, (Group 1), and 19 eyes of 19 patients with open angle glaucoma whose intra-ocular pressure had been successfully controlled either with medication or surgery, but that continued to show progression of the visual field loss and cupping, (Group 2), were included in this study.

All patients had a complete ophthalmological examination, and computerized visual field test (30-2 Humphrey analyzer). Blood flow evaluation of the retrobulbar vessels, (ophthalmic artery, short posterior ciliary arteries and central retinal artery and vein) was done with a Hewlett-Packard color triplex Doppler equipment, using a 7 Mhz transductor.

All eyes underwent the following surgical procedure:

- Decompression of the posterior scleral rim and canal, using a technique that has been previously described (Figure 1) [12–16].
- Pediculated muscle graft originated from the superior third of the anterior portion of the medial rectus, which was sutured to the epichoroidal layer at the laminar and prelaminar zone (Figure 2).

In 22 cases the procedure was done under general anesthesia and in 18 it was done under local. All cases were ambulatory. No significant complications were associated with the procedure. Visual acuity and blood flow evaluations were done 5 to 10 days after surgery. Visual field test was done 3–6 and 12 months after the procedure, and every year after the twelfth month.

The follow-up period covered one to four years.

## Results

Preoperative evaluation of the blood flow patterns at the retrobulbar vessels revealed an increased resistive index and a significant decrease in the diastolic velocity at the central retinal artery. The posterior ciliary arteries showed an increased systolic velocity, while the ophthalmic artery showed normal blood flow patterns in the two groups.

In Group 1, Table 1, the post-operative Doppler evaluation revealed normalization of the blood flow values in all cases at the central retinal artery. In 6

*Table 1.* Group 1: Normal pressure glaucoma. Post-operative Doppler values and visual results

| Normal pressure glaucoma with progression of the visual field loss | | | | |
|---|---|---|---|---|
| No. Cases | No. eyes | Doppler value normal | Visual field stability | Visual field improvement |
| 16 | 21 | 15 | 9 | 10 |

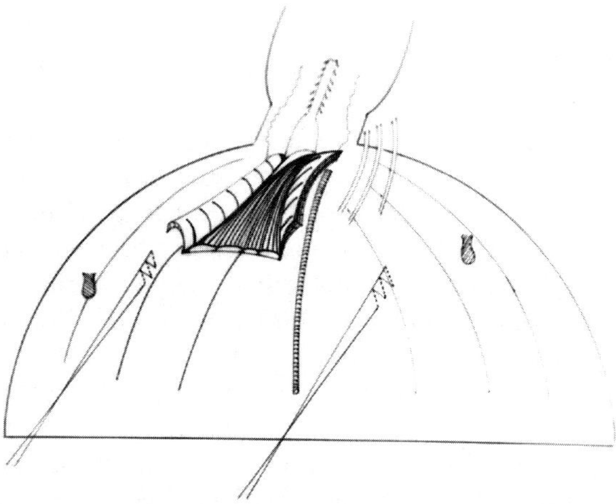

*Figure 1.* Decompression of the posterior rim and canal.

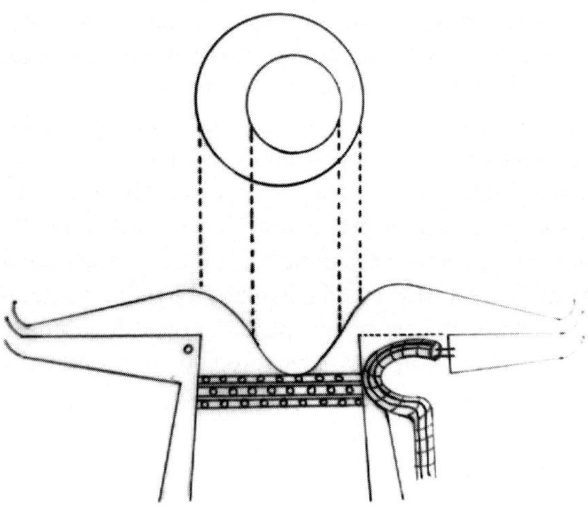

*Figure 2.* Pediculated muscle graft.

cases the flow velocities at the posterior ciliary arteries remained high.

During the follow-up 9 eyes showed stabilization of the visual field, while 10 showed a significant improvement (Figure 3). In two cases the visual loss continued. Visual acuity and IOP measurements did not show variation regarding the preoperative values.

In Group 2, Table 2 the Doppler evaluation revealed flow patterns similar to those found in Group one. Also in this group high velocity values were found at the short posterior ciliary arteries during the post-operative period in 8 cases. Regarding the visual field loss, it stopped in 10 cases, while 6 cases showed an improvement that persisted during the follow-up (Figure 4). In 3 cases the visual loss continued.

Two cases, a 58 and 25 years old who had dense central scotomas, recovered visual function in that area in the operated eye, and in both of them the visual acuity improved to 20/80. In the other cases of Group 2 visual acuity and IOP showed no change in their values with respect to the preoperative evaluation.

Observations during the surgical procedure revealed significant thickening of the scleral tissue around the neural-ocular junction.

Abundant transparent fluid was found in the subarachnoidal space, specially in the least advanced cases.

The medial rectus is one of the most vascularized and innervated muscles of the human body, containing sensitive and motor innervations. The ratio of muscle fibers to the motor nerves varies from 3 : 1 to 6 : 1, while in the biceps it is of approximately 100 : 1.

Between 30 and 50 sensory muscle spindles and other receptors occupy each rectus muscle [17].

In my opinion the muscle graft serves the following purposes:

- Prevents the closure of the scleral groove surgically created at the level of the neural-ocular junction.
- It may help to decrease the level of ischemia in this critical area because it may stimulate the formation of new capillaries that grow towards the optic nerve head.
- Its nerves contain large amounts of neural-trophic substances that might help revitalize axons that are not yet severely damaged. This might be the reason for the disappearance of the central scotomas and

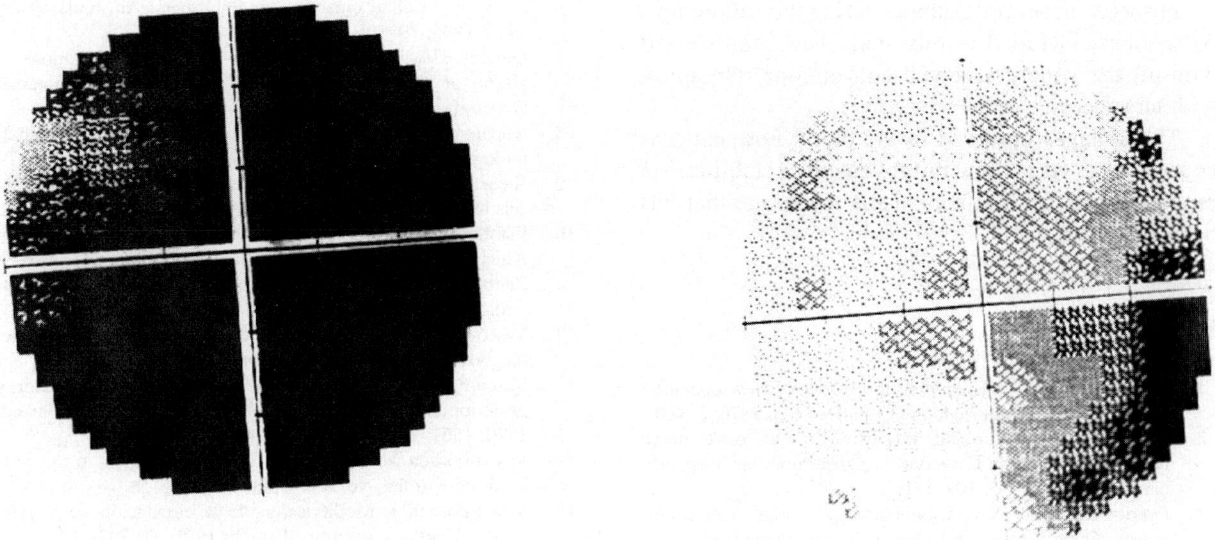

*Figure 3.* Normal pressure glaucoma. One year after the operation.

*Table 2.* Group 2: Open angle glaucoma. Post-operative Doppler values and visual results

| Open angle glaucoma. Controlled pressure and progression of the visual field loss | | | | | |
|---|---|---|---|---|---|
| No. cases | No. eyes | Doppler value normal | Visual field stability | Visual field improvement | Visual field loss control |
| 19 | 19 | 11 | 10 | 6 | 3 |

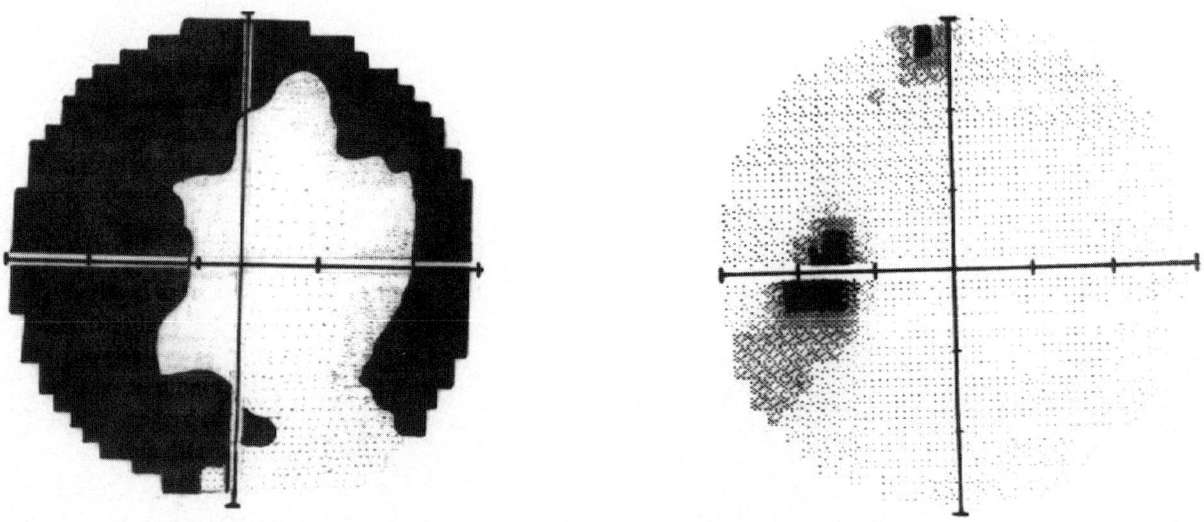

*Figure 4.* Open angle glaucoma. Four years after the operation.

the improvement in the peripheral visual field observed in several patients during the follow-up. All patients included in this study had been treated with all the widely accepted conventional therapies, without success.

The changes observed in the blood flow patterns as well as the visual field improvement or stabilization prove the beneficial effects of the technique that has been described.

## References


1. Argyl-Robertson D. Trephining the sclerotic. New operation for Glaucomas. Royal London Ophth, Hos Rep 1876; 8: 404.
2. Zeimer R. Biomechanical properties of the optic nerve head. Optic Nerve in Glaucoma. Kugler Publishers, Amsterdam/New York, 1995: 107–121.
3. Pyeritz RE. Biology and pathobiology of the extracellular matrix. Optic Nerve in Glaucoma. Kugler Publishers, Amsterdam/New York, 1995: 36–56.
4. Quigley HA. Overview and introduction to session on connective tissue of the optic nerve in glaucoma, in SM Drance DR Anderson (eds.), Optic Nerve in Glaucoma. Kugler Publishers, Amsterdam/New York, 1995: 15–36.
5. Morrison JC. The microanatomy of the optic nerve. Optic Nerve in Glaucoma. Kugler Publishers, Amsterdam/New York, 1995: 57–78.
6. Clark AF et al. Cellbiology of the human lamina cribrosa. Optic Nerve in Glaucoma. Kugler Publishers, Amsterdam/New York, 1995: 79–105.
7. Quigley HA, Brown A, Dorman-Pease ME. Alterations in elastin of the optic nerve head in human and experimental glaucoma. Brit J Ophthalmol 1991; 75: 552–557.
8. Kupersmith MJ. Berenstein. Neurovascular Neuro Ophthalmology. Publ. Springer-Verlag, 1993; 25.
9. Kaiser HJ et al. Blood-flow velocities in the extraocular vessels in normal volunteers. Amer J Ophth 1996; 122: 364–370.
10. Flammer J. Optic nerve circulation. Optic Nerve in Glaucoma. Kluger Publishers, Amsterdam/New York, 1995: 243–245.
11. Zeither A. Microcirculation. Optic Nerve in Glaucoma. Amsterdam/New York, 1995: 247–258.
12. Vasco-Posada J. Revascularización del segmento posterior del ojo. Arch Oftl Norte del Perú 1969; II(6): 7–18.
13. Vasco-Posada J. Revascularización del segmento anterior y posterior del ojo. Proceeding of XXI Intern Congress, México, 1970: 1561–1569.
14. Vasco-Posada J. Modification of the circulation in the posterior pole of the eye. Ann Ophth 1972; 48–56.
15. Vasco-Posada J. Modificación de la circulación en el polo posterior del ojo. Arch Soc Esp Oft 1971; 31: 797–806.
16. Vasco-Posada J. Glaucoma de presión normal y glaucoma crónico y pérdida progresiva del campo visual. Normalización de los valores del Doppler y estabilización del campo visual. Arch Soc Amer De Oftal Y Optom 1998; 26: 31–38.
17. Walsh FB, Hoyt WF. Clinical Neuro-Ophthalmology. Vol. 1. 3rd edn. 1969: 202.

*International Ophthalmology* **23:** 233–237, 2001.
*J.R. Sampaolesi (ed.), Laser Scanning: Update 1*, 51–55.




# Scanning Laser Polarimetry (SLP) for Optic Nerve Head Drusen

Andrea Mistlberger[1], S. Sitte[1], A. Hommer[2], M. Emesz[1], S. Dengg[1], W. Hitzl[3] & G. Grabner[1]
[1]*Department of Ophthalmology, County Hospital Salzburg, Austria;* [2]*Department of Ophthalmology, Sanatorium Hera, Vienna;* [3]*Office for Applied Statistics, Salzburg, Austria*

*Key words:* Optic Nerve Head Drusen, clinical visibility, Scanning Laser Polarimetry, visual field loss

## Abstract

Optic nerve head drusen (ONHD) are either clinically invisible or clearly protruding from the disc, in the later case leading to the condition of an irregular, indistinct disc margin or a swollen disc on biomicroscopy. They also may cause visual field defects, even with slow progression. Scanning laser polarimetry (SLP) has been proposed as a rapid, objective and reproducible technology for retinal nerve fiber layer (RNFL) assessment and clinical studies have demonstrated that SLP can help to distinguish between normal and glaucomatous eyes, identify glaucoma suspects and correlates well with visual field defects.

The purpose of this study was to evaluate the potential applicability of SLP in 20 consecutive patients with optic nerve head drusen (18 bilateral) that were clinically visible (22 eyes) and invisible (16 eyes). RNFL thickness was studied in patients with and without visual field defects.

Patients with visual field defects and ONHD were significantly older and had a small, but significant reduction of visual acuity. Some global SLP parameters (average thickness, ellipse average) were significantly different between subjects with normal and abnormal visual fields. The comparison of the groups with visible and invisible drusen showed that there was no difference in demographic or perimetric data. RNFL thickness measurements were also very similar in both groups.

Clinical visibility of drusen was not correlated with RNFL thinning as measured with the GDx™. SLP assessment, however, was well correlated with functional loss. This objective, non-invasive technology may be an additional option for RNFL evaluation in this condition and an especially useful tool for long-term follow-up.

## Introduction

Optic nerve head drusen (ONHD) are hyaline bodies in the optic nerve head either clinically invisible or clearly protruding from the disc, leading to the condition of an irregular, indistinct disc margin or a swollen disc on biomicroscopy [1, 2]. The pathogenesis is still under discussion. Spencer reported an axoplasmic transport alteration to be responsible for accumulation of material and formation of the drusen [3]. On the other hand congenitally dysplastic discs, abnormal disc vasculature and/or small scleral canals seem to play a role in the developement of ONHD [3–5].

Drusen may give rise to a pseudopapilledema appearance making it difficult to evaluate the standard disc parameters such as cupping or pallor. They also may cause visual field defects, even with slow progression [6]. It is therefore important to evaluate new technologies that are capable of measuring nerve fiber layer thickness or defects before visual field defects develop or progress [7].

Scanning laser polarimetry (SLP) has been proposed as a rapid, objective and reproducible technology for retinal nerve fiber layer (RNFL) assessment [8–11]. Clinical studies have demonstrated that this method can help to distinguish between normal and glaucomatous eyes, to identify glaucoma suspects and that it correlates well with visual field defects in glaucoma [10, 12, 13].

The purpose of this study was to evaluate the potential applicability of SLP in patients with optic nerve head drusen [14]. We studied RNFL thickness in patients with and without visual field defects as these parameters could prove to be of value in the diagnosis and follow-up of this uncommon disease.

## Patients and methods

Thirty-eight eyes of 20 consecutive patients with ONHD were examined at the Department of Ophthalmology at St. John's Hospital, Salzburg, Austria.

All patients underwent a complete ophthalmic examination including visual acuity testing, intraocular pressure (IOP) measurement, dilated slit-lamp biomicroscopy, optic nerve head stereophotography, B-scan echography, scanning laser polarimetry, and in some cases fluorescein angiography. Achromatic automated perimetry was performed with the Humphrey visual field analyzer (program 30-2 or 24-2, full threshold; Humphrey Systems, Inc, Dublin, CA). Any previous intraocular surgery, or anterior segment or retinal disorders that might cause visual field defects or thinning of the RNFL were considered to be exclusion criteria. None of the patients was on topical pressure lowering medication.

Visual field reliability criteria included fewer than 33% fixation losses and false-positive and false-negative rates. A normal visual field was based on a Glaucoma Hemifield Test (GHT) within normal limits, whereas abnormal visual fields had a GHT outside normal limits or borderline [15].

Further subdivisions in our cohort were: (a) eyes with either normal or abnormal visual fields, and (b) eyes with clinically visible or inapparent drusen that were diagnosed with ultrasound only.

Nerve fiber layer thickness was measured with the Nerve Fiber Analyzer (GDx$^{TM}$, Laser Diagnostic Technologies, Inc. San Diego, CA) with undilated pupils. The working principle of this technology is that polarized light passing through a form-birefringent medium will undergo a measurable phase shift directly proportional to the thickness of the nerve fiber layers [8,16]. SLP was performed in a standardized fashion by one operator (SD) and a baseline image was created using three of six scans. The placement ellipse at the disc margin was focused in comparison with both, the clinical appearance of the optic nerve head and the stereo disc photography. A second ellipse, the calculation ellipse, is then projected around the placement ellipse to a diameter of 1.75 disc diameter.

In this study, we investigated those SLP parameters designed to represent global indices of RNFL thickness. Ellipse average was calculated by averaging the RNFL thickness beneath the ellipse and ellipse modulation was defined as difference between the thickest and thinnest parts of the nerve fiber layer covered by the ellipse surrounding the optic nerve, whereas maximum modulation provided an indication of the difference between the thickest and thinnest parts using all points of the image. Average thickness was defined as the average thickness of all pixels in the image (excluding the optic disc) and total integral as the total area beneath the entire ellipse. The neural network number, an experimental number derived from a prototype neural network was generated by all values obtained from the image and is considered a possible indicator of glaucomatous damage.

*Table 1.* Summary of visibility and functional damage (number of eyes)

| Visual field (GHT) | Visible drusen | Invisible drusen | Total |
|---|---|---|---|
| Normal | 10 | 14 | 24 |
| Abnormal | 12 | 2 | 14 |
| Total | 22 | 16 | 38 |

GHT = Glaucoma Hemifield Test.

## Statistics

All computations and illustrations were done with Statistica 5.5. In order to compare the groups two sided t-tests were applied [17]. Multiple comparisons demand the use of a type I error rate adjustment in order to protect against an increase of the overall type I error rate. Assuming statistical independence, we computed the individual type I error rates to end up with a reasonable overall type I error rates for each table of 5% [18].

## Results

In all subjects the existence of drusen, even when clinically invisible, was confirmed by B-mode ultrasonography. Of the 20 patients with ONHD, 10 were male, 10 female, with a mean age of 42.1 years (range 13.6–72.6 years). Eighteen patients had bilateral ONHD and 2 had drusen in only one eye. In 22 eyes ONHD were visible, in the other 16 they were clinically indiscernible (Table 1).

Demographics and visual field indices of subjects with normal and abnormal visual field are listed in

*Table 2.* Demographics and visual function

| | Normal visual field $n = 24$ | Absnormal visual field $n = 14$ | $p$-value |
|---|---|---|---|
| Gender (male, female) | 11, 13 | 7, 7 | 0.8 |
| Eye (right, left) | 13, 11 | 7, 7 | 0.81 |
| Age (years) | 34.7 ± 19.0 (13.7, 72.6) | 56.2 ± 14.0 (31.0, 71.8) | 0.0015* |
| Visual acruity | 1.0 ± 0.1 (0.8, 1.0) | 0.8 ± 0.3 (0.1, 1.0) | 0.0012* |
| Refraction (diopter sphere) | 0.0 ± 1.1 (−5.4, 1.8) | −7.3 ± 6.7 (−6.3, 1.0) | 0.0337 |
| MD | −1.5 ± 1.8 (−5.4, 1.8) | −7.3 ± 6.7 (−16.7, 10.4) | 0.003* |
| CPSD | 1.7 ± 1.4 (0.0, 5.3) | 8.8 ± 3.3 (4.0, 13.6) | 0.0000* |

* Marked $p$-values are significant at 0.05/7 ≅ 0.007; two-sided $t$-test.
MD = mean deviation, CPSD = corrected pattern standard deviation.

*Table 3.* Description of SLP parameters

| | Normal visual field $n = 24$ | Abnormal visual field $n = 14$ | $p$-value |
|---|---|---|---|
| Number | 32.88 ± 23.7 (10, 90) | 53.36 ± 31.26 (9, 92) | 0.046 |
| Max modulation | 1.45 ± 0.38 (0.87, 2.11) | 1.41 ± 0.68 (0.44, 2.75) | 0.825 |
| Ellipse modulation | 2.36 ± 0.53 (1.25, 3.23) | 2.08 ± 0.94 (0.93, 3.75) | 0.319 |
| Average thickness ($\mu$m) | 69.6 ± 17.3 (48.0, 111.0) | 54.9 ± 6.34 (46.0, 65.0) | 0.0007* |
| Ellipse average ($\mu$m) | 71.42 ± 17.9 (47.7, 116.0) | 55.79 ± 6.85 (45.0, 66.0) | 0.0006* |
| Total polar integral (mm$^2$) | 0.66 ± 0.25 (0.07, 1.11) | 0.51 ± 0.07 (0.38, 0.62) | 0.02 |

* Marked $p$-values are significant at 0.05/6 ≅ 0.008; two-sided $t$-test.

*Table 4.* Correlation between visual field indices and GDx$^{TM}$ parameter

| | Normal visual field $n$ = 24 | | | | Abnormal visual field $n$ = 14 | | | |
|---|---|---|---|---|---|---|---|---|
| | MD | | CPSD | | MD | | CPSD | |
| Number | −0.82 | −0.03* | −0.23 | 0.71 | −0.45 | 0.68 | 0.26 | 0.94* |
| Max modulation | −0.17 | 0.75 | −0.65 | 0.34 | −0.68 | 0.45 | −0.76 | 0.45 |
| Ellipse modulation | −0.42 | 0.59 | −0.53 | 0.50 | −0.2 | 0.86 | −0.82 | 0.31 |
| Average thickness ($\mu$m) | −0.31 | 0.67 | −0.53 | 0.50 | −0.59 | 0.56 | −0.93 | −0.17* |
| Ellipse average ($\mu$m) | −0.24 | 0.71 | −0.57 | 0.45 | −0.62 | 0.52 | −0.91 | −0.05* |
| Total polar integral | −0.17 | 0.75 | −0.83 | −0.06* | −0.61 | 0.54 | −0.86 | 0.20 |

*Indicates a moderate correlation.

*Table 5.* Demographics and visibility of drusen

| | Visible drusen $n = 22$ | Invisible drusen $n = 16$ | $p$-value |
|---|---|---|---|
| Gender (male, female) | 8, 14 | 12, 4 | 0.029 |
| Eye (right, left) | 13, 9 | 7, 9 | 0.33 |
| Age (years) | 45.8 ± 19.7 (16.8, 72.6) | 36.5 ± 20.1 (13.7, 70.2) | 0.1897 |
| Visual acruity | 0.9 ± 0.2 (0.4, 1.0) | 0.9 ± 0.2 (0.1, 1.0) | 0.4029 |
| Refraction (diopter sphere) | −0.9 ± 1.9 (−6.3, 3.3) | 0.1 ± 0.8 (−1.5, 1.5) | 0.0351 |
| MD | −4.9 ± 5.9 (−16.7, 10.4) | −1.9 ± 3.0 (−10.6, 1.8) | 0.0489 |
| CPSD | 5.6 ± 4.4 (1.0, 13.6) | 2.2 ± 2.9 (0, 9.18) | 0.0298 |

* Marked $p$-values are significant at 0.05/7 ≅ 0.07; two-sided $t$-test.
MD = mean deviation, CPSD = corrected pattern standard deviation.

*Table 6.* Description of SLP parameter depending on drusen visibility

| | Visible drusen $n = 22$ | Invisible drusen $n = 16$ | $p$-value |
|---|---|---|---|
| Number | 50.9 ± 29.9 (9, 92) | 26.1 ± 18.1 (10, 67) | 0.0031 |
| Max modulation | 1.3 ± 0.5 (0.44, 2.4) | 1.6 ± 0.5 (0.9, 2.8) | 0.1094 |
| Ellipse modulation | 2.1 ± 0.7 (0.93, 3.3) | 2.5 ± 0.6 (1.5, 3.8) | 0.1019 |
| Average thickness ($\mu$m) | 60.8 ± 12.4 (48.0, 100.0) | 68.9 ± 19.1 (46.0, 111.0) | 0.1478 |
| Ellipse average ($\mu$m) | 61.7 ± 13.0 (47.0, 100.0) | 71.1 ± 19.7 (45.0, 116.0) | 0.1115 |
| Total polar integral (mm$^2$) | 0.6 ± 0.2 (0.07, 1.0) | 0.7 ± 0.2 (0.4, 1.1) | 0.1747 |

* Marked $p$-values are significant at 0.05/6 ≅ 0.008; two-sided $t$-test.

Table 2. Patients with visual field defects and ONHD were significantly older and had a small, but significant reduction of visual acuity.

Some global SLP parameters (average thickness, ellipse average) were significantly different between subjects with normal and abnormal visual fields (Table 3), there were correlations with the SLP parameters and the visual field indices (Table 4).

The comparison of the groups with visible and invisible drusen showed that there was no difference in demographic or perimetric data (Table 5). RNFL thickness measurements were also very similar in both groups (Table 6).

## Discussion

Drusen represent one of the major causes of pseudopapilledema. Their pathogenesis has not yet been elucidated, but alterations in axoplasmic transport seem to be one possible cause of drusen formation [3–5, 14, 19]. Because they are often not readily seen with biomicroscopy, imaging techniques have to be applied to confirm the diagnosis when suspected. Several studies of the recent past have discussed the pros and cons for the different imaging procedures [20–24]: Kheterpal et al performed computed tomography, magnetic resonance imaging, fluorescein angiography and B-mode ultrasonography suggesting the latest technique to be the imaging method of choice [20]. This recommendation was confirmed by Levin and co-workers, who found B-scan echography to be more reliable than computed tomography [22]. In all our patients ONHD, visible or deeply buried and thereby concealed, where confirmed and often detected by ultrasound only which is an objective, noninvasive, safe and reliable examination [24].

Nerve fiber layer thickness evaluation is important, because ONHD can go along with visual field defects that are sometimes slowly progressing [26]. The clinical appearance of the disc resembling pseudoneuritis with an abnormally small disc size makes it difficult to reliably interpret changes by clinical inspection alone [27]. Because thinning of RNFL procedes functional damage, the usefullness of nerve fiber layer evaluation has to be emphasized. Roh et al showed that optical coherence tomography is able to detect RNFL thinning in clinically visible ONHD and that it correlated with both, visual field defects and nerve fiber loss as seen by red-free photography [28].

In our study, SLP could not differentiate between visible and buried drusen, because of a wide overlap of the clinical parameters in each group. However, the more drusen become clinically visible, the higher the percentage of visual field loss (Table 1) [14].

The analysis of visual field defects showed the typical damage patterns in our series, such as enlargement of the blind spot, defects in the nasal quadrants and concentric restrictions [6, 26]. Global indices (average thickness and ellipse average) differed significantly in patients with and without field defects (Tables 3, 4).

Our data support some findings about ONHD. The alteration of axonal flow due to a small optic nerve scleral canal an/or abnormal disc vasulature probably leads to drusen formation that increases with age. This slow progression is sometimes followed by an appearance and steady increase in visual field defects (Table 2) [6, 14, 26].

In summary, clinical visibility of drusen was not correlated with RNFL thinning as measured with the GDx$^{TM}$ However, SLP assessment was well correlated with functional loss. This objective, non-invasive technology may be an additional option for RNFL evaluation in this condition and an especially useful tool for long-term follow-up.

## References


1. Boyce SW, Platia EV, Green WR: Drusen of the optic nerve head. Ann Ophthalmol 1978; 10: 695–704.
2. Friedman AH, Gartner S, Modi SS. Drusen of the optic disc. A retrospective study in cadaver eyes. Br J Ophthalmol 1975; 59, 413–421.
3. Spencer WH. Drusen of the optic disk and aberrant axoplasmic transport. The XXXIV Edward Jackson memorial lecture. Am J Ophthalmol 1978; 85: 1–12.
4. Sacks JG, Ogrady RB, Choromokos E, Leestma J. The pathogenesis of optic nerve drusen. A hypothesis. Arch Ophthalmol 1977; 95: 425–428.
5. Antcliff RJ, Spalton DJ. Are optic disc drusen inherited? Ophthalmology 1999; 106: 1278–1281.
6. Rossmann M, Rigal K, Harrer S, Partik G. Gesichtsfelddefekte bei Drusenpapillen. Spektrum Augenheilkd 1998; 12: 248–252.
7. Quigley HA, Addicks EM, Green WR. Optic nerve damage in human glaucoma, III: quantitative correlation of nerve fiber loss and visual field defect in glaucoma, ischemic neuropathy, papilledema, and toxic neuropathy. Arch Ophthalmol 1982; 100: 135–146.
8. Weinreb RN, Dreher AW, Coleman A, Quigley H, Shaw B, Reiter K. Histopathologic validation of Fourier-Ellipsometry measurements of the retinal nerve fiber layer thickness. Arch Ophthalmol 1990; 108: 557–525.
9. Zangwill L, Berry CA, Garden VS, Weinreb RN. Reproducibility of retardation measurements with the Nerve Fiber Analyzer II. J Glaucoma 1997; 6: 384—389.
10. Chen YY, Chen PP, Xu L, Ernst PK, Wang L, Mills RP. Correlation of peripapillary nerve fiber layer thickness by scanning laser polarimetry with visual field defect in patients with glaucoma. J Gaucoma 1998; 7: 312–316.
11. Essock EE, Sinai MJ, Fechtner RD. Interocular symmetry in nerve fiber layer thickness of normal eyes as determined by polarimetry. J Gaucoma 1999; 8: 90–98.
12. Weinreb RN, Shakiba S, Zangwill L. Scanning laser polarimetry to measure the nerve fiber layer of normal and glaucomatous eyes. Am J Ophthalmol 1995; 119: 627–636.
13. Choplin NT, Lundy DC, Dreher AW. Differentiating patients with glaucoma from glaucoma suspects and normal subjects by nerve fiber layer assessment with scanning laser polarimetry. Ophthalmology 1998; 105: 2068–2076.
14. Roh S, Noecker RJ, Schuman JS, Hedges TR, Weiter JJ, Mattox C. Effect of optic nerve head drusen on nerve fiber layer thickness. Ophthalmology 1998; 105: 878–885.
15. Asman P, Heijl A. Glaucoma Hemifield Test. Automated visual field evaluation. Arch Ophthalmol 1992; 110: 812–819.
16. Dreher AW, Reiter K. Retinal laser ellipsometry: A new method for measuring the retinal nerve fiber layer thickness distribution. Clin Vision Sci 1992; 7: 481–488.
17. Neter J, Wasserman W, Kuntner M. Applied Linear Statistical Models: Regression, Analysis of Variance and Experimental Design. Irwin Series in Statistics, 3rd edn. New York, 1990.
18. StatSoft, Inc. STATISTICA for Windows [Computer program manual]. StatSoft, Inc., Tulsa IK, 1999.
19. Friedman AH, Henkind P, Gartner S. Drusen of the optic disc. A histopathological study. Trans Ophthalmol Soc UK 1975; 95: 4–9.
20. Kheterpal S, Good PA, Beale DJ, Kritzinger EE. Imaging of optic disc drusen: A comparative study. Eye 1995; 9: 67–69.
21. Haynes RJ, Manivannan A, Walker S, Sharp PF, Forrester JV. Imaging of optic nerve head drusen with scanning laser ophthalmoscope. Br J Ophthalmol 1997; 81: 654–657.
22. Kurz Levin MM, Landau K. A comparison of imaging techniques for diagnosing drusen of the optic nerve head. Arch Ophthalmol 1999; 117: 1045–1049.
23. Bec P, Adam P, Mathis A, Alberge Y, Roulleau J, Arne JL. Optic nerve head drusen. High-resolution computed tomograhic approach. Arch Ophthalmol 1984; 102: 680–682.
24. Schön JKD, Nasemann JE, Boergen KP. Vergleichende Untersuchung tiefliegender Drusen der Papille mit Scanning Laser Ophthalmoskop und Funduskamera. Klin Monatsbl Augenheilkd 1992; 200: 175–177.
25. Boldt HC, Byrne SF, Di Bernardo C. Echographic evaluation of optic disc drusen. J Clin Neuroophthalmol 1991; 11: 85–91.
26. Savino PJ, Glaser JS, Rosenberg MA. A clinical analysis of pseudopapilledema.II. Visual field defects. Arch Ophthalmol 1979; 97: 71.
27. Jonas J, Gusek GC, Guggenmoos-Holzmann I , Naumann GOH. Optic nerve head drusen associated with abnormally small optic discs. Int Ophthalmol 1987; 11: 79–82.
28. Roh S, Noecker RJ, Schuman JS. Evaluation of coexisting optic nerve head drusen and glaucoma with optical coherence tomography. Ophthalmology 1997; 104: 1138–1144.

*International Ophthalmology* **23:** 239–244, 2001.
*J.R. Sampaolesi (ed.), Laser Scanning: Update 1*, 57–62.




# Evaluation and definition of physiologic macro cups with confocal optic nerve analysis (HRT)

Eduardo Arenas-Archila, Fernando Caycedo-Yunis & Mario Rodríguez Rodríguez
*Unidad de Optalmologia, Facultad de Medicina, Universidad Nacional de Colombia, Bogota, Colombia*

## Introduction

Glaucoma diagnosis and follow-up is based actually in three main pillars: Optic nerve appearance, visual field changes and intraocular pressure values. Optic nerve characteristics are the main clinical sign to include an eye as a glaucoma suspect. The vertical cup/disc ratio (CDR) has long been used in the assessment of the glaucoma suspect, though the wide range of CDR values in the normal population limits its use [1–3]. Cup size is related physiologically to disc size and pathologically to glaucomatous damage[4]. Large optical cupping it is easily found by the general ophthalmologist and is one of the features that shows to a glaucoma specialist that the patient could be loosing optic nerve axons. However there are many patients with large optical cupping that do not have or will never have glaucoma, while there are others with large cups that could have a clear disease. Patients with normal tension glaucoma may remain undetected, without pathological changes in the visual fields [5, 6]. Large cups tested with short-wavelength automated perimetry may show early visual field changes [7].

Embriologically the development of the physiologic cup is related to the size of the posterior scleral foramen and the regression of the hyaloid vessels of the Bergmeister's papilla [8]. Physiologic cupping of the optic nerve has been defined as a normal central escavation of the optic nerve where the retinal vessels and fibers form an infundibular depression. The normal size of this cupping has not been universally defined because among the normal population it is possible to find optical cups from almost zero to 4 or more CDR [9]. The size of the cup is generally similar in fellow eyes although it varies greatly among individuals and has a genetic determination [10]. It has been found that among races black population have larger CDR than whites A study made in the Micronesian population of the Marshall islands showed greater CDR than normals without signs of glaucoma even in cases with follow-ups as long as 18 years. Also the size of the CDR increases with age [2, 13].

In relationship with the refractive error it has been found that myopes have greater CDR than hyperopes [2, 14, 15]. however we have found that with the help of HRT which excludes the magnification or minification of the refractive media this assumption is not valid [16]. The term macro physiologic optic disk cupping (MC) has been introduced to include all those disks with cuppings larger than 3 DD but with not glaucomatous disease [17]. Several studies among normal population have defined as a normal range a CDR of maximum 0.30 [2–4].

Evaluation of the optic disc cupping can be achieved by several methods from the simple direct ophthalmoscopy, by ophthalmoscopy, using a slit lamp with the commonly used ophthalmoscopical lenses [18] to the high-resolution ocular ultrasound [19]. With the introduction of the Heidelberg Retina Tomograph (HRT) Heidelberg Engineering, the accuracy of these measurements has increased notoriously [20, 21].

## Methods

From a database of 621 consecutive patients, we studied 57 cases where a clinical diagnosis of Physiologic Macrocup (MC) was found. Inclusion criteria for (MC) were patients with a CDR greater than 0.3. Those cases were submitted to an HRT test because the unusual large cupping. All the tests to discard glaucoma including a complete eye examination, visual Fields, diurnal Intraocular pressure and at least one new ocular examination 6 months or later were performed. Some of our cases with physiological macro-cups were followed by several years without showing any mayor change in the HRT parameters. Exclusion criteria for (MC) were eyes with IOP higher than 18 mm Hg or a variance correlation in the HRT higher than 30. Patients classified as having (MC) were indexed and compared with two groups: a group of normal eyes NE and those with advanced glaucoma (AG).

Normal eyes (NE) were defined as those having a CDR lower than 0.3, with healthy neuretinal rim, normal visual fields, and not familiar or ocular signs of glaucoma. Thirty cases were included in this category. Advanced glaucomas (AG) were defined as those with a clear clinical diagnosis of glaucoma, visual fields with a standard deviation greater than −25 with the Humphrey Statpac 630 (San Leandro, CA) and with a variance correlation index in the HRT lower than 30. In this group we included 13 eyes.

All parameters found in the last software version of the Heidelberg Retina Tomograph (Heidelberg Engineering GmbH, Heidelberg, Germany) were included. For obtaining a numerical score we utilized an absolute unit combining the more indicative parameters of the HRT with a method published elsewhere [22]. Normal eyes being close to zero and advanced glaucomas close to 10.

All parameters were first described in the three different groups. The frequency distribution was analyzed with the Anova and the $t$ test performed when it found a significant difference. The kurtosis and skewness were also defined. Afterwards the correlation matrix of Spearman for categoric variables was done.[1]

In order to measure the neuroretinal rim width we utilized the HRT diagram and measure the four cardinal points where the lines of reference crossed the cup border. We measured with a mathematical scale ruler the superior, nasal, inferior and temporal distance between the disc border and the cup limit (Figure 1) We included this information in our database and filed as local width of the neuroretinal rim.

## Results

*Normal eyes.* All parameters were included in a table and the mean value of our series was obtained (Table 1) and compared them with other normal values that appeared in the literature (Table 2).

[1] The statistical analysis was performed by Klaus Mieth MD from the Centro de Estudios Epidemiólogicos de la Fundación (CEIS) Santa Fe de Bogotá.

The three groups showed different values in all mean parameters. MC was usually in the middle, but sometimes shifting more towards the normal group NE and in others approaching to the AG (Table 2). In order to understand better those tendencies we will include the data in 4 sub-groups.

*Similar values:* The only parameter with little differences between the three groups was the disc area. However the mean area in $mm^2$ was 2.26 for NE, 2.91 for MC and 2.41 for AG.

*Equidistant values:* In this group we included those parameters where the MC was separated from the other two groups in almost similar proportions. Those parameters were:

| | Normal patients | MC | Advanced glaucoma |
|---|---|---|---|
| Cup area | 0.4 | 1.26 | 1.713 |
| C/D area | 0.172 | 0.427 | 0.734 |
| Height variable contour | 0.41 | 0.35 | 0.285 |

*MC closer to normal:* There were six parameters in this category including also our score measure.

| | Normal patients | MC | Advanced glaucoma |
|---|---|---|---|
| Rim area | 1.86 | 1.65 | 0.777 |
| Rim volume | 0.522 | 0.368 | 0.118 |
| Cup shape measure | −0.239 | −0.117 | 0.022 |
| RNFL thickness | 0.282 | 0.25 | 0.087 |
| RNFL Ac. Sec. area | 1.502 | 1.356 | 0.469 |
| Scoring | −1.73 | 1.575 | 9.819 |

*MC closer to AG:* Only the cup volume showed a tendency to appear similar in MC with the group of real glaucoma.

| | Normal patients | MC | Advanced glaucoma |
|---|---|---|---|
| Cup volume | 0.074 | 0.437 | 0.734 |

When comparing the width of the optic rim between the three groups NE, MC and AG we found the following distribution (Table 3) Total rim widths was lower in AG than in the other two groups and MC showed slight lower values than NE. The sequence of

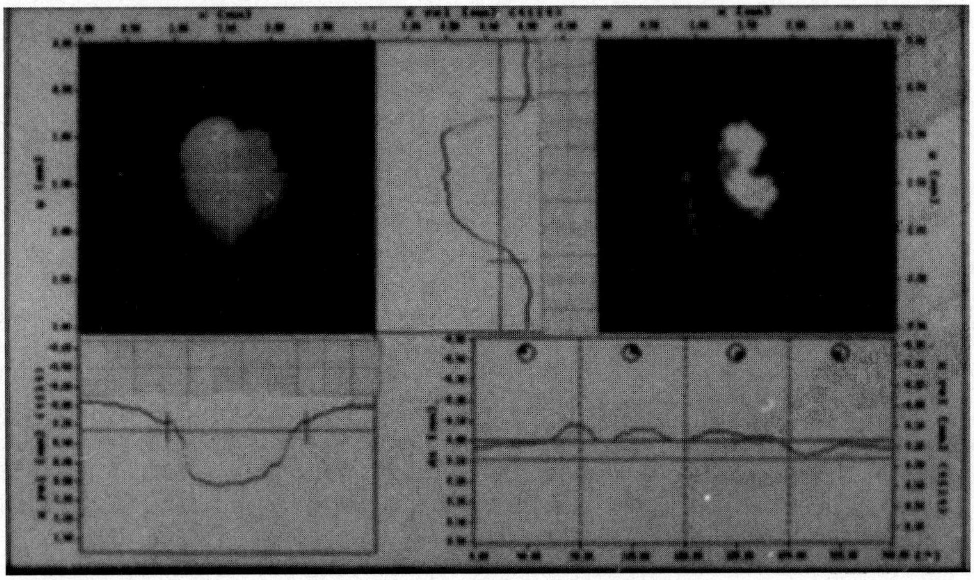

*Figure 1.*

*Table 1.*

| | Weinreb[23] | Capriolli[28] | Spaeth[24] | Drance[25] | Iester[29] | Flamer[26] | Arenas |
|---|---|---|---|---|---|---|---|
| Disk area | 2.02 | | 2.051 | 2.43 | 2.47 | 1.8 | 2.267 |
| Cup area | 0.46 | 0.47 | 0.418 | 0.71 | 0.73 | 0.406 | 0.405 |
| C/D area | 0.21 | 0.23 | 0.205 | 0.28 | 0.28 | 0.213 | 0.172 |
| Rim area | 1.57 | 1.53 | 1.646 | 1.73 | 1.76 | 1.4 | 1.863 |
| Cup volume | 0.1 | 0.1 | 0.074 | 0.23 | 0.23 | 0.095 | 0.074 |
| Rim volume | 0.43 | 0.37 | 0.476 | 0.48 | 0.49 | 0.388 | 0.522 |
| Mean cup depth | | | | 0.26 | 0.27 | 0.199 | 0.176 |
| Max. cup depth | 0.55 | | | 0.72 | 0.73 | 0.611 | 0.548 |
| Cup shape measure | 0.19 | 0.24 | | 0.2 | 0.24 | −0.235 | −0.239 |
| Height variable contour | 0.11 | 0.37 | | 0.4 | 0.4 | 0.391 | 0.419 |
| RNFL thickness | 0.24 | 0.23 | 0.276 | 0.26 | | 0.261 | 0.282 |
| RNFL Ac. S. A. | 1.22 | 1.17 | 1.367 | 1.41 | 1.43 | 1.234 | 1.502 |

thickness of the optical rim was identical in the NE and MC but coincidentally it was similar in the vertical plane in both NE and AG groups; the inferior being thicker than superior, but in the horizontal plane we found a greater difference between the temporal side, being almost three folds thicker the nasal side in NE when compared with AG. MC showed slight differences in all four cardinal measures.

## Discussion

Our group of NE has similar media values from others reported in the literature [27–30], However it is possible that other authors had included among the normals, those cases with MC, shifting some values to the left or to the right. It is probable that the normal value of the HRT parameters should be reviewed from time to time, until we can include only those cases that will never develop glaucoma after several years of observation.

It is interesting to observe that MC have a tendency to appear in large optical Discs, a finding already reported [3, 31], however, in our series the AG disk area was very similar with the NE.

From all HRT parameters great differences were found in those indicating loss of nerve fibers such as

*Table 2.*

| | Normal patients | MC | Advanced glaucoma |
|---|---|---|---|
| Disk area | 2.26 | 2.91 | 2.491 |
| Cup area | 0.4 | 1.26 | 1.713 |
| C/D area | 0.172 | 0.427 | 0.734 |
| Rim area | 1.86 | 1.65 | 0.777 |
| Cup volume | 0.074 | 0.437 | 0.734 |
| Rim volume | 0.522 | 0.368 | 0.118 |
| Mean cup depth | 0.17 | 0.34 | 0.452 |
| Maximum cup depth | 0.54 | 0.79 | 0.848 |
| Cup shape measure | −0.239 | −0.117 | 0.022 |
| Height variable contour | 0.41 | 0.35 | 0.285 |
| RNFL thickness | 0.282 | 0.25 | 0.087 |
| RNFL Ac. Sec. area | 1.502 | 1.356 | 0.469 |
| Points | −1.73 | 1.575 | 9.819 |

*Table 3.*

| | Normal patients | MC | Advanced glaucoma |
|---|---|---|---|
| Total horizontal disk | 26.433 | 30.018 | 24.769 |
| Total vertical disk | 28.433 | 30.473 | 27.462 |
| Horizontal cup | 9.267 | 19.929 | 20.923 |
| Vertical cup | 10.733 | 19.795 | 21.962 |
| Temporal rim | 4.683 | 3.277 | 1.038 |
| Nasal rim | 12.317 | 7.098 | 1.769 |
| Superior rim | 7.833 | 5.232 | 2.462 |
| Inferior rim | 9.183 | 6.054 | 2.962 |

rim area, rim volume, and the two RNLF values indicating a clear damage in the number of optic nerve fibers in the AG. Cup volume in MC was close similar to AG and should be a factor to discard when deciding if a MC has some features of a real glaucoma. Cup shape measure or third moment has been accepted as one of the most important parameters to diagnose glaucomatous changes in the optic nerve with the HRT [32–34]. The cup shape measure indicates the indentation of the layer of optic fibers against the papillary cup. This characteristic is difficult to analyze with opthalmoscopy or plannimetry and it is one of the advantages of the confocal examination with the HRT.

In our study we found that in spite that the cup measure has an intermediate value in MC, still shows some inclination similar to that found in less advanced glaucomas. As we have found in our previous study 22 this parameter has less value than those showing loss of fibers such as RNLF values. Other authors have found similar correlation [35].

We also compared the total value in units according to our scale from 0 to 10 and found that (MC) have HRT units of 1.5 units in comparison with the AG 9.1 and the NE minus 1.7.

Regarding the measurement of the neuroretinal rim: in AG it is thicker inferiorly followed by the superior, nasal and the thinnest is the temporal. In MC follows exactly the same sequence than the NE as follows: nasal, inferior, superior and also the thinnest temporal. A similar finding was reported in the Chinese literature [36]. The inferior zone should be always measured cautiously to differentiate (MC) from real normal tension glaucomas [37]. Planimetric measures are new tools that should be added to the HRT report to improve the information about neural rim sectorial thickness [38, 39]. Care should be taken to differentiate MC from optic nerve isquemic diseases [40] and from an incipient glaucoma controlling those cases with blue over yellow campimetries [41, 42]. It has also been found that, with growing optic disc size sensitivity of the retinal nerve fibre layer increased and third moment value decreased [43]. Recently a parallel involvement of both function and morphology-has been found, combining pattern electroretinogram amplitudes and optic disc cup measurements with the HRT [44].

## Conclusions

Confocal optic nerve analysis is an excellent tool to differentiate optic nerve cupping in patients with (MC) from those with real advanced glaucoma. Based on the HRT studies we can define (MC) as an escavation with a large CDR with a tendency to be larger in the horizontal plane but usually round, and with conservation of a thick neuroretinal rim in the inferior-nasal quadrant. The numerical scoring designed by us marks a clear limit between pathological and physiological MC. Further studies with a larger number of cases and longer follow-ups should be done to find out the real numerical parameters of this type of borderline cups.

## References

1. Armaly MF, Sayegh RE. The cup-disc ratio. The findings of tonometry and tonography in the normal eye. Arch Ophthalmol 1969; 82: 191–196.

2. Carpel EF, Engstrom PF. The normal cup-disk ratio. Am J Ophthalmol 1981; 91: 588–597.
3. Jonas JB, Budde WM, Panda Jonas S. Ophthalmoscopic evaluation of the optic nerve head. Surv Ophthalmol 1999; 43: 293–320.
4. Garway-Heath DF, Ruben ST, Viswanathan A, Hitchings RA. Vertical cup/disc ratio in relation to optic disc size: Its value in the assessment of the glaucoma suspect. Br J Ophthalmol 1998; 82: 1118–1124.
5. Schwartz B, Takamoto T. Measurement of retinal nerve fiber layer thickness and its functional correlation's with the visual field. Bull Soc Belge Ophtalmol 1992; 244: 61–72.
6. Schwartz B, Tomita G, Takamoto T. Glaucoma-like discs with subsequent increased ocular pressures. Ophthalmology 1991; 98: 41–49.
7. Mansberger SL, Sample PA, Zangwill L, Weinnreb RN. Achromatic and short-wavelength automated perimetry in patients with glaucomatous large cups. Arch Ophthalmol 1999; 117: 1473–1477.
8. Duke Elder SE, ed. System of Ophthalmology. Volume 3, Part 1. Normal and Abnormal Development. Embryology. CV Mosby, St Louis, 1963: 109–110.
9. Jonas JB, Gusek GC, Naumann GO. Optic disc, cup and neuroretinal rim size, configuration and correlation in normal eyes [published errata appear in Invest Ophthalmol Vis Sci 1991; 32: 1893 and 1992; 32: 474–475.
10. Armaly MF. Genetic determination of cup/disc ratio of the optic nerve. Arch Ophthalmol 1967; 78: 5–43.
11. Beck RW, Messner DK, Musch DC, Martonyi CL, Lichter PR. Is there a racial difference in physiologic cup size? Ophthalmology 1985; 92: 873–876.
12. Maisel JM, Pearlstein CS, Adams WH, Heotis PM. Large optic disks in the Marshallese population. Am J Ophthalmol 1989; 107: 145–150.
13. Schwartz JT, Reuling FH, Garrison RJ. Acquired cupping of the optic nerve head in normotensive eyes. Br J Ophthalmol 1975; 59: 216–222.
14. Tomlinson A, Phillips CI. Ratio of optic cup to optic disc. In relation to axial length of eyeball and refraction. Br J Ophthalmol 1969; 53: 765–768.
15. Jonas JB, Dichtl A. Optic disc morphology in myopic primary open-angle glaucoma. Graefes Arch Clin Exp Ophthalmol 1997; 235: 627–633.
16. Forero S, Arenas E. Scanning optic nerve diameter analysis in different ametropias with the optical Coherence Heilderberg retina Tomograph (HRT) Opthalmol. Invest Ophthal & Visual Sciences 1998; S39 abs 4342.
17. Jonas JB, Gusek G, Naumann GO. Makropaillen mit physiologischer Makroexkavation (Pseudo-Glaukompapillen). Papillometrischer Charakteristika in 17 Augen. Klin Monatsbl Augenheilkd, 1987; 191: 452–457.
18. Papastathopoulos KI, Jonas JB. Ophthalmoskopische Abschätzung der Größe der Papilla N.Optici. Klin Monatsbl Augenheilkd 1997; 211: 291–295.
19. Beatty S, Good PA, McLaughlin J, Tsaloumas M, ONeill EC. Evaluation of optic disc cupping using high-resolution ocular ultrasound. Eye 1998; 12: 54–60.
20. Saurian A, Orgül S, Koçak I, Prünte C, Flammer J. Descriptive information of topographic parameters computed at the optic nerve head with the Heidelberg retina tomograph. J Glaucoma 1998; 7: 420–479.
21. Sampaolesi R, Sampaolesi JR. Tomografąa confocal del nervio óptico y de la retina. Archivos de Oftalmología de Buenos Aires. Relato anual 1995.
22. Arenas E, Rodriguez M. Diseño de una nueva unidad cuantitativa en la interpretacion del HRT. Revista Peruana de Oftalmologia 1998; 22: 25–31.
23. Zangwill LM, van Horn S, de Souza Lima M, Sample PA, Weinreb RN. Optic nerve head topography in ocular hypertensive eyes using confocal scanning laser ophthalmoscopy. Am J Ophthalmol 1996; 122: 520–525.
24. Eid TM, Spaeth GL, Katz LJ, Azuara Blanco A, Agusburger J, Nicholl J. Quantitative estimation of retinal nerve fiber layer height in glaucoma and the relationship with optic nerve head topography and visual field. J Glaucoma 1997; 6: 221–230.
25. Iester M, Mikelberg FS, Courtright P, Drance SM. Correlation between the visual field indices and Heidelberg retina tomograph parameters. J Glaucoma 1997; 6: 78–82.
26. Saruhan A, Orgül S, Koçak I, Prünte C, Flammer J. Descriptive information of topographic parameters computed at the optic nerve head with the Heidelberg retina tomograph. J Glaucoma 1998; 7: 420–429.
27. Asawaphureekorn S, Zangwill L, Weinreb RN. Ranked-segment distribution curve for interpretation of optic nerve topography. J Glaucoma 1996; 5: 79–90.
28. Uchida H, Brigatti L, Caprioli J. Detection of structural damage from glaucoma with confocal laser image analysis. Invest Ophthalmol Vis Sci 1996; 37: 2393–2401.
29. Iester M, Broadway DC, Mikelberg FS, Drance SM. A comparison of healthy, ocular hypertensive, and glaucomatous optic disc topographic parameters. J Glaucoma 1997; 6: 363–370.
30. Eid TM, Spaeth GL, Katz LJ, Azuara Blanco A, Agusburger J, Nicholl J. Quantitative estimation of retinal nerve fiber layer height in glaucoma and the relationship with optic nerve head topography and visual field. J Glaucoma 1997; 6(4): 221–230.
31. Jonas JB, Gusek GC, Naumann GO. Optic disc, cup and neuroretinal rim size, configuration and correlations in normal eyes [published errata appear in Invest Ophthalmol Vis Sci 1991; 32: 1893 and 1992; 32(2): 474–475] Invest Ophthalmol Vis Sci 1988; 29: 1151–1158.
32. Iester M, Mikelberg FS, Swindale NV, Drance SM. ROC analysis of Heidelberg Retina Tomograph optic disc shape measures in glaucoma. Can J Ophthalmol 1997; 32: 382–388.
33. Bathija R, Zangwill L, Berry CC, Sample PA, Weinreb RN. Detection of early glaucomatous structural damage with confocal scanning laser tomography. J Glaucoma 1998; 7: 121–127.
34. Teesalu P, Vihanninjoki K, Airaksinen PJ, Tuulonen A, Läärä E. Correlation of blue-on-yellow visual fields with scanning confocal laser optic disc measurements. Invest Ophthalmol Vis Sci 1997; 38: 2452–2459.
35. Mistlberger A, Liebmann JM, Greenfield DS, Pons ME, Hoh ST, Ishikawa H, Ritch R. Heidelberg retina tomography and optical coherence tomography in normal, ocular-hypertensive, and glaucomatous eyes. Ophthalmology 1999; 106: 2027–2032.
36. Xu L, Liu L, Yang H. [A study of neuroretinal rim morphology in physiologic large cups and early glaucoma] Chung Hua Yen Ko Tsa Chih 1996; 32: 114–117.
37. Tomita G, Yamada T, Matsubara K, Kitazawa Y. Glaucoma-like disc without increased intraocular pressure or visual field loss and early stage normal-tension glaucoma disc] Nippon Ganka Gakkai Zasshi 1993; 97: 225–231.
38. Weber J, Dannheim F, Dannheim D. The topographical relationship between optic disc and visual field in glaucoma. Acta Ophthalmol (Copenh) 1990; 68: 568–574.

39. Vihanninjoki K, Tuulonen A, Burk RO, Airaksinen PJ. Comparison of optic disc measurements by Heidelberg Retina Tomograph and manual planimetric techniques. Acta Ophthalmol Scand 1997; 75: 512–515.
40. Orgül S, Gass A, Flammer J. Optic disc cupping in arteritic anterior ischemic optic neuropathy. Ophthalmologica 1994; 208(6): 336–338.
41. Johnson CA, Adams AJ, Casson EJ, Brandt JD. Blue-on-yellow perimetry can predict the development of glaucomatous visual field loss. Arch Ophthalmol 1993; 111: 645–650.
42. Kwon YH, Park HJ, Jap A, Ugurlu S, Caprioli J. Test-retest variability of blue-on-yellow perimetry is greater than white-on-white perimetry in normal subjects. Am J Ophthalmol 1998; 126: 29–36.
43. Mardin CY, Horn FK. Influence of optic disc size on the sensitivity of the Heidelberg Retina Tomograph. Graefes Arch Clin Exp Ophthalmol 1998; 236: 641–645.
44. Salgarello T, Colotto A, Falsini B, Buzzonetti L, Cesari L, Iarossi G; Scullica L. Correlation of pattern electroretinogram with optic disc cup shape in ocular hypertension. Invest Ophthalmol Vis Sci 1999; 40: 1989–1997.

*International Ophthalmology* **23:** 245–250, 2001.
*J.R. Sampaolesi (ed.), Laser Scanning: Update 1*, 63–68.




# Detecting AMD with Multiply Scattered Light Tomography

A.E. Elsner[1], Q. Zhou[2], F. Beck[2], P.E. Tornambe[3], S.A. Burns[1], J.J. Weiter[1,4] & A.W. Dreher[2]
[1]*Schepens Eye Research Institute;* [2]*Laser Diagnostic Technologies;* [3]*Retina Consultants, San Diego;* [4]*Retina Specialists of Boston*

## Abstract

*Purpose:* to use a novel technique, Multiply Scattered Light Tomography (MSLT), to provide a comfortable, rapid, and noninvasive method for detection and management of Age-related Macualar Degeneration. *Methods:* two patient groups were studied in clinical settings with MSLT and confocal scanning laser tomography. In Poway, CA, 21 retinal patients underwent tomography, and the 17 patients with suspicion of exudation also had ICG. An Angio-Scan (Laser Diagnostic Technologies, Inc.) was used to provide simultaneous fundus reflectance and ICG imaging. In Methuen, MA, 20 retinal patients underwent tomography with fluorescein angiography for suspicion of exudation. The MSLT was based on the TopSS (Laser Diagnostic Technologies, Inc.), with a Vertical Cavity Surface Emitting Laser array at 850 mm as the illumination source. The central laser produced confocal images. The surrounding lasers produced multiply scattered light images. *Results:* MSLT emphasized structures beneath the retina such as drusen, choroidal new vessel membranes, and pigment epithelial detachments. Exudation seen on angiography was visualized by MSLT as topographical structures with distinct borders. Superficial structures, e.g., cysts and epiretinal membranes, were visualized in 850 nm images. *Discussion:* confocal tomography and MSLT provided a rapid, noninvasive method to detect and localize macular degeneration and pathological structures found in eyes of older patients.

## Introduction

Age-related macular degeneration (AMD) is the leading cause of blindness in adults in industrialized countries (Liebowitz et al., 1980). The widespread methodology for the detection and localization of pathological changes, as well as the management of patients, remains based upon technology that was never optimized for either AMD or use in older patients. Primary examples include ophthalmoscopy, color photography, and fluorescein poor ocular media, particulary with short wavelength light necessary for fluorescein excitation.

We performed three main areas of technical research and development for the early detection of AMD and its exudative complications: (1) Confocal Imaging Tomography (CT), (2) Multiply Scattered Light Tomography (MSLT), and (3) interpretation of the novel image data and quantification. The technical development centered around new technology, MSLT.

The term MSLT denotes the use of light that is scattered multiple times, rather than singly scattered. By systematically changing the plane of focus, images acquired with this mode of imaging form a series of tomographic images of structures. The MSLT imaging methology builds on the indirect mode imaging work with a Scanning Laser Ophthalmoscope (Elsner, 1992, 1996; Hartnett, 1996a, b; Remky, 1999), and is related to dark field microscopy and to a lesser extent the view obtained with a slit lamp. Examples of posterior pole structures that singly scatter light are those with a high index of refraction change, situated roughly perpendicular to the optical axis of the eye and imaging device, such as the inner limiting membrane/retinal nerve fiber layer interface. In MSLT, the light from the single scattering sources is blocked, and the remaining multiply scattered light reveals structures not imaged with other techniques. This novel technique can help elucidate structures beneath strong single scatters, which are optimally imaged with confocal, polarization, or coherence methodology. Not only is the MSLT instrumentation described here novel (Elsner et al., 1998; Zhou et al., 1999), but also it goes beyond indirect mode imaging since there is a stack of images that permits new software for computations in three dimensions (Dreher et al., 1999).

The use of confocal tomographic (CT) methods have been available commercially for most of this decade. For example, the TopSS$^{TM}$ (Laser Diagnostic Technologies, San Diego, CA), performs measurements on structures that are mainly single scatterers, especially the retinal nerve fiber layer. CT has had greater penetration into clinical use with optic nerve head diseases rather than macular diseases. Clinical studies have shown that CT provides accurate (Dreher and Weinreb, 1991) and reproducible (Dreher et al., 1991; Fechner, 1992) quantification of the optic nerve head. Height measurements with the Topographic Scanning System TopSS$^{TM}$ (Laser Diagnostic Technologies, Inc., San Diego, CA; Fig. 2) are reproducible to within less than 30 $\mu$m for a pixel size of 11 $\mu$m (Fechner, 1992). The TopSS$^{TM}$ uses near infrared illumination, penetrating moderate Cataract or cloudy media, thereby being useful in a large segment of the aging population, since these patients often have cloudy media, intraocular lenses, or small pupils. The illumination is readily tolerated, and the instrument is optimized for a 3 mm pupil.

Macular use of CT instrumentation has primarily been confined to superficial structures, such as macular holes, cysts, and edema (Bartsch, 1989, 1994; Beausencourt 1998, in press; Hudson, 1998). Recently, using CT, we have shown that CT is useful in the management of one of the serious exudative complications of AMD: pigment epithelial detachment (PED). The relative height to width ratio of a PED can be readily quantified. The ratio, height, and shape are related to severity of vision loss (Kunze, 1998, 1999).

The new MSLT methodology allows the use of CT, MSLT, or the two imaging modes nearly simultaneously via line-by-line alternation. The rapid alternation provides direct comparison of information from both modes of imaging. We performed the initial clinical studies to test the feasibility of diagnosing subretinal changes associated with AMD with MSLT, using the modified TopSS$^{TM}$ system.

## Methods

We performed 3 sets of clinical feasibility studies. To demonstrate the ease of portability of the MSLT device, all but one retinal patient was tested in a clinical office rather than a laboratory. Patients signed an approved consent form, and there was human studies approval of the protocol by the boards of LDT and SERI for subjects from California and Massachusetts, respectively for all sites.

First, we tested 21 patients with retinal disease in an ophthalmology practice in Poway, CA. Of these, 17 had AMD, and 15 patients were suspicious for current exudative AMD prior to study entry, by either examination by a retinal specialist or fluorescein angiography (FA). These patients also underwent confocal indocyanine green angiography with an Angioscan$^{TM}$ (Laser Diagnostic Technologies, San Diego, CA), under the direction of their physician, Dr. Paul Tornambe. In many cases the photographic fluorescein angiography (FA) was of poor quality in these older eyes with media problems, and the reason for the return visit for to perform improved imaging prior to potential treatment. Next, in Methuen, MA, 20 retinal patients underwent tomography. There was photographic fluorescein angiography for patients with suspicion of exudation, under the direciton of Dr. John Weiter. In addition, over 40 normal subjects have been tested to date, aged 20-69 y.

## MSLT Apparatus

A prototype instrument was based on the TopSS$^{TM}$. A 3 $\times$ 3 square array of vertical cavity surface emitting lasers (VCSELs) (VIXEL Corp., Broomfield, CO) with a physical spacing of 125 $\mu$m between lasers in a row or column was implemented as the illumination source. The VCSEL array, which is a solid state laser device capable of GHz bandwidth, was mounted on a microchip. The array consisted of one central and eight peripheral lasers, each individually controlled. The lasers were used in continuous mode, although the center laser could be alternated with the surrounding lasers, switching at a line frequency of 4 kHz.

VCSELs have radially symmetric beam profiles, unlike the edge-emitting laser diodes previously used in scanning laser devices. The lateral and axial transfer functions are potentially benefitted by a well-focused spot on the retina. The result is the possibility of higher contrast for retinal structures. The VCSELs used were 850 nm, slightly longer wavelength than in commercially available devices, which is also an advantage (Elsner et al., 1996). The total power at the cornea from the VCSEL array did not exceed 1 mW.

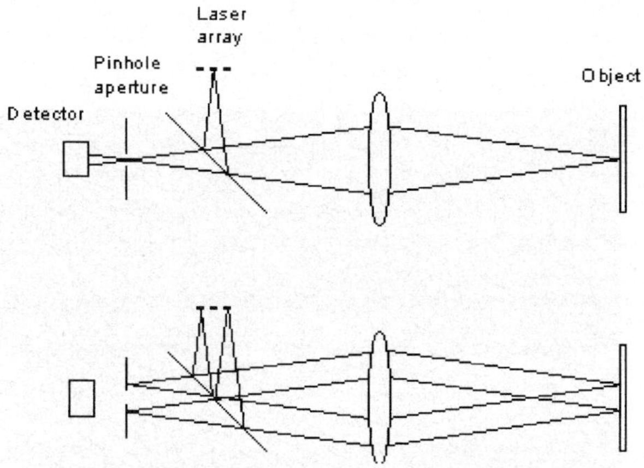


*Figure 1.* Diagram of the basic optical principal for multiply scattered light tomography. A scanning laser device is modified to have a VCSEL array as the illumination source. (a) Top: When the center laser in the array, which is on axis with the confocal aperture, is illuminated, then images are dominated by directly backscattered light. Confocal imaging results. (b) Bottom: When the outer lasers in the array, which are offset from the confocal aperture, are illuminated, light that is directly backscattered is blocked. The multiply scattered light can pass through the aperture, and the images are dominated by structures that can scatter light multiple times.

Additional optics were introduced so that the VCSEL beam was collimated, and the laser spacing on the retina was minified to 47 $\mu$m. The pinhole size was 40 $\mu$m in diameter with respect to the retina, while the commercial TopSS$^{TM}$ is 24 $\mu$m. Pilot data indicated that larger pinhole apertures, while providing more light to the detector from the eye, broadened the axial profile and led to decreased contrast in the confocal images. The confocal aperture was on axis with the center laser. Illuminating the center laser produced directly backscattered light images for CT (Figure 1a). The surrounding lasers were off axis with respect to the confocal aperture. Illuminating the surrounding lasers produced multiply scattered light images for MSLT (Figure 1b). Scanning in depth with the center laser illuminated resulted in CT data. Scanning in depth with the surrounding lasers resulted in MSLT data.

For each image series, 32 images were acquired in 0.9 sec while the focusing mirror was scanned in an axial dimension. The central and surrounding lasers could be illuminated alternatively, line-by-line, to provide an image in which the two modes were in axial and lateral registration. Each image had 256 $\times$ 256 pixel resolution and 8 bits of grayscale. The images were typically 10 $\times$ 10, 15 $\times$ 15, or 20 $\times$ 20 deg visual angle, but for the Poway study we used 10 $\times$ 10, 20 $\times$ 20, and 30 $\times$ 30 deg to match the Angioscan field size and resolution.

A new software package, developed at LDT, separated and processed data from the two imaging modes. To provide the lateral location of the optic nerve head borders, the multiply scattered light images were aligned using the landmarks from the matching confocal sections. This permitted use of the commercial TopSS$^{TM}$ software, including plotting axial profiles.

To obtain full information for the three-dimensionality of structures, a cross-sectional approach was taken. Intensity values, or the intensity difference values, were plotted to form a cross-sectional image from the 32 sections in the $x$, $z$ plane, using a Matlab (MathWorks, Natick, MA) program specially written by one of the authors (SAB). Any $x$, $z$ or $y$, $z$ plane can be selected by the operator, according to the feature of interest.

## Results

Most Poway patients were clearly imaged without the instillation of mydriatic drops, although a few patients with pupils $<$2 mm diameter had reduced image quality and brightness. In no case was the FA, typically of poor quality due to lens opacities and in one patient a nonmydriatic pupil, equal in quality to the either the MSLT or the ICGA images.

Figure 2 shows a comparison of ICGA and MSLT, with both identifying the choroidal neovascularization components readily. All types of imaging agreed as to the location and extent and location of the choroidal neovascularization and the fluid accumulation. The MSLT device readily detected exudation without the injection of dye. As expected from our previous work (Elsner et al., 1996), the accumulation of fluid in exudation appeared dark with respect to the surrounding retina in both CT and MSLT. The MSLT images showed the borders of exudative lesions, beneath the retina. In contrast, CT emphasized the retinal features, including the elevation and deformation of the retinal nerve fiber layer by subretinal exudation. CT also emphasized small exudates, typically indicating exudation beneath the retina at that location. The live images using MSLT provide the added cue of focusing and motion parallax, enabling the operator to localize the lesion in seconds. A typical example that drusen are readily seen with MSLT is also demonstrated by this figure, appearing as round structures that are brighter than the surrounding retina. This pa-

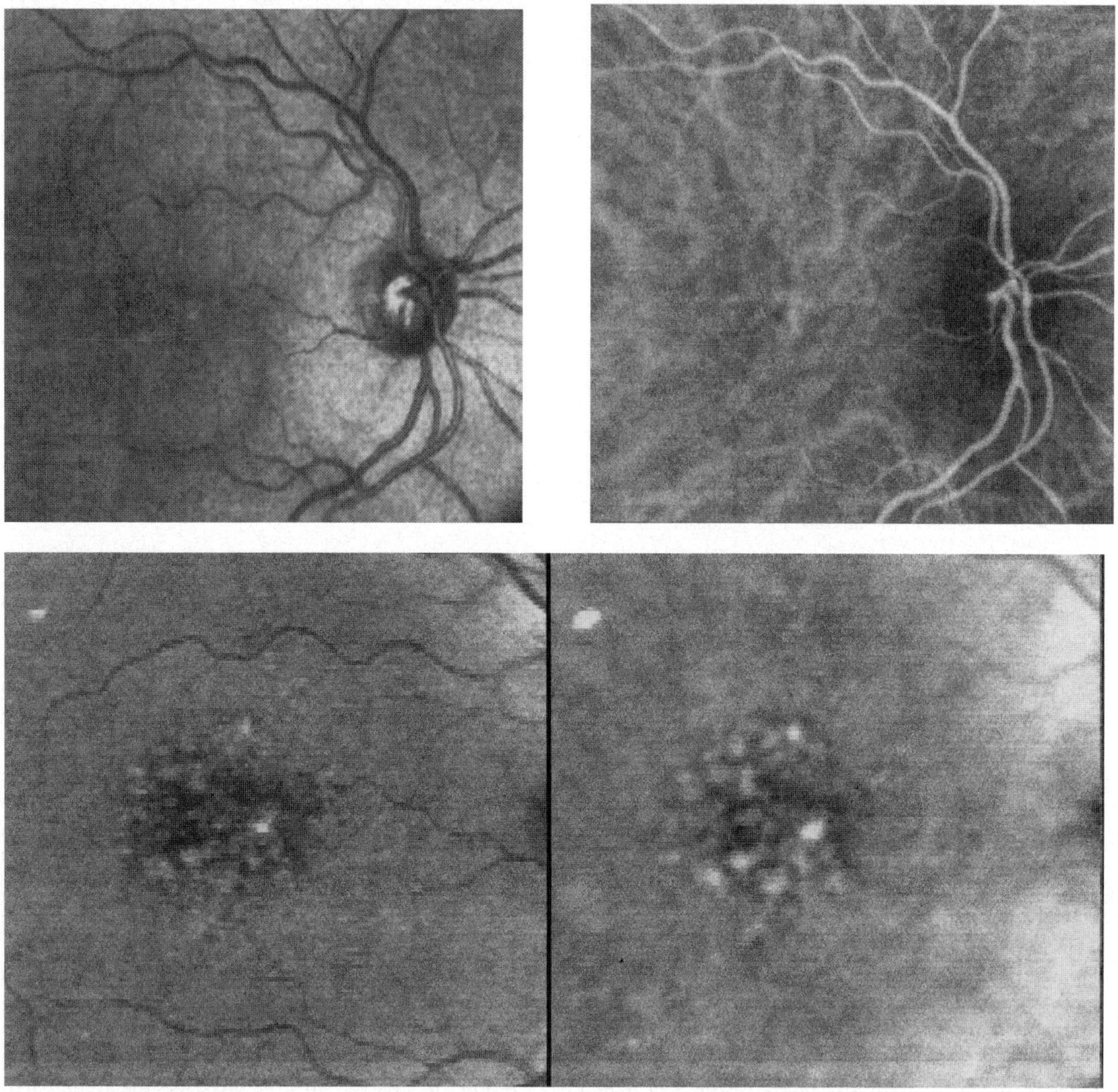

*Figure 2.* Top: 30 deg intensity and ICGA images of the posterior pole of a patient with exudative age-related macular degeneration, centered on the macula. Top left: summary intensity image at 790 nm, showing a shallow pigment epithelial detachment involving the fovea and inferior macula. Surrounding the fovea are bright exudates. Top right: mid-phase ICGA, taken at exactly the same time as the intensity image, and thus in registration. There is a subfoveal, choroidal neovascular membrane that is less than one disk diameter, surrounded by fluid, lying underneath the bright exudates. Bottom: 20 deg image of the posterior pole of the same patient, centered on the bright exudates and taken on the same day with the MSLT device. Bottom left: Confocal Tomography (CT) emphasizing superficial features such as retinal blood vessels, from the average of only two sections. The fluid exudation is clearly seen as darker than the surrounding retina, but the entire exudative region, i.e., the inferior retina, is too extensive to be imaged at this field size. Right: Multiply Scattered Light Tomography (MSLT) image, acquired as interlaced images with the CT image at left, emphasizing deeper structures scuh as choroidal vessels.

tient also has reticular pseudodrusen peripheral to the fovea in both the inferior and superior retina.

While performing the Poway study, it became clear that the confocal imaging provided unexpectedly clear information concerning retinal abnormalities. This is despite the relatively long wavelength used, 850 nm. Relatively small retinal features, such as retinal blood vessels and small exudates, are seen in the CT image. The CT and MSLT techniques combined to reveal neovascularization that is both small and subtle, less than one disk diameter, necessary if treatments are going to be applied to small lesions early on in the disease. Adequate depth resolution is also necessarily if there is to be evaluation of potential fluid detachment of the neurosensory retina. There were clear-cut differences among images from different focal planes. When the summary image of all 32 sections is shown, as is the case in commercial tomographic software, the superficial and flatter tissue layers typically dominate the confocal image. The bright retinal nerve fiber layer/inner limiting membrane interface particularly dominates the summary image. The structures beneath this layer, often crucial for retinal disease, can be detected better with confocal images in the deeper focal plane than a summary image. MSLT often provides even better images if the target features are not extremely small.

The Methuen patients were also clearly imaged, although these patients had received mydriatic drops prior to imaging, as part of their ophthalmological examination. For these patients, the main source of degradation of image quality was abnormal corneal tear film. This was primarily a factor for the most anterior or posterior image planes. The image acquisition time is rapid enough to avoid most of the effects of vitreous opacities. Again, patients preferred the comfort of CT and MSLT to FA. Findings were similar to those with the Poway patients, in that deeper and exudative structures were well-visualized by MSLT, although sometimes seen in images from CT at the appropriate depth. Small retinal features were best imaged with individual sections or the average of 2–3 sections, rather than the summary image, using CT.

## Discussion

The MSLT device readily detected exudation, and in some patients the abnormal new choroidal vessels themselves, without the injection of dye. Many patients did not need mydriatic drops, although the instrument worked well if these had already been instilled. Dry corneas degraded image quality in some patients, but not to the extent that measurements were precluded. Patients preferred this instrument to the conventional photography, primarily because of the uncomfortably bright lights commonly used in photography.

With the standard TopSS, a CT instrument, we used quantitative height measurements and confocal imaging to quantify exudative structures in pigment epithelial detachments and occult choroidal neovascularization (Kunze et al., 1999). This furthered our earlier work on retinal vascular anomalous complexes (Hartnett et al., 1996). Patients with exudation that involves the retinal structures is especially high and has a very poor prognosis compared with shallower lesions. The MSLT instrumentation combines this capability with scattered light imaging.

Although we have previously published high contrast images images at wavelength ranges including and even exceeding 850 nm (Elsner et al., 1996), it is often erroneously assumed that a wavelength of 850 nm is too long for good confocal images. The lateral transfer and axial transfer functions both increase with wavelength. However, the minimal decrease in resolution compared with 633 or 790 nm is more than compensated by the improved contrast due to less absorption anterior to the retina. In addition, the excellent beam quality and circular symmetry of the 850 nm VCSEL permit a small point spread function in the plane of focus compared with edge-emitting diodes that diverge more in one direction than another. This led to better than anticipated CT images, capable of showing abnormal retinal vessels in the central macula when the highest magnification was used.

## References


1. Bartsch D-U, Intaglietta M, Bille JF, et al. Confocal laser tomographic analysis of the retina in eyes with macular hole formation and other focal macular diseases. Am J Ophthalmol 1989; 108: 277–287.
2. Bartsch D-U, Freeman WR. Axial intensity distribution analysis of the human retina with a confocal scanning laser tomograph. Exp Eye Res 1994; 58: 161–173.
3. Beausencourt E, Elsner AE, Hartnett ME, Trempe CL. Quantitative analysis of macular holes with scanning laser tomography. Ophthalmol 1997; 104: 2018–2029.
4. Beausencourt E, Remky R, Elsner AE, Hartnett ME, Trempe CL. Infrared scanning laser tomography of macular cysts. Ophthalmology (In press).
5. Dreher AW, Tso PC, Weinreb RN: Reproducibility of topographic measurements of the normal and glaucomatous

optic nerve head with the laser tomographic scanner. Am J Ophthalmol 1991; 111: 221.
6. Dreher AW and Weinreb RN. Accuracy of topographic measurements in a model eye with the laser tomographic scanner. Invest Ophthalmol Vis Sci 1991; 32(11): 2991.
7. Dreher AW, Zhou Q, Elsner AE, Beck FW, Tornambe PE, Burns SA. A novel image processing technique to detect and map retinal and choroidal lesions. [ARVO Abstract]. Invest Ophthalmol Vis Sci 1999; 40(4): S119. Abstract No. 642.
8. Elsner AE, Burns SA, Hughes GW, Webb RH. Reflectometry with a scanning laser ophthalmoscope. Applied Optics 1992; 31: 3697–3710.
9. Elsner AE, Burns SA, Weiter JJ, Delori F. Infrared imaging of sub-retinal structures in the human ocular fundus. Vision Res 1996; 36: 191–205.
10. Elsner AE, Dreher AW, Zhou Q, Beausencourt E, Burns SA, Webb RH. Multiply scattered light tomography: vertical cavity surface emitting laser array used for imaging subretinal subretinal structures. Lasers and Light in Ophthalmology 1998; 8: 193–202.
11. Elsner AE, Bartsch D-U, Weiter JJ, Hartnett ME. New devices in retinal imaging and functional evaluation, in W Freeman (ed.), Practical Atlas of Retinal Disease and Therapy, 2nd edn. Lippincott-Raven, New York, 1998: 19–35.
12. Fechtner RD, Dreher AW, Weinreb RN. The laser scanning ophthalmoscope, in R Varma and G Spaeth (eds.), Optic Nerve. JB Lippincott Company, 1992: 269–277.
13. Hartnett EM, Elsner AE. Characteristics of exudative age-related macular degeneration determined *in vivo* with confocal direct and indirect infrared imaging. Ophthalmology 1996; 103: 58–71.
14. Hartnett ME, Weiter JJ, Staurenghi G, Elsner AE. Deep retinal vascular anomalous complexes in advanced age-related macular degeneration. Ophthalmol 1996; 103: 2042-2053.
15. Hudson C, Flanagan JG, Turner GS, McLeod D. Scanning laser tomography Z profile signal width as an objective index of macular retinal thickening. Br J Ophthalmol 1998; 82: 121–130.
16. Kunze C, Moraes L, Elsner AE, Hartnett ME. Alterations in the foveas of aging subjects and patients with ARM, and AMD. Retinal sensitivity in pigment epithelial detachments correlates to three-dimensional extent, American Academy of Ophthalmology, 1998.
17. Kunze C, Elsner AE, Beausencourt E, Moraes L, Hartnett ME, Trempe CL. Spatial Extent of Pigment Epithelial Detachments in Age-related Macular Degeneration. Ophthalmol 1999; 9: 1830–1840.
18. Leibowitz H, Krueger DE, Maunder LR, et al. The Framingham Eye Study Monograph: An ophthalmological and epidemiological study of cataract, glaucoma, diabetic retinopathy, macular degeneration and visual acuity in a general population of 2631 adults, 1973–1975. Surv Ophthalmol 1980; 335–610.
19. Remky A, Beausencourt E, Hartnett ME, Trempe CL, Arend O, Elsner AE. Infrared imaging of cystoid macular edema. Graefe's. Arch Clin Exp Ophthalmol 1999; 237: 897-901.
20. Zhou Q, Elsner AE, Beck FW, Tornambe PE, Burns SA, Dreher AW. Macular exudation imaged with muliply scattered light compared with confocal imaging and Indocyanine Green angiography. [ARVO Abstract]. Invest Ophthalmol Vis Sci 1999; 40(4): S123. Abstract No. 666.

*International Ophthalmology* **23:** 251–257, 2001.
*J.R. Sampaolesi (ed.), Laser Scanning: Update 1*, 69–75.




# Large optic nerve heads: megalopapilla or megalodiscs

Roberto Sampaolesi & Juan Roberto Sampaolesi
*University of Buenos Aires, School of Medicine, Department of Ophthalmology, Buenos Aires, Argentina*

## Abstract

*Purpose*: To evaluate megalopapillas in order to determine whether they constitute a different population, according to their specific features. *Material*: Evaluation of the optic nerve head of 405 eyes divided into 172 normal eyes, 168 pre-perimetric glaucomas (phase 3 or 4), 30 primary congenital glaucomas (children with pure congenital glaucomas operated more than once and diagnosed within the first year of age), and 35 megalopapillas. *Method*: The examinations were performed with the HRT (Heidelberg Retina Tomograph), using a wavelength of 680 nm and the new standard reference plane. Each examination results from the mean of 3 image acquisitions with a SD lower than n = 20. Each optic nerve head was studied in 360 degrees (segments), as well as in quadrants and octants separately (predefined segments). The visual fields were assessed with either the Octopus 1-2-3 or the Octopus 101, programs G2 and G2x (three complete phases). The intraocular pressure was measured by means of daily pressure curves including 7 measurements with applanation tonometry at the office and the first one at 6/7 a.m. with the patient still in bed. Gonioscopic examinations were also performed for classification purposes. *Results*: By comparing all the stereometric parameters with the Total Area (disc area) and the Rim Volume it has been concluded that the megalopapilla group is different from both the normal control group and the glaucoma group. The disc area of the congenital glaucomas was not statistically different from the megalopapillas, but it was in terms of rim volume and other parameters. *Conclusion*: megalopapilla is an entity characterized by a large optic nerve head which may appear abnormal, with an increased cup, but associated with a normal rim volume, normal visual field and normal IOP. Special attention should be given to its differentiation from pseudoglaucomatous diseases. *Discussion*: The frequency of megalopapillas seems to be quite higher than in congenital optic nerve head anomalies, but is has been virtually ignored by the literature worldwide so far.

## Background

Bock and Franceschetti [1] originally coined the term megalopapilla for enlarged optic discs not associated with any other morphological anomalies.

Megalopapilla may present with either of two phenotypic features:

1. Normal configuration with an abnormally large optic disc, high cup/disc ratio, disc surface pallor, neuroretinal rim pallor due to the presence of axons spread over a large surface, usually round or horizontally oval-shaped optic disc cup and absence of nasalization of vessels at their point of origin (usually bilateral and congenital).
2. (Less common) The cup is displaced towards the top and thus obliterating the adjacent neuroretinal rim [2].

The appearance of the megalopapilla thus resembles that of glaucomatous neuropathy. It is indeed, a pseudoglaucomatous disc. Most cases of megalopapilla may be merely a statistical variant of normality.

Megalopapillas may occasionally result form altered optic axonal migration, early in embryogenesis, in children with basal encephalocele [3]. Basal encephalocele is often associated with other midline anomalies such as hypertelorism, broad nasal root, cleft lip and cleft palate. Optic disc anomalies, such as pallor, dysplasia, optic disc pits, colobomas and megalopapillas have been reported [4]. Caprioli has described an association between basal encephalocele and Morning Glory Syndrome [5]

According to our experience, megalopapillas may be divided into congenital and acquired ones. Acquired megalopapillas are associated with type II congenital glaucomas (reoperated congenital glaucomas) [6]. Grimson [7] described an orbital glioma case which, in children, can lead to progressive enlargement of a previously normal-sized optic disc.

Collier [8] describes four cases of megalopapilla associated with pulverulent cataract and suggest that this finding may be explained by the development of the epithelial primitive disc during the second month of gestation and of the pulverulent cataract in the third month of gestation.

Hirokane [9] described megalopapilla in eight eyes of four children. The disc, with glaucomatous-like cupping, was so large that the macular diameter (measured from the center of the disc to the macular edge) to disc diameter ratio (DM/DD) was lower than 2.2. Although the cup/disc ratio and the appearance of the disc margin and the disc vessels in these cases were the same as those of glaucomatous eyes, the population they reported showed absence of rim pallor, notch and nerve fiber layer defects typically present in glaucomatous eyes. These findings demonstrate the importance of determining the DM/DD ratio for the differentiation of megalopapillas from glaucomas in children with large optic cups.

Jonas [10] published an interesting paper on "pseudoglautomatous" physiologic large cups in megalopapillas with the use of planimetric analysis of stereoscopic optic disc photographs. Upon studying twenty-one megalopapillas he concluded that "here are individuals with abnormally high cup/disc ratio but no pathologic findings". It was possible to identify intrapapillary and juxtapapillary characteristics that may be helpful in the differentiation of glaucomatous eyes and normal eyes with high cup/disc ratios.

Maisel [11] studied 141 native adults from the Marshall Island. In this genetically isolated population he found 22 eyes of 15 subjects with megalopapilla and normal IOP. Three large discs with an 18-year photographic follow-up showed no change.

## Purpose

The purpose of this presentation is to evaluate megalopapillas in order to determine whether they constitute a different population, by showing their specific features, as well as to find the differences between this entity and other optic nerve head neuropathies with enlarged disc area, like congenital glaucomas, goniodysgenesis with megalopapilla and advanced optic nerve head glaucomatous damage, and to perform differential diagnosis between megalopapilla and pure congenital glaucomas in children.

## Material

The population studied was divided into 4 groups:

- Group 1: 33 congenital megalopapillas, 13 of which were not associated with any other disease, 11 were associated with goniodysgenesis, 6 with open-angle glaucoma, 2 with hypophysis non-functional tumors and 1 with exfoliation syndrome.
- Group 2: 30 optic neuropathies in pure congenital glaucomas diagnosed within the first year of age and operated from 2 to 7 times,[1] which became acquired megalopapillas.
- Group 3: 168 open-angle glaucomas in the perimetric stage (with visual field defects characteristic of phase 3 or 4).
- Group 4: 172 normal subjects as controls.

## Method

The optic nerve was evaluated with the HRT, software version 2.12 (wavelength: 680 nm). All the stereometric parameters were measured. The visual fields were examined with Octopus perimetry. Slit-lamp examination of the posterior segment, as well as daily pressure curves, by which mean and standard deviation were obtained, were also performed.

## Classification

By definition, a megalopapilla is an optic nerve head disc with a surface greater than 2.5 $mm^2$.

Type 1: normal configuration, bilateral, high cup/disc ratio, pale disc surface, round or oval cup and congenital.

[1] This group belongs to a previous series divided into three groups:
Group 1: primary or pure congenital glaucomas diagnosed within the first year of age successfully operated once.
Group 2: with more than one surgery required and which is the group we have further studied here.
Group 3: late congenital glaucomas or goniodysgenesis.
Groups 1 and 2 have a significantly enlarged axial length according to echometry, while in group 3, since diagnosis was made after 5 years of age, when the eye was no longer elastic, the axial length is normal [12].

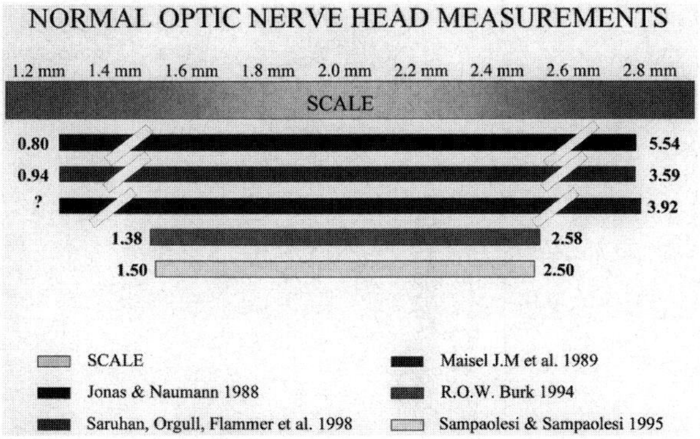


*Figure 1.*

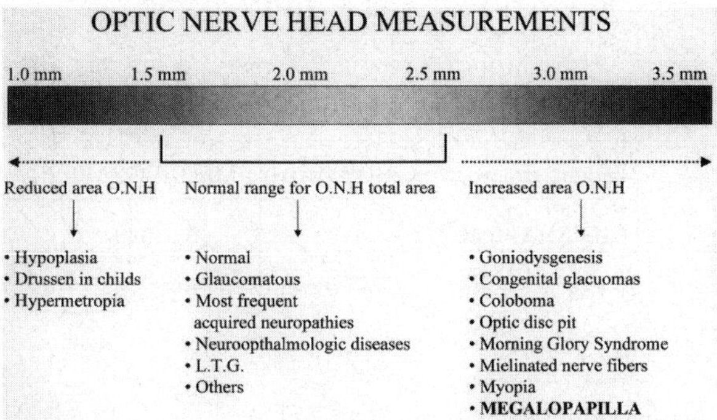


*Figure 2.*

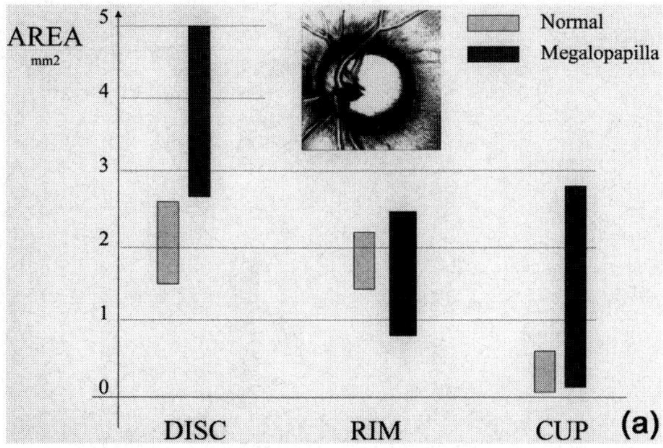


RESULTS

| Parameter/ Group | NORMAL | MEGALOPAPILLA |
|---|---|---|
| DISC AREA | 2.05 | 3.07 *** |
| RIM AREA | 1.74 | 1.50 * |
| CUP AREA | 0.32 | 1.57 *** |

(b)

ns: not significant *: $p < 0.01$ ***: $p < 0.0001$

*Figure 3.*

Type 2: decentrated to top cup, unilateral, higher frequency of cilioretinal arteries, neuroretinal rim reduced in top, round cup and congenital.

*Normal optic nerve head measurements*

Figure 1 shows different measurements made by Jonas and Naumann (1989) [10], Saruhan et al. (1998) [13], Maisel et al. (1989) [11], Burk (1992) [14] and Sampaolesi and Sampaolesi (1995) [15].

For this study we have used the values determined by us, which are very similar to those determined by Burk. This narrower range enables a better statistical differentiation.

Figure 2 demonstrates that normal disc areas may be present in the glaucomas, as well as in the most frequent acquired neuropathies, neurophthalmologic disorders and low-tension glaucomas, while reduced disc areas may be found in optic nerve hypoplasias, children with drusen and hyperopic cases. Increased disc areas may be found in goniodysgenesis, congenital glaucomas, optic disc colobomas, optic disc pits, Morning Glory Syndrome, myelinated nerve fibers, myopias and megalopapillas.

## Results

We will now compare the optic nerve head parameters of the megalopapilla with those of normal discs, first in terms of area, then in terms of volume and lastly, in terms of the third moment.

The ordinates of Figure 3a show the area in $mm^2$, while in the abscissas, the first group compares the disc area, the second group, the rim area and the third one, the cup area.

Figure 3b shows the statistical values of these comparisons, with a highly significant difference for disc area and cup area. The difference is poorly significant for rim area.

Figure 4a compares volumes. The abscissa shows volume values in $mm^3$ and there are two groups represented in the ordinate, the first one showing cup volume and the second one, rim volume.

The statistical analysis of Figure 4b shows a significant difference in cup volume and no significant difference in rim volume.

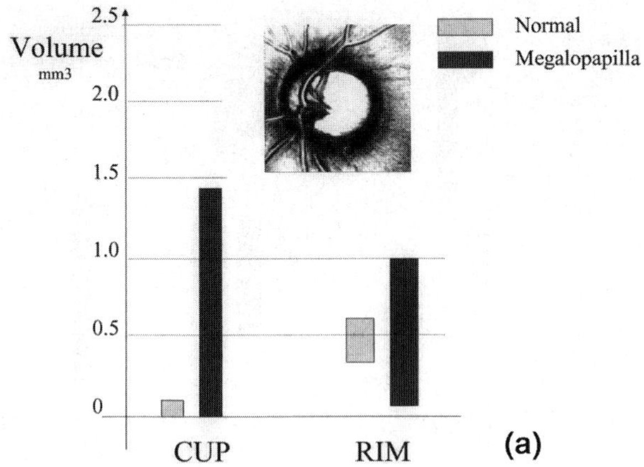


=NORMAL OPTIC NERVE HEAD PARAMETERS

(b)

| Parameter/ Group | NORMAL | MEGALOPAPILLA |
|---|---|---|
| RIM VOLUME | 0.48 | 0.36 ns |
| CUP VOLUME | 0.06 | 0.59 *** |

ns: not significant *: $p < 0.01$ ***: $p < 0.0001$

*Figure 4.*

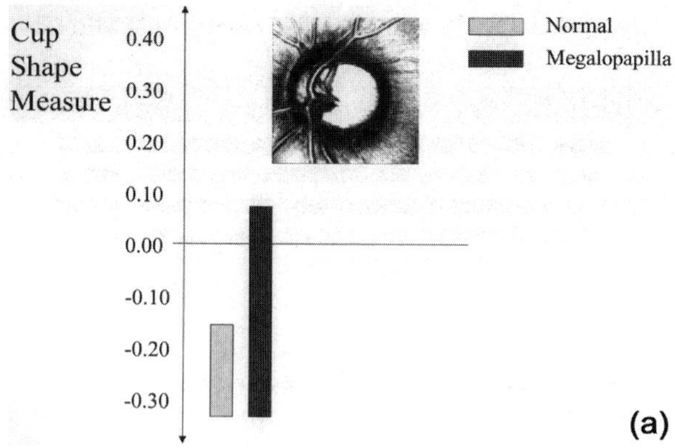


=NORMAL OPTIC NERVE HEAD PARAMETERS

(b)

| Parameter/ Group | NORMAL | MEGALOPAPILLA |
|---|---|---|
| CUP SHAPE M. | -0.22 | 0.06 *** |

ns: not significant *: $p < 0.01$ ***: $p < 0.0001$

*Figure 5.*

The ordinates in Figure 5a show the values for cup shape measure and the abscissa represents the comparison between the groups studied.

There is a statistically high difference which makes the image of megalopapillas always seem very pathological (Figure 5b).

The following figures compare the advanced glaucoma group (with visual field defects in phase 3 or 4) with megalopapillas.

The abscissa in Figure 6a shows area values in $mm^2$ and in the ordinate there are 6 groups represented in order to make a comparison in terms of disc area, rim area, rim volume, cup area, cup volume and cup shape measure.

When comparing the megalopapillas with advanced glaucomas, there is a statistically highly significant difference in terms of disc area, rim area and rim volume, while the difference is less significant in terms of cup area and cup volume and there is no significant difference as regards cup shape measure (Figure 6b).

When comparing the megalopapillas with congenital glaucomas:

In Figure 7 the ordinates show parameter values either in $mm^2$ or in $mm^3$, and the different optic nerve head parameters are represented in the abscissa: disc area, rim area, rim volume, cup area, cup volume and cup shape measure.

The statistical difference is highly significant as regards cup shape measure, poorly significant as regards rim area and non significant in terms of disc area, rim volume, cup area and cup volume (Figure 7b). It should be kept in mind that the cup area has the same value in the megalopapillas as in reoperated congenital glaucomas and that the mean disc area is 3 $mm^2$ in both groups. In congenital glaucomas the axial length of the eye is large, thus enlarging the Elschnig's ring, and the disc in turn [6].

The appearance of the megalopapilla in the right eye illustrated by the HRT shown by Figure 8 is almost identical to that of advanced glaucoma (left eye, same figure), thus leading to mistaken diagnosis, since upon observing a megalopapilla the ophthalmologist will tend to think of a congenital glaucoma first. As seen in the figure, the area of the megalopapilla is even larger.

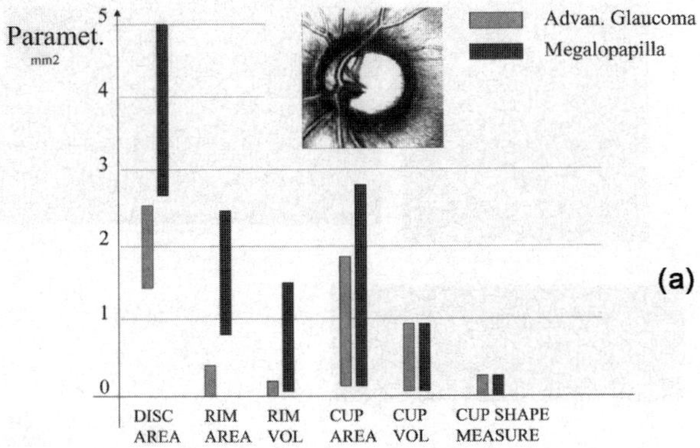


(b)

| Parameter/ Group | ADV.GLA. | MEGALOPAPILLA |
|---|---|---|
| DISC AREA | 2.05 | 3.07 *** |
| RIM AREA | 0.20 | 1.50 *** |
| RIM VOLUME | 0.05 | 0.36 *** |
| CUP AREA | 1.80 | 1.57 * |
| CUP VOLUME | 0.96 | 0.59 * |
| CUP SHAPE M. | 0.00 | 0.06 ns |

ns: not significant *: $p < 0.01$ ***: $p < 0.0001$

*Figure 6.*

Figure 9 shows three different discs:

a) a normal disc with a very small cup

b) a glaucomatous disc with a very big cup and a lower number of fibers because many of them have died and disappeared.

c) A megalopapilla without glaucoma. It is a disc with a very big surface compared with a and b, and a very big cup. The most remarkable thing is that the number of nerve fibers that we drew in the neuroretinal rim of the megalopapilla is the same as that drawn in the normal disc.

At the bottom of a, b and c there is a comparison between a normal optic disc with and a glaucomatous one (a and b). The area of the disc is the same but the area of the cup is bigger in the glaucomatous one, and the area of the neuroretinal rim is smaller. In the normal disc (a), the slope (third moment or cup shape measure) is smooth, but in the glaucomatous one (b), the slope is steeper, and in the megalopapilla (c) the slope is equally steeper. The cup shape measure, which represents the slope of the fibers, is positive in glaucomatous discs and megalopapillas, while in normal discs it is negative.

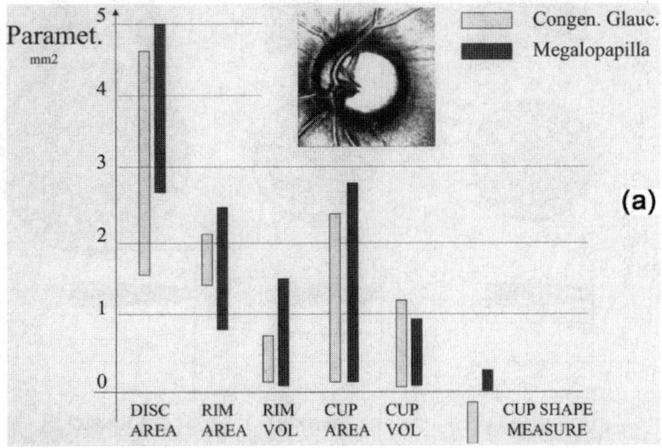


(b)

| Parameter/ Group | CONG.GLAUC. | MEGALOPAPILLA |
|---|---|---|
| DISC AREA | 3.07 | 3.07 ns |
| RIM AREA | 1.70 | 1.50 * |
| RIM VOLUME | 0.42 | 0.36 ns |
| CUP AREA | 1.37 | 1.57 ns |
| CUP VOLUME | 0.56 | 0.59 ns |
| CUP SHAPE M. | -0.14 | 0.06 *** |

ns: not significant *: $p < 0.01$ ***: $p < 0.0001$

*Figure 7.*

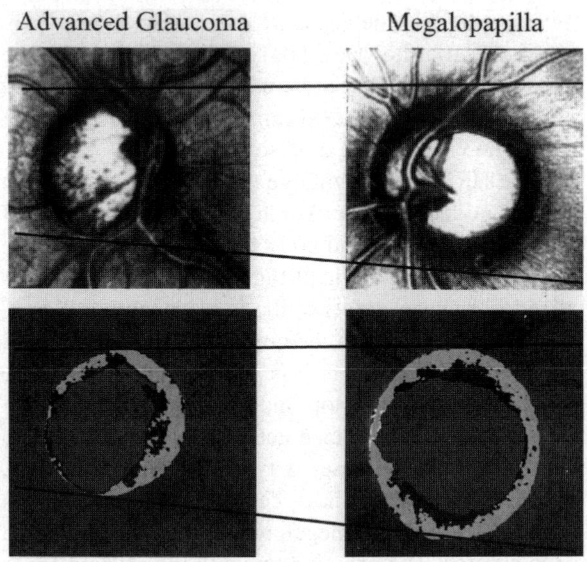


*Figure 8.*

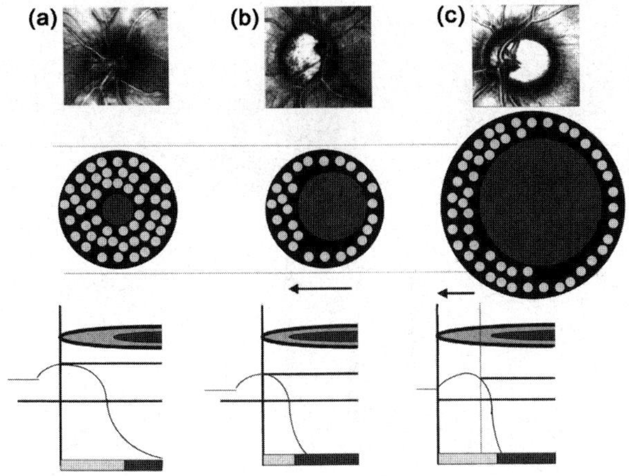


*Figure 9.*

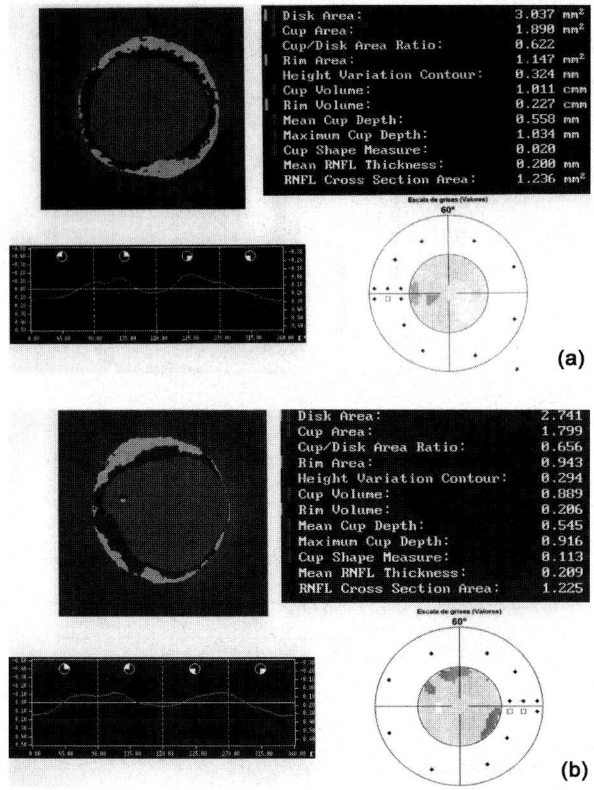


*Figure 10.*

We will now describe a very interesting clinical case of megalopapilla coexisting with glaucoma.

C.H No. 6400, male, 34 years of age.

Diagnosis: megalopapilla with open-angle glaucoma.
IOP: OD: 26 mmHg, OS: 32 mmHg.
BCVA: OD: 20/20, OS: 20/30
Gonioscopy: BE: goniodysgenesis in 360 degrees with a prominent Schwalbe's line.
Visual field: BE: borderline.

Figure 10a shows the HRT of the right eye, with the following values: Disc area: 3.037 $Mm^2$; cup volume: 1.011 $mm^3$; rim volume: 0.227 $mm^3$; cup shape measure: 0.020. The visual fields are also shown.

Figure 10b shows the HRT of the left eye, with the following values: Disc area: 2.741; cup volume: 0.889 $mm^3$; rim volume: 0.206 $mm^3$; cup shape measure: 0.11. The visual fields are also shown.

The HRF of the right eye (Figure 11a) shows that there is no localized perfusion defect but an increased intercapillary space and no ischemic localized zone.

In the megalopapilla of Figure 11b there is no localized fiber layer defect at the inferior temporal area, as clearly seen both in the color and in the black and white photograph where it is signaled by two arrows, one coming from the top and the other one from the bottom. The absence of a notch in this localized defect, as observed in this case, is typical of megalopapilla. However, should this have been a glaucomatous localized defect with no megalopapilla, there would have been a notch. The examination with the HRF in these cases shows a clear localized nerve fiber layer defect with localized reduced perfusion, increased intercapillary space (black squares) and an ischemic localized zone (bright).

## Conclusions

Megalopapillas may be confused with advanced glaucomatous optic nerve head damage and other neuropathies, but they actually represent a different entity, characterized by a pseudoglaucomatous disc, with normal visual field and normal daily pressure curve.

The characteristics shared with glaucoma are the following: increased Cup Area, Cup Volume and Cup Shape Measure. However, the following parameters are completely different: Disc Area, Rim Volume and Rim Area.

No significant differences were found upon comparing the megalopapilla group with the congenital glaucoma group in all the optic nerve head parameters but one, Cup Shape Measure, which is almost preserved in the congenital glaucoma group.

The increase in the total area leads to an increase in the Cup Area, though with no decrease in the Rim Area, unlike what happens in adulthood glaucoma.

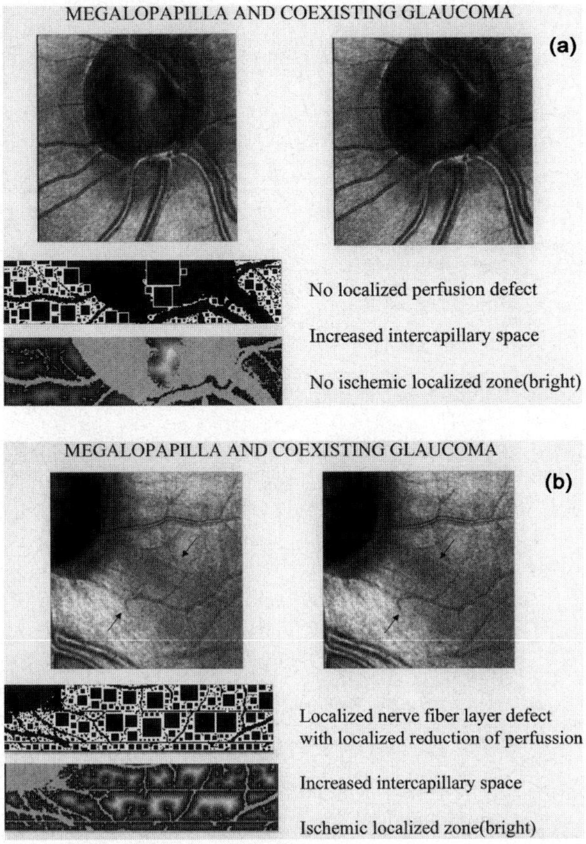


*Figure 11.*

The increase of the papillary border, cause the fibers to enter almost vertically, a fact that turns the cup shape measure values into positive ones (pathological values).

*Megalopapilla is an entity which seems to have been ignored by the literature worldwide, and it is not even named in most well known ophthalmology textbooks.*

## References

1. Franceschetti A, Bock R. Megalopapilla. A new congenital anomaly. Amer J Ophthalmol 1950; 33: 227–235.
2. Brodsky CM. Congenital optic disk anomalies. Survey of Ophthalmol 1994; 39: 89–112.
3. Goldhammer Y. Optic nerve anomalies in basal encephalocele. Arch Ophthalmol 1975; 2: 115–118.
4. Theodossiadis GP, Kollia AK, Theodossiadis PG. Cilioretinal arteries in conjunction witha pit of the optic disc. Ophthalmologica 1992; 204: 115–121.
5. Caprioli J. Basal encephalocele and Morning Glory syndrome. Br J Ophthalmol 1989; 107: 145–150.
6. Sampaolesi R, Sampaolesi JR. Etude du nerf optique dans le glaucome congénital par la tomographie confocale au laser. Ophtalmologie 1998; 12: 205–213.
7. Grimson BS, Perry DD. Enlargement of the optic disk in childhood optic nerve tumors. Amer J Ophthalmol 1984; 97: 627–631.
8. Collier M. Megalopapilla and central pulverulent cataract. Bull Soc Franc Ophtalmologie 1965; 9: 719–724.
9. Hirokane K, Kimura T, Kimura W, Sawada T, Ohte A, Kobayashi M. Megalopapilla in four children. Folia Ophthalmol JPN 1995; 731–735.
10. Jonas JB, Zäch FM, Gusek GC and Naumann GOH. Pseudoglaucomatous physiologic large cups. Amer J Ophthalmol 1989; 107: 137–144.
11. Maisel JM, Pearlstein CS, Adams WH, Heotis PM. Large optic disks in the Marshallese Population. Amer J Ophthalmol 1989; 107: 145-150.
12. Sampaolesi R. Congenital glaucoma. The importance of echometry in its diagnosis, treatment and functional outcome. In Cennamo G and Rosa N (eds) Ultrasonography in Ophthalmology XV, Proceedings of the 15th SIDUO Congress, Cortina, Italy, 1994, Kluwer Academic Publishers, Dordrecht, 1997: 1–47.
13. Saruhan A, Orgül S, Koçak I, Prünte C, Flammer J: Descriptive information on topographic parameters computed at optic nerve head with Heidelberg Retina Tomograph. J Glaucoma 1998; 7: 420–429.
14. Burk ROW, Rohrschneider K, Noack H, Völcker HE: Are large optic nerve heads susceptible to glaucomatous damage at normal intraocular pressure? Graefes Arch Ophthalmol 1992; 230: 552–560.
15. Sampaolesi R, Sampaolesi JR: Tomografía confocal del nervio óptico y de la retina (Relato Anual). Arch Oftalmol B Aires 1995; 70: 1–566.

*International Ophthalmology* **23:** 259–261, 2001.
*J.R. Sampaolesi (ed.), Laser Scanning: Update 1*, 77–79.



# Optic nerve head damage progression in patients with glaucoma

Roberto Sampaolesi & Juan Roberto Sampaolesi
*Fundación Argentina Oftalmológica Juan Sampaolesi and Department of Ophthalmology, School of Medicine, University of Buenos Aires, Argentina*

## Abstract

*Purpose:* To steady the optic nerve head damage progression in glaucoma. *Material and methods:* 178 eyes with open angle glaucoma were studied in order to classify them into different stages in correlation with their visual fields. 172 normal eyes were studied and compared as a control group. Examinations were performed with the Heidelberg Retina Tomograph (longwave: 680 nm), using the new standard reference plane. Each examination is a mean of 3 acquisitions with 5 STD lower than $N = 20$. Each optic nerve head was studied in 360 degrees (segments) and in quadrants and octants separately (predefined segments). Visual fields were performed with Octopus 23 or 101, programs G2 and 62x, with 3 phases completed. Diurnal pressure curve with applanation tonometry and gonioscopy (for classification) were also performed. *Results:* Glaucomatous optic neuropathy was classified into 6 stages: NORMAL, BORDERLINE, PHASE 1, PHASE 2, PHASE 3 and PHASE 4. The group with normal optic nerve head was the only one included in the HYPERTENSIVE PERIOD of the disease. Borderline, Phase 1 and Phase 2 were included in the PREPERIMETRIC PERIOD, and Phase 3 and Phase 4 were included in the PERIMETRIC PERIOD of this disease. *Conclusion:* Confocal tomography of the optic nerve head is helpful in showing glaucomatous damage, specially in the preperimetric period, when visual field remains normal. It also helps to differentiate hypertensive eyes from early preperimetrics eyes. *Discussion:* Confocal tomography of the optic nerve head seems to detect glaucomatous optic nerve head damage much earlier than visual field does.

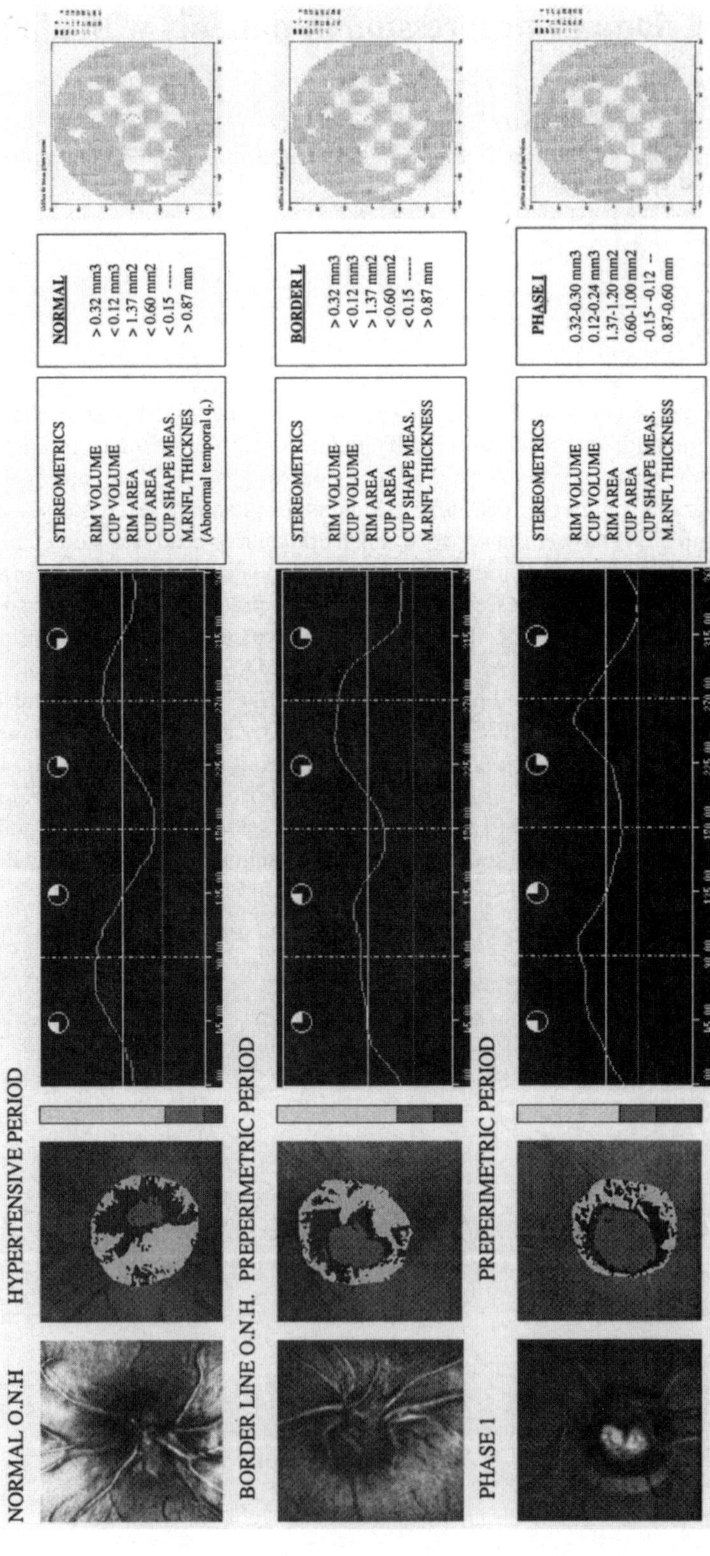
NORMAL O.N.H
HYPERTENSIVE PERIOD
STEREOMETRICS
RIM VOLUME
CUP VOLUME
RIM AREA
CUP AREA
CUP SHAPE MEAS.
M.RNFL THICKNES
(Abnormal temporal q.)
NORMAL
> 0.32 mm3
< 0.12 mm3
> 1.37 mm2
< 0.60 mm2
< 0.15 -----
> 0.87 mm
BORDER LINE O.N.H.
PREPERIMETRIC PERIOD
STEREOMETRICS
RIM VOLUME
CUP VOLUME
RIM AREA
CUP AREA
CUP SHAPE MEAS.
M.RNFL THICKNESS
BORDER L
> 0.32 mm3
< 0.12 mm3
> 1.37 mm2
< 0.60 mm2
< 0.15 -----
> 0.87 mm
PHASE 1
PREPERIMETRIC PERIOD
STEREOMETRICS
RIM VOLUME
CUP VOLUME
RIM AREA
CUP AREA
CUP SHAPE MEAS.
M.RNFL THICKNESS
PHASE I
0.32-0.30 mm3
0.12-0.24 mm3
1.37-1.20 mm2
0.60-1.00 mm2
-0.15- -0.12 --
0.87-0.60 mm

PHASE 2 PERIMETRIC PERIOD STARTS

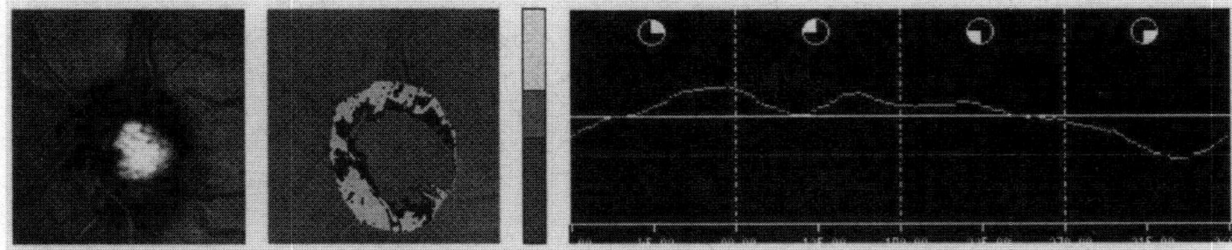

| STEREOMETRICS | PHASE II |
|---|---|
| RIM VOLUME | 0.30-0.20 mm3 |
| CUP VOLUME | 0.24-0.48 mm3 |
| RIM AREA | 1.20-0.80 mm2 |
| CUP AREA | 1.00-1.50 mm2 |
| CUP SHAPE MEAS. | -0.12- -0.07 -- |
| M.RNFL THICKNESS | 0.60-0.40 mm |

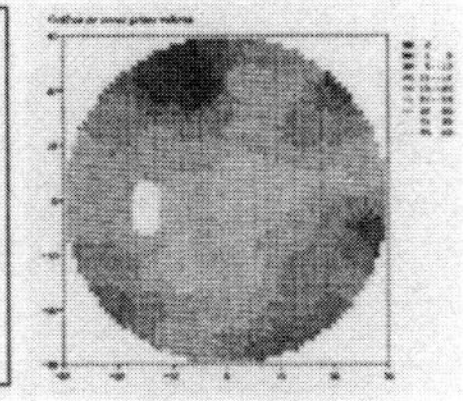

PHASE 3 PERIMETRIC PERIOD

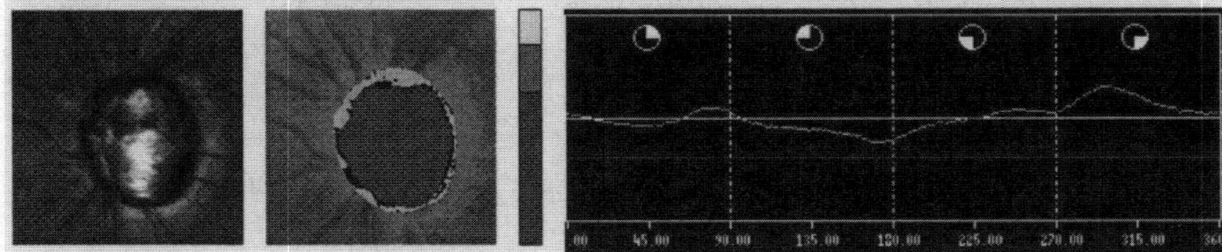

| STEREOMETRICS | PHASE III |
|---|---|
| RIM VOLUME | 0.20-0.10 mm3 |
| CUP VOLUME | 0.48-0.96 mm3 |
| RIM AREA | 0.80-0.40 mm2 |
| CUP AREA | 1.50-1.80 mm2 |
| CUP SHAPE MEAS. | -0.07- -0.02 -- |
| M.RNFL THICKNESS | 0.40-0.20 mm |

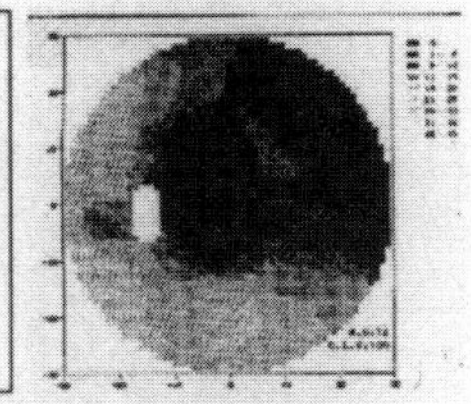

PHASE 4 PERIMETRIC PERIOD

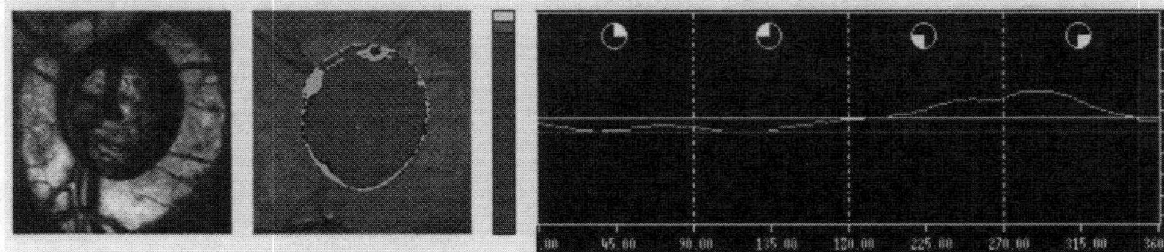

| STEREOMETRICS | PHASE IV |
|---|---|
| RIM VOLUME | 0.10-0.00 mm3 |
| CUP VOLUME | 0.96 o > mm3 |
| RIM AREA | 0.40-0.00 mm2 |
| CUP AREA | 1.80 o > mm2 |
| CUP SHAPE MEAS. | 0.00 o > -- |
| M.RNFL THICKNESS | 0.20-0.00 mm |

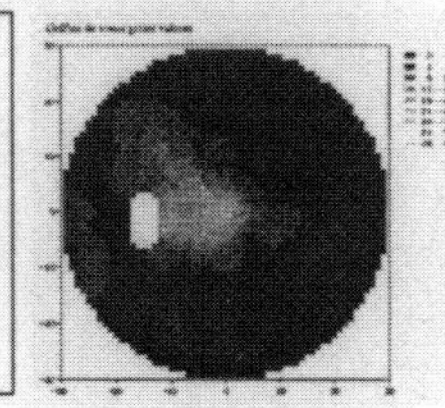

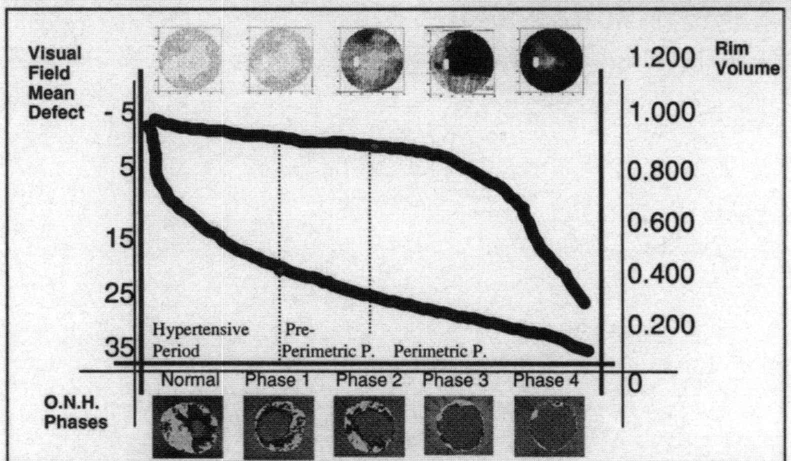


| I.O.P. | OPTIC NERVE HEAD | | | | | | | | |
|---|---|---|---|---|---|---|---|---|---|
| 50 40 30 20 10 | N | BL | PI | PII | PIII | PIV | | | |
| | VISUAL FIELD | | | | | | | | |
| | N | | | | | BL | S1 | SII | SIII |
| HYPERTENSIVE | PRE-PERIMETRIC | | | | | PERIMETRIC | | | |

Supported by Fundación Argentina Oftalmológica Juan Sampaolesi

International Ophthalmology **23:** 263–265, 2001.
J.R. Sampaolesi (ed.), Laser Scanning: Update 1, 81–83.




# Congenital anomalies of the optic nerve head – review

R. Sampaolesi, J.R. Sampaolesi & R.A. Ucha
*Fundación Argentina Oftalmológica Juan Sampaolesi and Instituto de la Visión, Buenos Aires, Argentina*

## Abstract

*Purpose:* To show the topographic features and specific measurements of the different optic nerve head congenital anomalies. *Material:* Confocal tomography was performed to the following optic nerve head congenital anomalies: Morning Glory Syndrome, Optic Disc Coloboma, Peripapillary Staphylomas, Megalopapilla, Optic Disc Pit, Tilted Disc Syndrome and Pseudopapilla. *Methods:* Examinations were performed with the Heidelberg Retina Tomograph (long wave: 680 nm), using the new standard reference plane. Each eye was also examined with fundus biomicroscopy. *Results:* The examinations with confocal tomography show great differences and specific features in each anomaly. Topographic data and stereometric measurements are showed for each case. *Conclusion:* Confocal tomography is able to demonstrate and separate the different optic nerve head anomalies, and the measure O. N. H. parameters.

## MEGALOPAPILLA

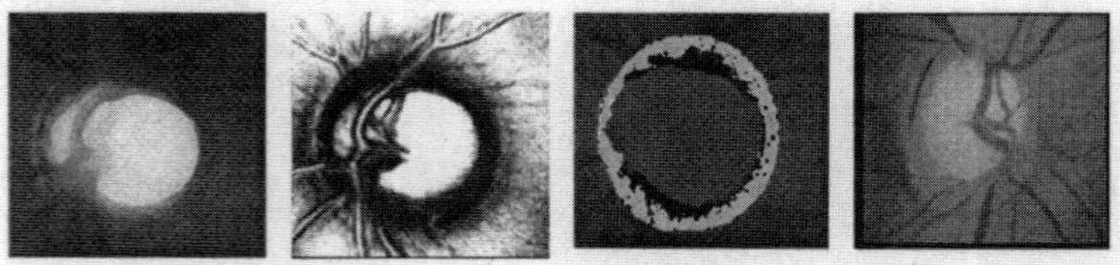

**TYPE I :**
- Surface > 2 .5 mm2.
- Normal configuration / higher cup/disc ratio
- Pale neuroretinal rim
- Bilateral

**TYPE II :**
- Surface > 2 .5 mm2.
- SUPERIOR DECENTERED CUPPING.
- Bigger NRR at the inferior part.
- Unilateral

## OPTIC DISC PITS

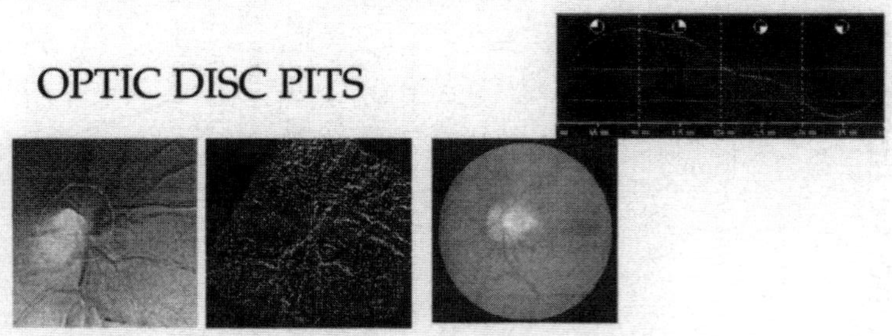

**OPHTHALMOSCOPY**
- Round or oval depression, grey, white or yellow.
- Temporal location.
- Peripapillary pigmentary changes.
- One or two cilioretinal arteries start at the pits.
- Generally unilateral, ( 15 % bilateral ).

**VISUAL FIELD**
- The defects correlate with the pit's position.
- Arcuate scotoma, central scotoma.

**RETINAL DETACHMENT**
- The macular serous detachment is related to the pit.
- Macular hole, retinal detachment surrounding it.

## MORNING GLORY

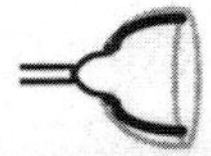

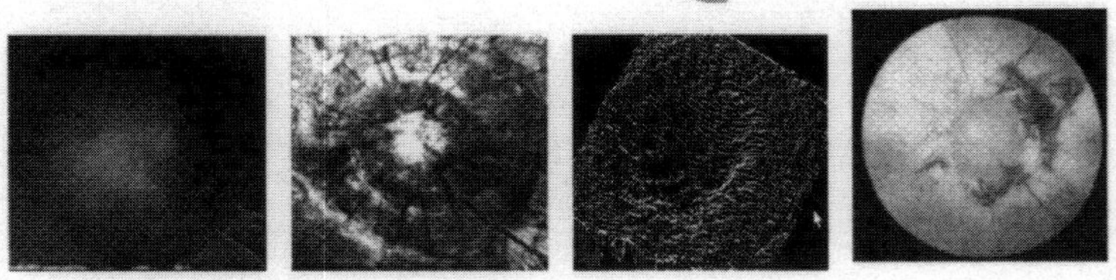

- Increased number of vessels and difficult to distinguish
- Veins from arteries.
- White central veil ( glia ).
- Macular capture / retinal detachment, small holes near the O.N.
- Communicacion between vitreous body and subarachnoid space.

## PERIPAPILLARY STAPHYLOMA

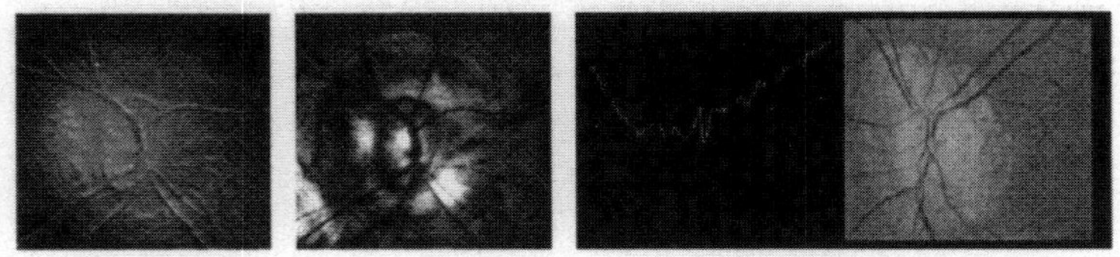

**OPHTHALMOSCOPY:**
- Deep cupping surrounding the O.N.H. ( specially inferior )
  Surrounded by a pigment hale, ( = M.Glory ).
  There is no central glical white veil.
  Unfrequent / Unilateral

**VISUAL ACUITY :**
- Very reduced, though normal sometimes.
- Usually associated with iris, retinal and ciliary body colobomas.

## OPTIC NERVE COLOBOMAS

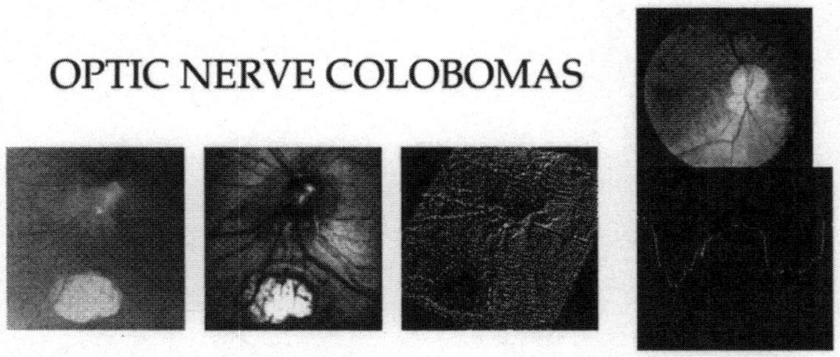

- Abnormal coaptation of the proximal part, optical slit.
- Big O.N., definite edges, white, deep cupping.
- CUP DECENTERED TOWARDS THE INFERIOR PART.
- Absent NRR at the inferior part.
- Inferior extension : 'CHOROIDAL COLOBOMA" and/or iris coloboma.
- Posterior pole cupping ( = M. Glory ),
- bilateral ( = M. Glory )
- The V.A. depends on the integrity of the papillo-macular bundle.
- Serous macular detachment / Normal retinal vessels

## TILTED DISC SYNDROME

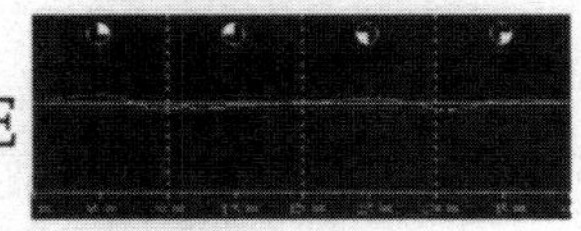

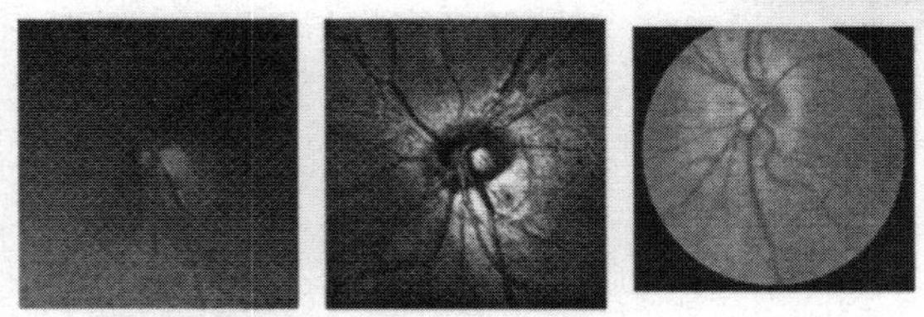

**OPHTHALMOSCOPY**
- Superonasal overelevated O.N.
- Inferonasal O.N. with posterior displacement.
- Oblique major axis.
- Situs Inversus of the papillary vessels..
- Congenital inferonasal conus.
- Retinal epithelium and inferonasal choroid thinning.
- Inferonasal albinism

## PSEUDOPAPILLA

HERVOUET, 1958 : HYPOTHESIS: IT IS FEASIBLE FOR A DOUBLE OPTIC NERVE TO DEVELOP IF THERE IS A DOUBLING OF THE OPTICAL PEDICULUM, WITH THE DICHOTOMIZATION OF THE HYALOID ARTERY.

**OCULAR MANIFESTATIONS**

THERE ARE 2 CLINICO-OPHTHALMOSCOPIC FORMS:

A) OPTIC DISCS JOINED BY THE VERTICAL MERIDIAN: where the supernumeric optic disc joins the inferior part of the main optic disc. Vascularization is shared.

B) OPTIC DISC DOUBLING: the requirements are :
- existence of a double optic disc contour.
- existence of double vascularization of a central type.
- existence of two joined discs without pigmentary elements between them.

C) VASCULAR PSEUDOPAPILLA: there is no actual second optic disc, but just a colobomatous scar through which retinal veins enter and choroidal arteries emerge. *

## VASCULAR PSEUDOPAPILLA

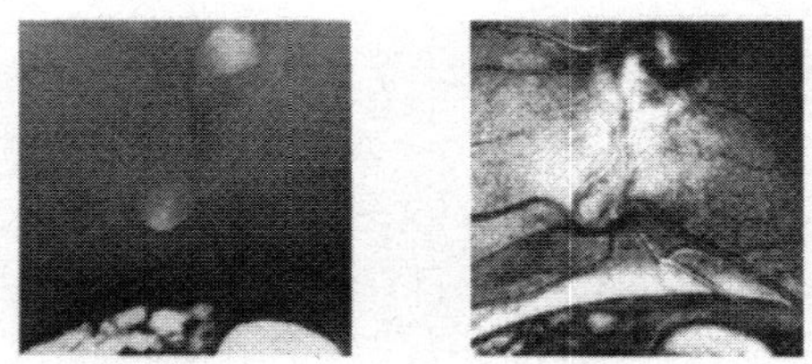

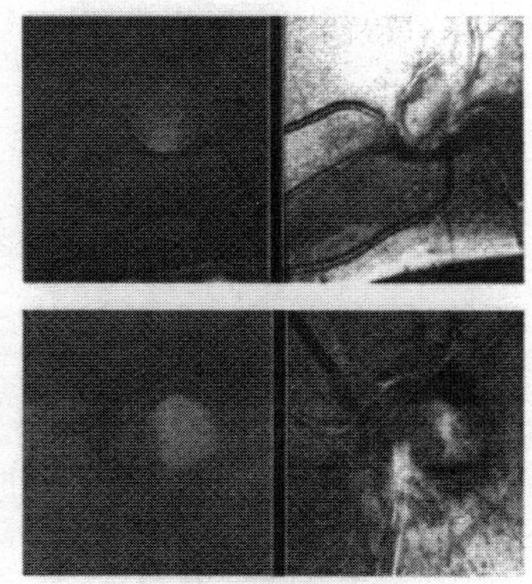

Supported by Fundación Argentina Oftalmólogica Juan Sampaolesi

International Ophthalmology **23:** 267–269, 2001.
J.R. Sampaolesi (ed.), Laser Scanning: Update 1, 85–87.



# The pseudoglaucomas

Juan Sampaolesi & Roberto Sampaolesi
*Fundación Argentina Oftalmológica Juan Sampaolesi and Department of Ophthalmology, School of Medicine, University of Buenos Aires, Argentina*

## Abstract

*Purpose:* To study the differences between different pseudoglaucomas and advanced glaucomatous optic nerve head damage. *Materials and methods:* Confocal tomography was performed in patients with advanced glaucoma (phase 4, perimetric period), Megalopapilla, optic nerve head atrophy produced by Anterior Ischemic Optic Neuropathy, with Meningioma of the optic nerve sheaths and with non functional adenoma. Examinations were performed with the Heidelberg Retina Tomography (longwave; 680 nm) using the new standard reference plane. Each eye was also examined with fundus biomicroscopy. O. N. H. parameters were measured and compared between the different cases. *Results:* While the 4 different diseases are characterized by a great cupping and apparently important neuroretinal rim reduction, the analysis with confocal tomography detected significant differences and particular features for each of these diseases. *Conclusion:* Confocal tomography is very useful in order to distinguish between true glaucomas and pseudoglaucomas.

ADVANCED GLAUCOMA (PERIMETRIC PERIOD)

Increased cup volume and area. Reduced rim volume and area. Normal disc area.

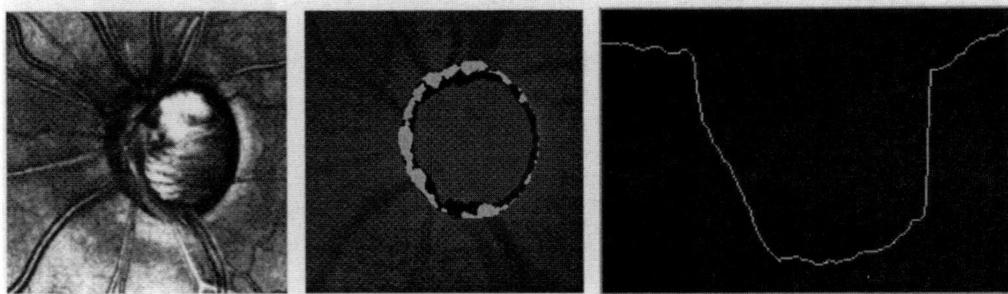

| | |
|---|---|
| Disk Area: | 1.788 mm² |
| Cup Area: | 1.228 mm² |
| Cup/Disk Area Ratio: | 0.687 |
| Rim Area: | 0.560 mm² |
| Height Variation Contour: | 0.136 mm |
| Cup Volume: | 0.838 cmm |
| Rim Volume: | 0.061 cmm |
| Mean Cup Depth: | 0.636 mm |
| Maximum Cup Depth: | 1.094 mm |
| Cup Shape Measure: | 0.059 |
| Mean RNFL Thickness: | 0.106 mm |
| RNFL Cross Section Area: | 0.504 mm² |

MENINGIOMA OF THE OPTIC NERVE SHEATHS

Aparently increased cup, with ectasic of the venous vessels, protruding towards the vitreous body.

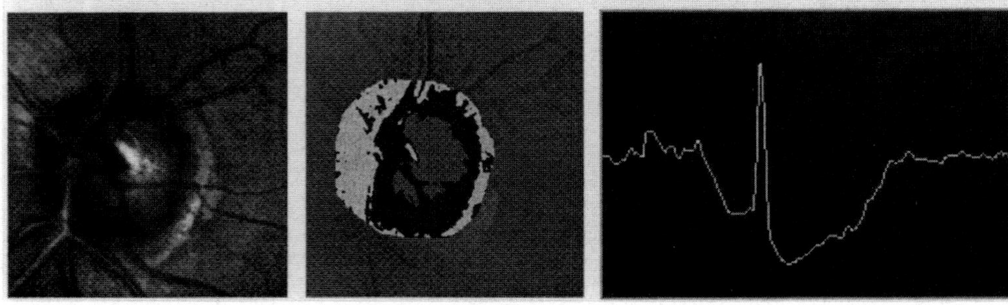

| | |
|---|---|
| Disk Area: | 1.876 mm² |
| Cup Area: | 0.342 mm² |
| Cup/Disk Area Ratio: | 0.182 |
| Rim Area: | 1.534 mm² |
| Height Variation Contour: | 0.201 mm |
| Cup Volume: | 0.022 cmm |
| Rim Volume: | 0.379 cmm |
| Mean Cup Depth: | 0.198 mm |
| Maximum Cup Depth: | 0.453 mm |
| Cup Shape Measure: | -0.105 |

## NON FUNCTIONAL ADENOMA

Increased disc area, abnormal cup volume and area. AT shows non functional adenoma.

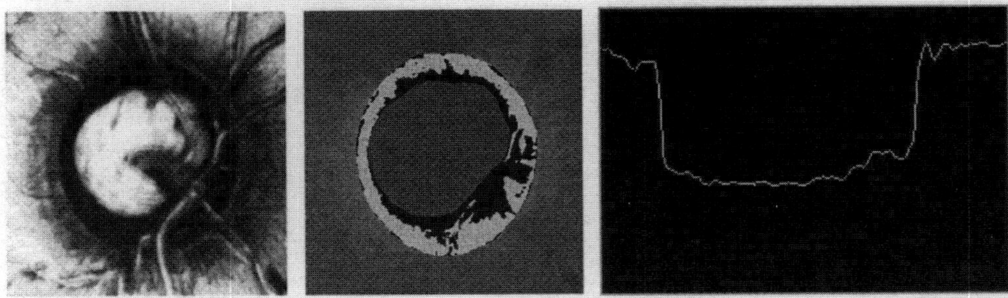

| | |
|---|---|
| Disk Area: | 3.029 $mm^2$ |
| Cup Area: | 1.525 $mm^2$ |
| Cup/Disk Area Ratio: | 0.504 |
| Rim Area: | 1.504 $mm^2$ |
| Height Variation Contour: | 0.275 mm |
| Cup Volume: | 0.840 cmm |
| Rim Volume: | 0.278 cmm |
| Mean Cup Depth: | 0.514 mm |
| Maximum Cup Depth: | 0.892 mm |
| Cup Shape Measure: | 0.051 |
| Mean RNFL Thickness: | 0.172 mm |
| RNFL Cross Section Area: | 1.064 $mm^2$ |

## MEGALOPAPILLA

Increased disc area, abnormally high cup volume and area, with normal rim volume and normal visual field.

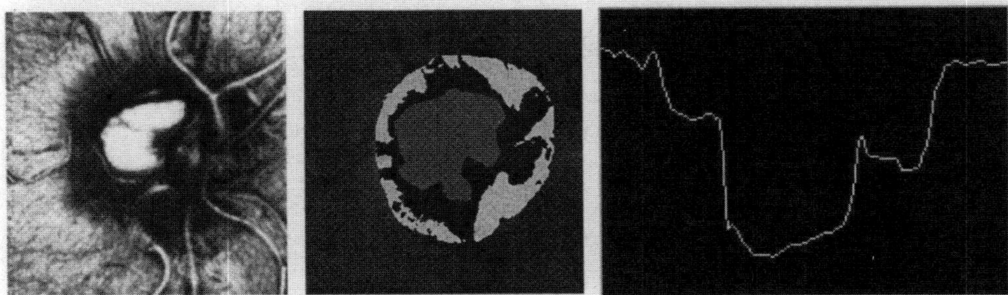

| | |
|---|---|
| Disk Area: | 2.644 $mm^2$ |
| Cup Area: | 0.830 $mm^2$ |
| Cup/Disk Area Ratio: | 0.314 |
| Rim Area: | 1.815 $mm^2$ |
| Height Variation Contour: | 0.344 mm |
| Cup Volume: | 0.258 cmm |
| Rim Volume: | 0.441 cmm |
| Mean Cup Depth: | 0.289 mm |
| Maximum Cup Depth: | 0.807 mm |
| Cup Shape Measure: | -0.216 |
| Mean RNFL Thickness: | 0.256 mm |
| RNFL Cross Section Area: | 1.479 $mm^2$ |

## O.N.H. ATROPHY PRODUCED BY A.I.O.N.

Hemi cupped disc with hemi disc edema. Visual field doesn't match with glaucomatous damage.

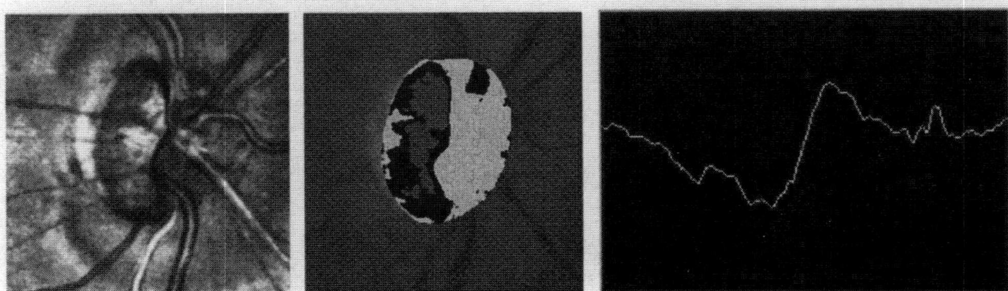

| | |
|---|---|
| Disk Area: | 1.605 $mm^2$ |
| Cup Area: | 0.280 $mm^2$ |
| Cup/Disk Area Ratio: | 0.175 |
| Rim Area: | 1.324 $mm^2$ |
| Height Variation Contour: | 0.405 mm |
| Cup Volume: | 0.014 cmm |
| Rim Volume: | 0.391 cmm |
| Mean Cup Depth: | 0.140 mm |
| Maximum Cup Depth: | 0.392 mm |
| Cup Shape Measure: | -0.167 |
| Mean RNFL Thickness: | 0.245 mm |
| RNFL Cross Section Area: | 1.107 $mm^2$ |

International Ophthalmology **23:** 271–274, 2001.
J.R. Sampaolesi (ed.), Laser Scanning: Update 1, 89–92.




# Arterial narrowing as a predictive factor in glaucoma

Moussalli M. Angelica, A. Sanseau & Carlos Argento*
**Author for correspondence: Instituto de la Visión, Buenos Aires, Argentina*

## Abstract

*Purpose*: To evaluate whether retinal arteriolar widths change in normal patients and in different glaucoma types. We measured the arteriolar width at the merge in the optic nerve (MD), at the edge of the optic nerve (ED) and at the peripapillary area (1 disc diameter distance) (disc distance D). HRT software 1.11, Interactive Means program was used.

One hundred forty three eyes of seventy-two patients divided in four subgroups normal, low-tension glaucoma (LTG), primary open angle glaucoma (POAG) and ocular hypertensive (OH). No statistically significant was found in each studied group. The normal patients showed no changes in the retinal arteriolar width compared with all the glaucoma groups.

## Introduction

The authors wished to determine the importance of focal or diffuse retinal arteriolar narrowing as a predictor factor in glaucoma. There is no clear evidence about the real importance of this sign in previous reports. Severe optic neuropathy with or without glaucoma shows focal narrowing of retinal arterioles at the peripapillary area [1–3]. There is no correlation between the changes in the arteriole width and the different glaucoma subgroups. Although no difference between the moderate glaucomatous damage and normal group of patients [2, 3, 5, 6].

*Purpose*: To evaluate the changes in retinal vessels width in relationship as a predictor factor in glaucoma.

## Conclusion

The differences in the arteriolar width between all the groups were not statistically significant ($p = 0.75$)

The arteriolar width was bigger in the normal group of patients

The less width was found at the peripapillary area (D) in all the examined groups, this was considered as a focal narrowing.

There were no differences between the measures taken at the superior and inferior pole of the optic nerve.

The ocular hypertensive subgroup showed the smallest width.

These changes at the retinal arterioles are important but we can not consider it as a predictor factor in glaucoma or a diagnostic sign. In the clinical management of patients with glaucoma, in some eyes, focal arteriolar narrowing may suggest progression of disease.

*Methods*: we retrospectively reviewed 143 normal, ocular hypertensive, glaucoma damage and low tension glaucomatous eyes with the HRT software 1.11 Interactive Means program.

*Results*: we measure the width of the retinal vessels in inferior and superior areas of the optic nerve at 3 different reference levels. The changes were not statistically significant between groups.

*Conclusion*: narrowing of the retinal vessels is not a predictor factor in glaucoma. Focal narrowing could be related to others optic nerve disease. More studies must be carried out to know the importance of these changes.

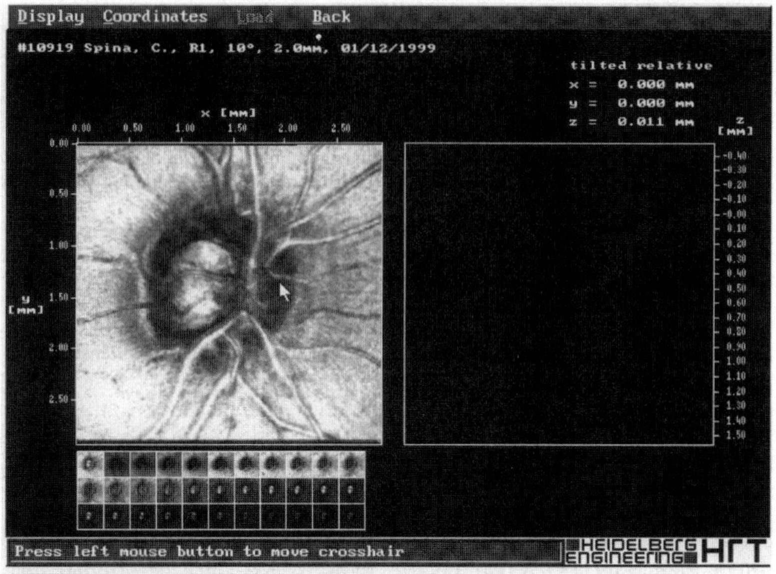


EYES: 26
AGE: 36 (14-70)
ENTRANCE (D) X: 0.068 mm
DISC EDGE (ED) X: 0.072 mm
DISTANCE DISC X: 0.068 mm

*Figure 1.* HRT and measures in a normal patient.

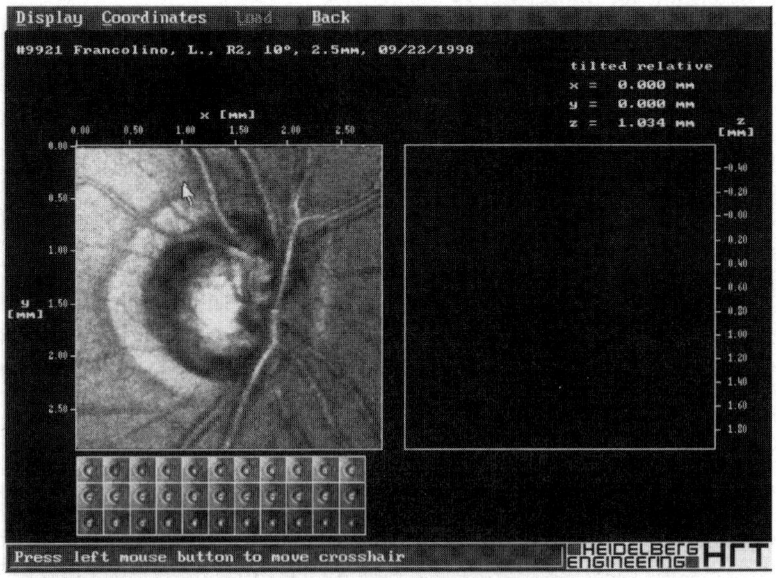


N: 20 EYES: 40
ENTRANCE X: 0.072 mm
DISC EDGE X: 0.053 mm
DISTANCE DISC X: 0.046 mm

*Figure 2.* HRT and measures in an ocular hypertensive patient.

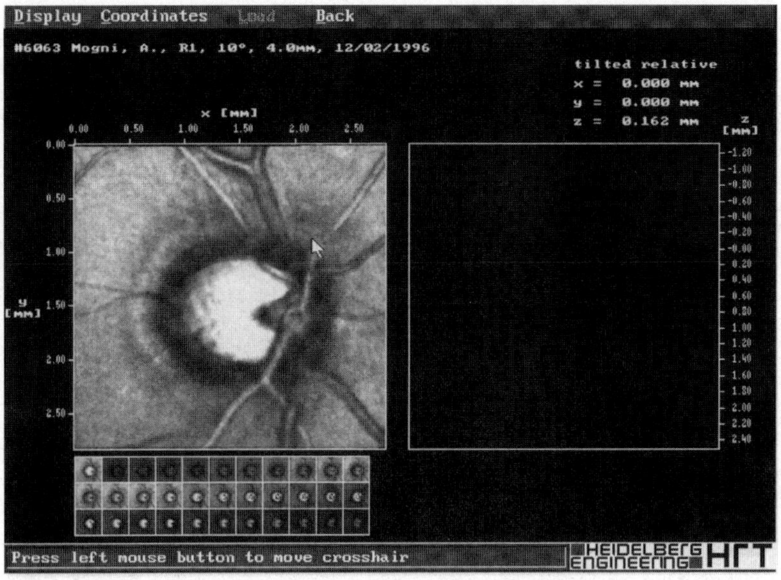


N: 3 EYES: 6

| | |
|---|---|
| ENTRANCE | X: 0.071 mm |
| DISC EDGE | X: 0.054 mm |
| DISTANCE DISC | X: 0.055 mm |

*Figure 3.* HRT and measures in a low-tension glaucoma patient.

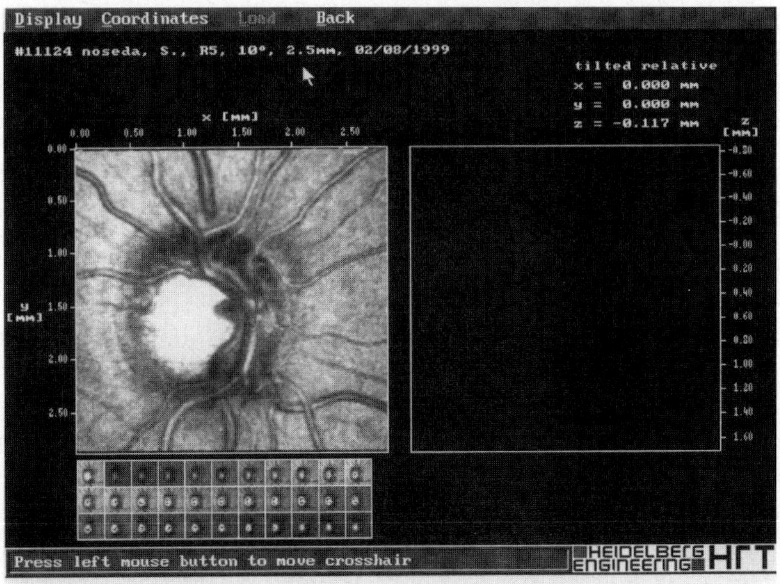


N: 36 EYES: 71

| | |
|---|---|
| ENTRANCE | X: 0.073 mm |
| DISC EDGE | X: 0.055 mm |
| DISTANCE DISC | X: 0.050 mm |

*Figure 4.* HRT and measures in a primary open glaucoma patient.

## References


1. Honrubia FM,García Sanchez J, Pastor JC. Diagnostico Precoz del Glaucoma. LXXII Ponencia de la Sociedad Española de oftalmología. Analisis de la papila Optica. Evaluación de la papila en el diagnostico precoz del glaucoma. 192. LAb. Alcon Cusí 1997.
2. Papastathopoulos KI, Jonas JB. Follow up of focal narrowing of retinal arterioles in glaucoma. British Journal of Ophthalmology 1999; 83/3: 285–289.
3. Papastathopoulos KI, Jonas JB. Fluorescein angiographic correlation of focal narrowing of retinal arterioles in glaucoma. British Journal of Ophthalmology 1998; 82/1: 48–50.
4. Rankin-SJA, Drance-SM. Peripapillary focal retinal arteriolar narrowing in open angle glaucoma. Journal of Glaucoma 1996; 5/1: 22–28.
5. Sampaolesi. Glaucoma. Biometria papilar V. 30 (319). Ed. Med. Panamericana Bs. As. 1991.
6. Tezel-G, Kass-MA, Kolker-AE, Wax-MB. Comparative optic disc analysis in normal pressure glaucoma, primary open-angle glaucoma and ocular hipertension. Ophthalmology 1996; 103/12: 2105–2113.

International Ophthalmology **23:** 275–278, 2001.
J.R. Sampaolesi (ed.), Laser Scanning: Update 1, 93–96.




# Papillary drusen and ocular hypertension

M. Angelica Moussalli, Ana Sanseau & Roberto Ebner
*Instituto de la Visión, Hospital Britanico, Buenos Aires, Argentina*


## Abstract

*Purpose:* To evaluate if the changes in the peripapillary and papillary retinal nerve fiber layer, in a young girl who presents papillary drusen and ocular hypertension in both eyes. *Methods:* We studied this case with retinography, Humphrey Visual Field, HRT, GDx, and diary curve tonometry. *Results:* After three years of follow up, no changes were observed in the drusen at the peripapillary and papillary retinal nerve fiber layer. *Conclusion:* In a well controlled ocular hypertensive patient there is no evidence of changes at the optic nerve head and RFNL related to the drusen or to the high pressure. All the diagnosis methods were correlationated with the clinical evolution over time.


## Introduction

Nowadays there are many imaging technologies to evaluate the optic nerve and the retinal nerve fiber layer [1, 2]. The accuracy, predictability is still being discussed in many reports. Optic nerve drusen mask changes in the RNFL [2], secondary to ocular hypertension (OHT) [3, 4].

## Methods

We present a female patient, 21 years old, visual acuty in OU 20 /20 with sph. −1.50 in both eyes, and all the studies, that we used to evaluate this case. The examination with direct oftalmoscope and retinographies (Figure 1) showed as a bilateral papillary drusen. The follow up was three years, with diary curve tonometry (×21 mmHG) (Figure 2), refraction, Humphrey visual field, HRT, and GDx. The GDx was performed at the time of the second examination.

## Results

The refraction and the tonometry were stable in this patient. The visual fields shows a moderate peripapillary depression, stable along the time (Figures 3–5). HRT (Figures 6–8) did not show variations in the height of the drusen nor in the estimation of the RFNL width. Data obtained with GDx differs from data obtained by HRT images. Values obtained with the GDx (Figures 9–10) for the RE was outside normal figures, while from LE, most of them were within normal figures. Curves corresponding to NFL were abnormally in OU.

## Conclusions

The correlation between tonometry, visual field, and HRT did not show variations in the height of the drusen nor in the estimation of the RFNL width. We think that the lack of appropriate arrangement of microtubes (parallelism disruption)in the surrounding of the disc with the drusen, impaired the perpendicularity of RFNL with the scan beam of the GDx instrument as the reason for the difference between HRT and GDx methods.

Optic nerve changes secondary to OHT are difficult to detect in patients with optic nerve drusen. Image technology is helpful to carry out a careful follow up over time, but there is no unique diagnostic method indicated as a best image to use as a diagnostic tool. In our case, HRT provided images in a 3 year period demonstrating stability of parameters related to the disc.

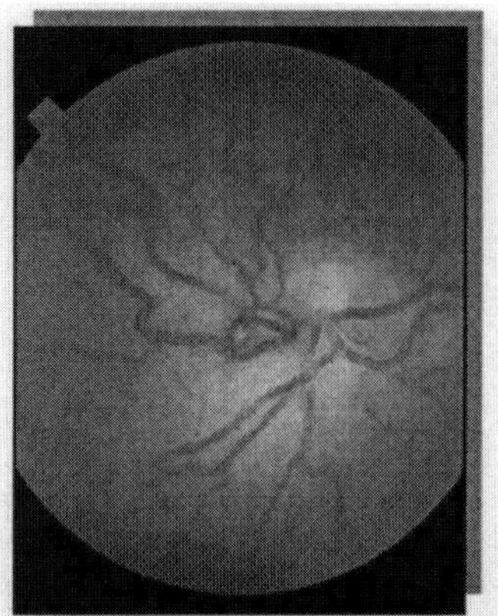

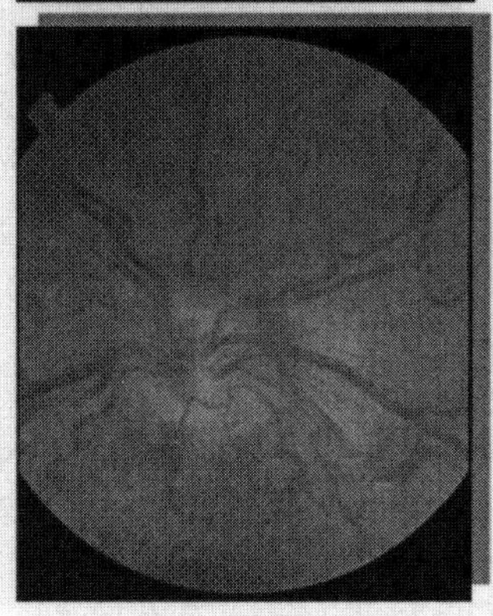

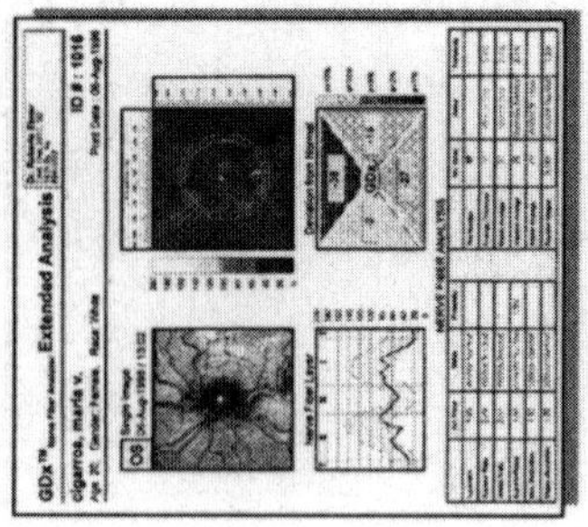

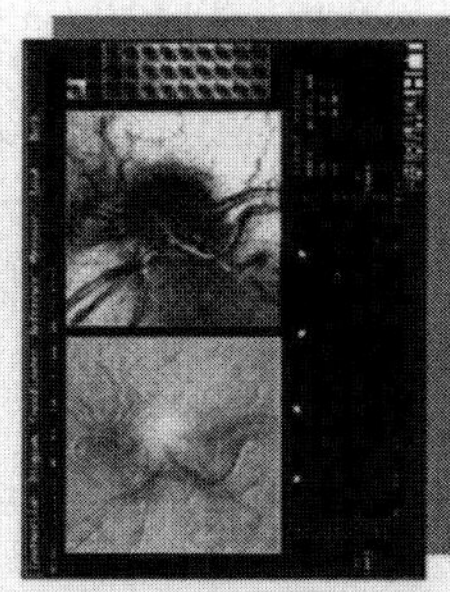

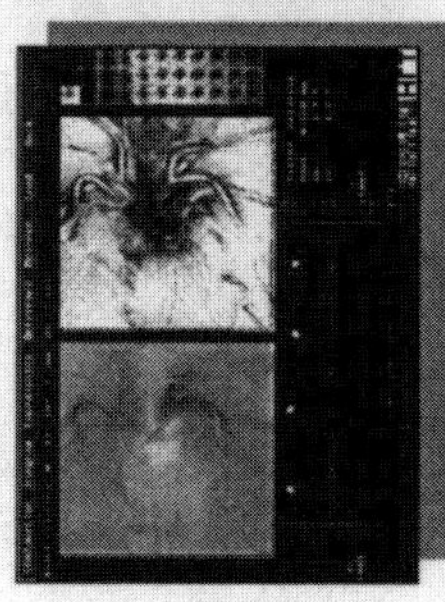

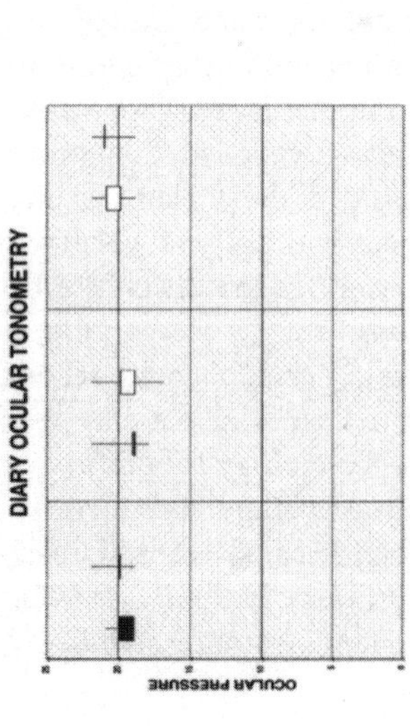
DIARY OCULAR TONOMETRY
OCULAR PRESSURE
OD OI
1997
OD OI
1998
OD OI
1999

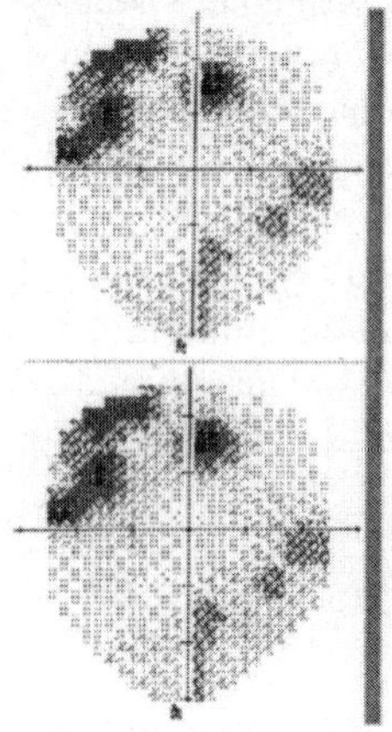

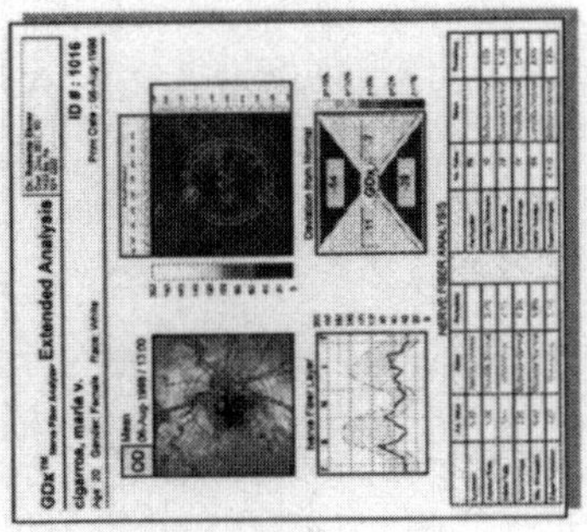

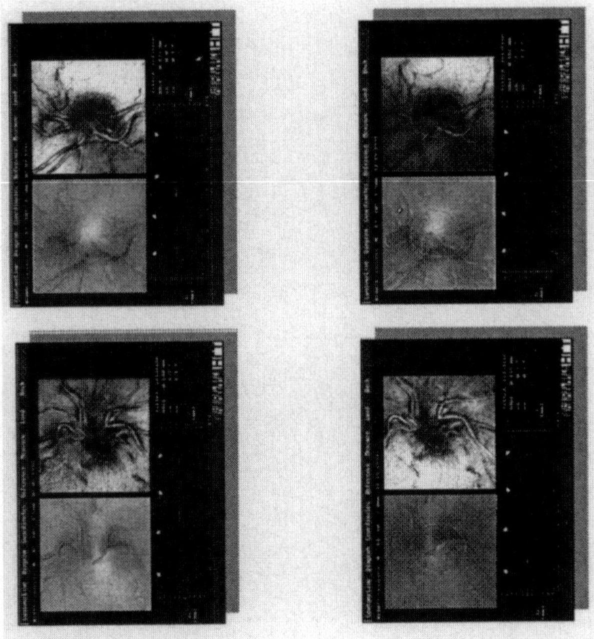

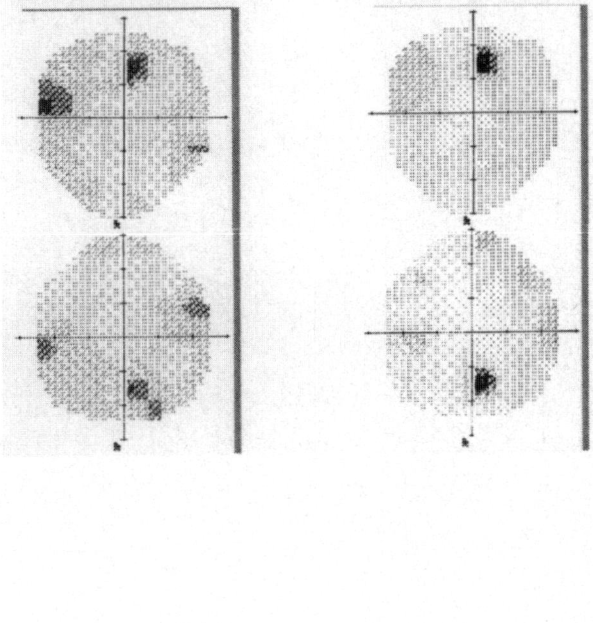

## References


1. Jonas JB, Dichtl A. Evaluation of the retinal nerve fiber layer. Surv Ophthalmol 1996; 40(5): 369–378.
2. Kline LB. Developmental and hereditary optic nerve disorders. Ophthalmol 1996.
3. Piccone MR, Piltz-Seymour J, Shin D. Coexisting optic nerve head – drusen and glaucoma. Lett Ophthalmol 1998; 105(5): 761–762.
4. Shiyoung R, Noecker R, Schuman Joel. Evaluation of coexisting optic nerve head – drusen and glaucoma with optical coherence tomography. Ophthalmol 1997; 104(7): 1138–1144.

International Ophthalmology **23:** 279–289, 2001.
J.R. Sampaolesi (ed.), Laser Scanning: Update 1, 97–107.




# Non-invasive measurement of the concentration of melanin, xanthophyll, and hemoglobin in single fundus layers *in vivo* by fundus reflectometry

Martin Hammer, Dietrich Schweitzer, Eike Thamm & Achim Kolb
*University of Jena, Department of Ophthalmology, Jena, Germany*

**Abstract**

A new model of the reflection of the human ocular fundus on the basis of the adding-doubling method, an approximate solution of the radiative transport equation, is described. This model enables the calculation of the concentration of xanthophyll in the retina, of melanin in the retinal pigment epithelium and in the choroid, and of hemoglobin in the choroid from fundus reflectance spectra. The concentration values found in 12 healthy subjects are in excellent agreement with literature data. In single cases of pathologic fundus alterations possible benefits to the ophthalmologic diagnostics is demonstrated.

## 1. Introduction

The reflection spectra of the human ocular fundus are characterized by the absorption of melanin, hemoglobin, and xanthophyll. The knowledge of the concentration of these pigments may be of clinical interest for diagnostic purposes as well as for the planing of laser treatment. Thus, it was the goal of this investigation to determine these concentrations separately in the retina, the retinal pigment epithelium (RPE), and in the choroid from *in vivo* measured ocular fundus reflection spectra.

A first attempt to model the reflection spectra by a model on the basis of Lambert–Beer's law were published by van Norren and Tiemeijer [1]. This model yields stable results and is very useful for relative measurements. Comparison of the obtained concentrations with physiological data, however, is difficult. Firstly, concentrations are given as optical densities only, which had to be corrected for the thickness of the respective fundus layer. And secondly, there is a deviation from the values known from psychopysical or *in vitro* measurements. The reason of this deviation is clearly the disregard of scattering in the tissue. In order to overcome this disadvantage, Delori and Pflibsen [2] as well as Schweitzer et al. [3] proposed fundus reflection models on the basis of the two photon-flux theory by Kubelka and Munk [4]. Unfortunately, the results of these models are in disagreement with measurements by Hammer et al. [5], because the requirements of the theory are not fulfilled by experimental conditions: The scattering in the ocular fundus tissues is neither isotropic, nor is a diffuse illumination possible at the fundus *in situ*.

Because of the transparency of the cornea, the aqueous humor, the lens, and the vitreous, the propagation of light in these media is described by geometrical optics. Inside the ocular fundus, however, the spread of the light is governed by the law of radiation transport. Furthermore, the fundus has to be considered as a structure of thin layers, obeying very different optical properties [6]. Unfortunately, an analytic solution of the radiation transport equation is not available for this geometry [7]. Thus, the problem has to be solved by a numerical solution or by an approximation. First, a Monte-Carlo simulation [8] of light transport in a stack of different layers was considered. This simulation, certainly, gives a correct description of the light backscattered from the fundus into the pupil of the eye. But it is not useful for the inverse problem of the calculation of the optical properties of single layers from the reflection spectrum of the stack because of its computational expenditure. Another solution widely used in biomedical optics is the diffusion approximation [7]. But application of this approximation requires almost isotropic radiation fields and scattering coefficients much greater than the absorption coefficients. Both conditions are violated

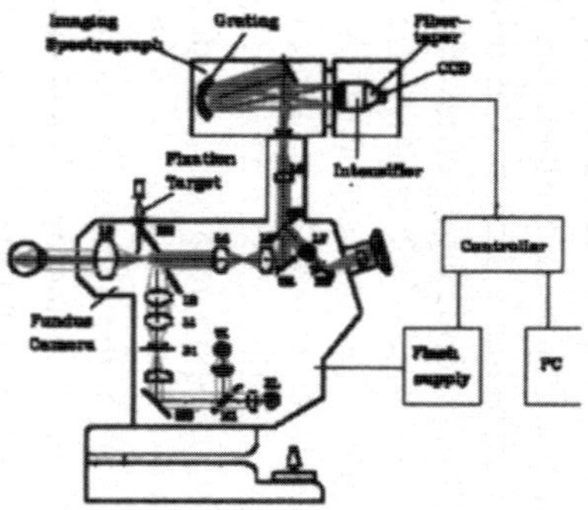


Scheme of the imaging ophthalmospectrometer

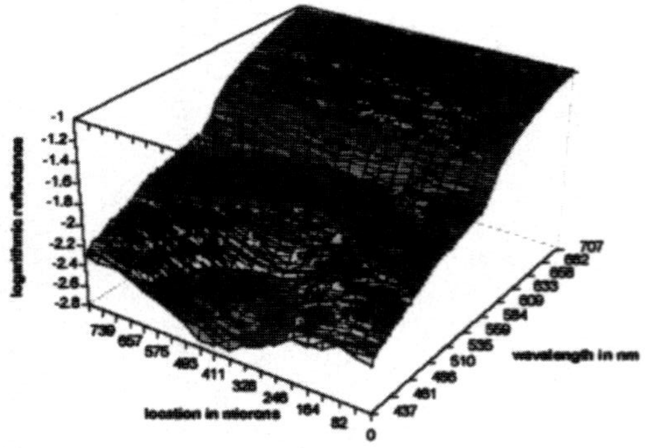


Fundus reflection spectra of the macular region

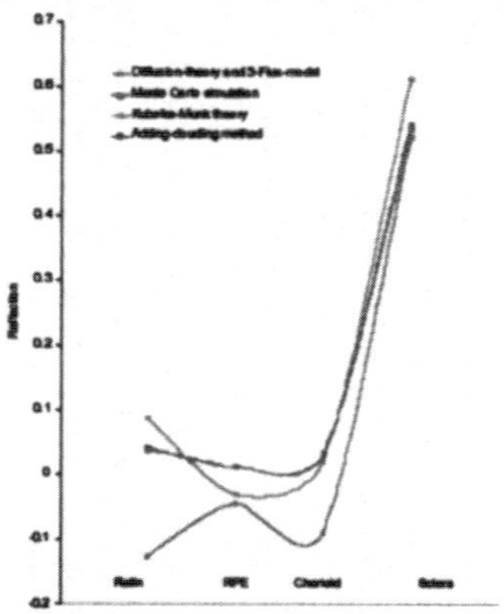

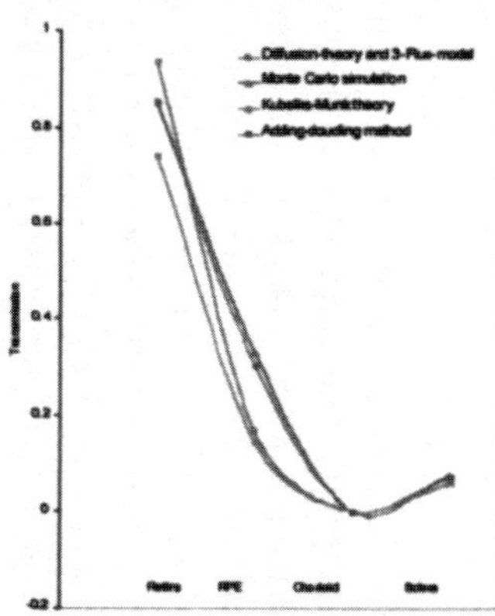


Comparison of the reflection and transmission of ocular fundus tissues calculated from different radiation transport models. These calculations are based on optical properties of the tissues determined in vitro[1]. Remarkable deviations of the results of Kubelka-Munk theory, diffusion approximation, and a 3-flux model from that of the Monte Carlo simu-lation, serving as a reference, were found. A good agreement, however, was obtained by the adding-doubling method.

The adding-doubling method[2] calculates the reflected intensity $I_{ref}$ of a plane parallel absorbing and scattering layer as the product of the Reflection Operator **R** and the incident intensity $I_0$. If the incident intensity is symmetric to the normal of the tissue slab, all intensities and operators are functions of the cosine $\eta$ of the angle to the normal. Thus, the reflected intensity is given by

$$I_{ref}(\eta) = \int_0^1 R(\eta,\eta^0) I_0(\eta^0) 2\eta^0 d\eta^0$$

The transmitted intensity is described in the same way by the transmission operator **T**. Considering a double layer, the reflection and transmission are described by convergent infinite series of reflections and transmissions of the single layers:

$$R_{double} = R + TRT + TRRRT + \ldots = R + T(1 - RR)^{-1} RT$$

$$T_{double} = TT + TRRT + TRRRRT + \ldots = T(1 - RR)^{-1} T$$

Here, the product of two operators is an integral over the cosine $\eta$ too, which can be solved by Gaussian quadrature, e.g.

$$\int_0^1 R(\eta,\eta')T(\eta',\eta'')2\eta' d\eta' = \sum_{j=1}^{n} R_{ij} 2\eta_j w_j T_{jk} = \mathbf{R} \bullet \mathbf{T},$$

resulting in a discretisation of the cosine. By this means, the operators R and T are transformed into matrices. Their *ij*-element gives the intensity, which is reflected or transmitted from angle-element *i* into angle-element *j*. Using the matrix multiplication derived above in the series describing the double reflection and transmission respectively, the intensities reflected and transmitted by a layer of arbitrary thickness can be calculated by iterative doubling an infinitesimal thin layer. For this infinitesimal layer, the reflection and the transmission operator can be obtained from an approximate solution of the discretised radiation transport equation, assuming the intensity inside the thin layer to be the mean value of the intensities at the surfaces, which are known from the boundary conditions.

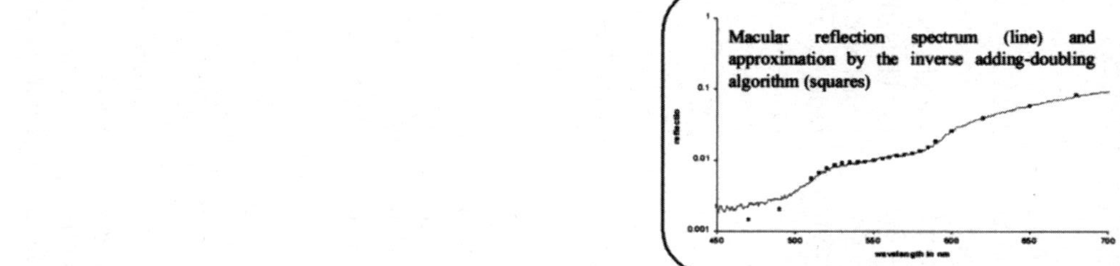


Macular reflection spectrum (line) and approximation by the inverse adding-doubling algorithm (squares)

An inverse algorithm was set up in order to calculate the pigment concentrations from the measured reflection spectra on the basis of the known absorption spectra of melanin, hemoglobin, and xanthophyll using the adding-doubling method. Further information needed are the scattering properties of the tissues, which were obtained from in vitro measurements[1], and the thicknesses of the single layers, which are to be obtained from OCT- or ultrasound measurements. In principle, the inverse algorithm is a least square fit procedure, varying the concentrations till the deviation of the model spectrum (squares in the figure) from the measurement (line) is minimal. Since the model was not given analytically, Powell's conjugate direction set method[3] was used as a derivative-free least square fit procedure.

**Results**

| | *xanthophyll* | *melanin (RPE)* | *melanin (chorioid)* | *hemoglobin* |
|---|---|---|---|---|
| makula | 1.01±0.23 | 0.99±0.71 | 0.020±0.036 | 1.71±1.49 |
| paramakular | 0.03±0.14 | 0.88±0.74 | 0.054±0.070 | 1.73±1.80 |

The concentrations, determined from the fundus reflection spectra by inverse adding-doubling method, are given in mmol/l (hemoglobin) or arbitrary units (xanthophyll: 1=value, given by Wyszecki and Stiles[4], melanin: 1=value given by Gabel et al.[5] for RPE).

**Discussion**

The xanthophyll concentration in the fovea is exactly that reported by Wyszecki and Stiles[4] from psychophysical measurements. The concentration of melanin in the RPE is equal to that measured by Gabel et al.[5] microspectrophotometrically. For the choroid we determined melanin concentrations which are much smaller as in the RPE. The hemoglobin concentration we found results in a blood content of the choroid of 20%.

## Conclusion:

**A new developed model of the ocular fundus reflection on the basis of the inverse adding-doubling method enables the calculation of the concentrations of xanthophyll in the retina, of melanin in the RPE as well as in the choroid, and of hemoglobin in the choroid from fundus reflection measurements in vivo.**


**References** 1. Hammer M, Roggan A, Schweitzer D and Müller G 1995 *Optical properties of ocular fundus tissues - an in vitro study using the double -integrating-sphere technique and inverse Monte Carlo Simulation* Phys. Med. Biol **40** 963-978
2. van de Hulst H C 1980 *Multiple light scattering*. Vol 1; Academic, New York
3. Brent R P 1973 *Algorithms for Minimization without Derivatives*. Prentice-Hall, Englewood Cliffs
4. Wyszecki G, Stiles W S 1967 *Colour science*. Wiley & Sons, New York
5. Gabel V-P, Birngruber R and Hillenkamp F 1978 *Visible and near infrared light absorption in pigment epithelium and choroid*. Proc. XXIII concilium ophthalmologicum, Kyoto, pp 658-662; ed. K Shimizu and J A Oosterhuis (Amsterdam Oxford: Excerpta Medica)

in our case: Due to the high concentrations of melanin and hemoglobin in the retinal pigment epithelium and in the choroid respectively, the scattering and absorption coefficients of these layers are in the same order. Furthermore, an isotropic field can not establish in very thin layers under nearly collimated illumination. Therefore, the application of the diffusion approximation for the interpretation of ocular fundus spectra did not produce reasonable results. The drawbacks of the two photon flux theory were already mentioned, the extension of this theory to four fluxes with appropriate boundary conditions is equivalent to the diffusion approximation [9].

An elegant solution of the problem, however, was found using the adding-doubling method developed by van de Hulst for the description of the reflection of planetary atmospheres [10, 11] and introduced to tissue optics by Prahl [12, 13].

## 2. Theory

The radiance reflected from ($I_{\text{ref}}$) or transmitted through ($I_{\text{tr}}$) a thin absorbing and scattering layer is given by $I_{\text{ref}} = \mathbf{R} \cdot I_0$ and $I_{\text{tr}} = \mathbf{T} \cdot I_0$ respectively, where $I_0$ is the incident radiance and **R** and **T** are reflection and transmission operators. Than, the reflection and transmission of a double layer is described by infinite series converging to the operators [11] (doubling)

$$\begin{aligned}\mathbf{R}_{\text{double}} &= \mathbf{R} + \mathbf{TRT} + \mathbf{TRRRT} + \cdots \\ &= \mathbf{R} + \mathbf{T}(\mathbf{1} - \mathbf{RR})^{-1}\mathbf{RT} \\ \mathbf{T}_{\text{double}} &= \mathbf{TT} + \mathbf{TRRT} + \mathbf{TRRRRT} + \cdots \\ &= \mathbf{T}(\mathbf{1} - \mathbf{RR})^{-1}\mathbf{T}\end{aligned} \quad . \qquad (1)$$

Once having the operators for an infinitesimal thin layer, that for a layer of an arbitrary thickness can be calculated by iterative doubling of the initial thin layer. In the same way, two layers of different optical properties can be combined [14] (adding)

$$\begin{aligned}\mathbf{R}_{\text{adding}} &= \mathbf{T}^{10}(1 - \mathbf{R}^{12}\mathbf{R}^{10})^{-1}\mathbf{R}^{12}\mathbf{T}^{01} + \mathbf{R}^{01} \\ \mathbf{T}_{\text{adding}} &= \mathbf{T}^{12}(1 - \mathbf{R}^{10}\mathbf{R}^{12})^{-1}\mathbf{T}^{01}\end{aligned} \quad , \qquad (2)$$

where the superscripts denote the consecutive layer interfaces. Now we consider the following geometry suitable for the treatment of ocular fundus reflection: Let $I_0$ be constant within the illuminated spot at the tissue surface with an angular distribution symmetric around an axis vertical to the infinitely extended stack of tissue layers. This geometry eliminates the azimuth dependence and the reflected radiance can be given in terms of the cosines $\eta_0$ and $\eta$ of the elevation angles of illumination and observation:

$$I_{\text{ref}}(\eta) = \int_0^1 R(\eta, \eta_0) I_0(\eta_0) 2\eta_0 d\eta_0. \qquad (3)$$

The same holds for the transmission as well as for the products of reflection and transmission operators in Eqs. (1) and (2). All the appearing integrals can be solved by numerical quadrature resulting in the following expression for the product of two operators **A** and B:

$$\begin{aligned}\int_0^1 A(\eta, \eta')B(\eta', \eta'')2\eta' d\eta' &= \sum_{j=1}^{n} A_{ij} 2\eta_j w_j B_{jk} \\ &= \mathbf{A} * \mathbf{B}.\end{aligned} \qquad (4)$$

This is a special type of matrix multiplication with the identity matrix $\mathbf{E} = \frac{1}{2\eta_j w_j}\delta_{ij}$ where $w_j$ is the weight of the $j$-th quadrature angle which cosine is $\eta_i$. Thus, the reflection and transmission operators are transformed into matrices which $ij$-element gives the radiance which is reflected or transmitted from the $i$-th to the $j$-th angular element respectively. Though gaussian quadrature is the most popular one, we followed Prahl [13] and used Radau quadrature [15] since this gives the opportunity to chose the cosine $\eta_0 = 1$ (normal incidence of the illumination) as a quadrature point. The $\eta_i$ and $w_i$ are for $n$ quadrature angles given by [13]

$$\eta_I = \frac{1}{2} - \frac{1}{2}x_i \quad \text{and} \quad w_I = \frac{1}{2(1 - x_i)[P'_{n-1}(x_i)]^2} \qquad (5)$$

where $x_i$ with $1 < I < n - 1$ are the solutions of

$$P_{n-1}(x) + \frac{x - 1}{n} P'_{n-1}(x) = 0 \qquad (6)$$

and $P_{n-1}(x)$ and $P'_{n-1}(x)$ are the $n - 1$-th Legendre polynomial of zero order and its first derivative respectively, whereas $x_n = 1$ and $w_n = w_n(-1)$.

The reflection and transmission matrices for an infinitesimal thin layer are obtained from the solution of the one-dimensional radiation transfer equation using discretization of the angle:

$$\pm\eta_i \frac{dI(\tau,\pm\eta_i)}{d\tau} + I(\tau,\pm\eta_i) = \frac{W_0}{2}\sum_{j=1}^{n}$$

$$w_j[h(\eta,\eta_j)I(\tau,\pm\eta_j) + h(\eta_i,-\eta_j)I(\tau,\mp\eta_j)] \quad (7)$$

where $\tau$ is the optical depth. The redistribution function $h(\eta_i,\eta_j)$ is the phase function integrated over the azimuth angle $\phi$:

$$h(\eta_I,\eta_j) = \frac{1}{2\pi}\int_0^{2\pi} \times p\left(\eta_i\eta_j + \sqrt{1-\eta_i^2}\sqrt{1-\eta_j^2} \times \cos\phi\right) d\phi. \quad (8)$$

Approximating the phase function by a sum of a Dirac delta function and a series on $n-1$ Legendre polynomials and substituting

$$\tau' = (1 - W_0 \cdot g^n)\cdot\tau \quad \text{and} \quad W_o' = \frac{1-g^n}{1-W_0\cdot g^n} W_0, \quad (9)$$

where $W_0$ is the albedo and $g$ is the scattering anisotropy yields in

$$h(\eta_i,\eta_j) = \sum_{k=0}^{n-1}(2n+1)\chi_k P_k(\eta_i)P_k(\eta_j). \quad (10)$$

Assuming the Henyey-Greenstein phase function [16], we find $\chi_k = \frac{g^k - g^n}{1-g^n}$. Using substitution Eq. (9) the integration of Eq. (7) over an infinitesimal layer thickness

$$\Delta\tau' = \tau_1' - \tau_0' < \eta_1 \quad (11)$$

then yields in

$$\pm\eta_i[I(\tau_1',\pm\eta_i) - I(\tau_0',\pm\eta_i)] + \Delta\tau' I_{0/1}(\pm\eta_i)$$

$$= \frac{W_0'}{2}\sum_{j=1}^{n} w_j \Delta\tau'$$

$$[h(\eta_i,\eta_j)I_{0/1}(\pm\eta_i) + h(\eta_i,-\eta_j)I_{0/1}(\mp\eta_i)]$$

with

$$I_{0/1}(\eta) = \frac{1}{\Delta\tau'}\int_{\tau_1'}^{\tau_2'} I(\tau',\eta)d\tau'. \quad (12)$$

Following Wiscombe [17], we assumed the radiance in this layer to be the average of that at the top and at the bottom of the layer:

$$I_{0/1}(\eta) = \frac{1}{2}[I(\tau_0',\eta) + I(\tau_1',\eta)]. \quad (13)$$

Thus, Eq. (12) is an approximate solution of the one dimensional discretized radiative transfer equation resulting in the following expressions for the reflection and transmission matrices of a thin layer:

$$\mathbf{R}_{\Delta\delta'} = 2\mathbf{G}\cdot\mathbf{B}\cdot([\delta_{ij}] + \mathbf{A})^{-1}$$

and

$$\mathbf{T}_{\Delta\tau'} = 2\mathbf{G} - [\delta_{ij}]$$

with

$$\mathbf{A} = \left[\frac{\delta_{ij}}{\eta_i}\right]\cdot\left([\delta_{ij}] - \frac{W_0'}{2}[h(\eta_i,\eta_j)]\cdot[w_i\delta_{ij}]\right)\frac{\Delta\tau'}{2}$$

$$\mathbf{B} = \frac{W_0'}{2}\left[\frac{\delta_{ij}}{\eta_i}\right]\cdot[h(\eta_i,-\eta_j)]\cdot[w_i\delta_{ij}]\frac{\Delta\tau'}{2}$$

and

$$\mathbf{G} = ([\delta_{ij}] + \mathbf{A} - \mathbf{B}\cdot([\delta_{ij}] + \mathbf{A})^{-1}\cdot\mathbf{B})^{-1}. \quad (14)$$

The calculation of the reflection of a layer with finite thickness needs the following steps: First, choosing the number of quadrature points and calculating the cosines of quadrature angles and their weights from Eq. (5). Using the transformations Eq. (9), the optical thickness of the infinitesimal layer is estimated by $2^m \cdot \Delta\tau' = \tau'$ where $\tau'$ is the transformed optical thickness of the layer under investigation and m is chosen to fulfill the condition Eq. (11).

Now, the reflection and transmission matrices can be calculated from Eq. (14) and an $m$-fold doubling according to Eq. (1) results in the reflection and transmission of the finite layer under consideration. By multiplication of that matrices with the incident radiance $\mathbf{I}_0(\eta_i)$ we obtain the reflected and transmitted radiances respectively. Summing up these radiances over all angular elements the total reflection and transmission was found.

Applying this formalism to the single layers of the ocular fundus (optical properties of the layers from Hammer et al. [6], thicknesses from Rohen [18]), we found an excellent agreement with the results of a Monte Carlo simulation using 8 quadrature angles only. Assuming collimated illumination ($\mathbf{I}_0 = \{0, \ldots 0, I\}$), the largest difference between the results of adding-doubling method and Monte Carlo simulation was found to be 2.6% for the retinal pigment epithelium.

## 3. Inverse adding-doubling model for the calculation of chromophore concentrations from reflection spectra

As the reflection of the ocular fundus representing a stack of plane parallel layers can be obtained from the optical parameter and the thicknesses of this layers by the adding-doubling method, a complete reflection spectrum can be calculated if the spectra of the absorption and scattering coefficients as well as that of the scattering anisotropy of all layers is known. Our problem, however, is the reverse one: We have to determine the chromophore concentrations from a measured spectrum. We assume the ocular fundus to consist of four layers: the retina containing oxygenated and reduced hemoglobin in their capillaries and xanthophyll in the macular region, the RPE heavily pigmented with melanin, the choroid containing hemoglobin and melanin, and the sclera. Thus, the absorption coefficients $\mu_a$ are related to the concentrations of this pigments in the following way:

$$\mu_{a_{\text{Retina}}}(\lambda) = c_x \epsilon_x(\lambda) + c_{\text{Hb}} \epsilon_{\text{Hb}}(\lambda) + c_{\text{HbO}_2} \epsilon_{\text{HbO}_2}(\lambda)$$

$$\mu_{a_{\text{RPE}}}(\lambda) = c_{\text{Me,RPE}} \epsilon_{\text{Me}}(\lambda) \tag{15}$$

$$\mu_{a_{\text{Chorioidea}}}(\lambda) = c_{\text{Me.Ch}} \epsilon_{\text{Me}}(\lambda) + c_{\text{HbO}_2\text{,Ch}} \epsilon_{\text{HbO}_2}(\lambda)$$

where $c$ are the chromophore concentrations and $\epsilon$ the extinction coefficients. The indices Hb and $\text{HbO}_2$ denote reduced and oxygenated hemoglobin respectively, X and Me denote xanthophyll and melanin, and Ch denotes the choroid. Reduced choroidal blood was neglected since the arterio-venous difference in oxygen saturation is only 3% for the choroids [19]. The molar extinction coefficients for hemoglobin were obtained from van Assendelft [20]. Melanin and xanthophyll, however, are not distinct chemical compounds but composite substances. Therefore, we obtained the melanin extinction spectrum from microspectrophotometric measurements by Gabel et al. [21] and that of xanthophyll from a survey of psychophysical data given by Wyszecki and Stiles [22]. Thus, the concentrations of hemoglobin are given in mmol/l whereas that of melanin and xanthophyll can be given in relative units only, which are factors to the physiological values found in the literature [21, 22]. The scattering coefficient $\mu_s$ and the anisotropy of scattering g for each layer were obtained from *in vitro* measurements of Hammer et al. [6]. Since post mortem changes of the tissue scattering are possible, the scattering coefficients were allowed to vary linearly. The scattering anisotropy values were kept constant, i.e. the number of scatterers was assumed to be variable at constant scatterer size and geometry. Also the absorption and scattering coefficient of the sclera were kept constant at the literature data [6] since a variation of these values showed negligible effect on the reflection of the whole fundus (<1% at variations of $\mu - a$ and $\mu_s$ by a factor of 8). In order to take the transmission of the anterior ocular media into account, we multiplied the calculated reflection of the fundus with the double ocular media transmission given by van Norren and Vos [23] corrected for age according to the formula by Pokorny et a1. [24].

Thus, we established a model of ocular fundus reflection $f$ on the basis of the adding-doubling theory that had to be fitted to the measured spectra $R(\lambda)$ using a least squares procedure:

$$\begin{aligned} &\sum_\lambda (R(\lambda) - f(c_X \epsilon_X(\lambda), c_{\text{Hb}} \epsilon_{\text{Hb}}(\lambda), c_{\text{HbO}_2} \epsilon_{\text{HbO}_2}(\lambda), \\ &c_{\text{Me,RPE}} \epsilon_{\text{Me}}(\lambda), c_{\text{Me,Ch}} \epsilon_{\text{Me}}(\lambda), c_{\text{HbO}_2\text{,Ch}} \epsilon_{\text{HbO}_2}(\lambda), \\ &c_{s,\text{R}} \mu_{s,\text{R}}(\lambda), c_{s,\text{RPE}} \mu_{s,\text{RPE}}(\lambda), c_{s,\text{Ch}^{\mu}_{\text{s}},\text{Ch}}(\lambda)))^2 \rightarrow \text{Min} \end{aligned} \tag{16}$$

where the $c_s$ are factors to the scattering coefficients and the index $R$ denotes the Retina. Eq. (16) is a nonlinear optimization problem with the handicap, however, that the reflection model $f$ can not be given analytically but is the result of the iterative adding-doubling process. This makes the solution by a Gauss–Newton method very expensive since the Jacobi-matrix had to be determined numerically performing small variations of each parameter at any iteration step. In this case, a derivative-free method is more comfortable. We used a self-learning procedure on the basis of Powell's conjugate direction set method [25].

In order to optimize the model, we fixed some of the parameter: First, we neglected the hemoglobin absorption in the retina since this showed to be far below that of the choroids in preliminary fits. This is in agreement with the anatomy: The capillary density of the retina which is much smaller than that of the choroid. Further, we fixed the factors $c_s$ of the scattering coefficients in the following way: We ran the program according to Eq. (16) for three macular reflection spectra of normal volunteers using six different sets of initial parameter values. From this calculation, we found the mean values and standard deviations $c_{s,\mathrm{R}} = 0.24 \pm 0.019$, $c_{s,\mathrm{RPE}} = 0.59 \pm 0.14$, and $c_{s,\mathrm{Ch}} = 1.64 \pm 1.32$ which were used in all further fits. An interesting alternative would be the *in vivo* measurement of the scattering parameters by optical coherence tomography (OCT) [26].

Thus, the remaining parameters to be determined from ocular fundus reflection spectra are the retinal xanthophyll concentration $c_X$, the concentration of melanin in the RPE and in the choroid $c_{\mathrm{Me,RPE}}$ and $c_{\mathrm{Me,Ch}}$, and the choroidal hemoglobin concentration $c_{\mathrm{HbO_2,Ch}}$.

## 4. Validation of the reflection model in clinical examples

### A. Methods

The model for the determination of ocular fundus chromophore concentrations introduced in section 3 was tested in first experiments. The fundus reflection spectra of 12 normal caucasians, as well as patients suffering from macular holes, age related an juvenile macular degeneration (AMD), and high myopia were measured by imaging fundus reflectometry. The technique is described in detail elsewhere [27]. In principle, the light of the xenon flash of a Zeiss fundus camera (Carl Zeiss Jena, Germany) reflected from the ocular fundus was dispersed by an imaging spectrograph (Jobin Yvon, France) and detected by an intensified slow scan CCD camera (Princeton Instruments, New Jersey). The reflected intensity was measured spectrally and locally resolved in relative units dividing the signal detected from the fundus by that detected from an artificial eye with a diffuse white reflector (Spectralon, Labsphere Inc., New Hampshire) in its focal plane.

The thickness of the retina was measured by an optical coherence tomography scanner (Humphrey Instruments Inc., California). Unfortunately, the depth resolution and the sensitivity of this device were not sufficient to measure the thickness of the other layers. Therefore, we used literature values [18] instead: RPE: 10 $\mu$m, choroid: 250 $\mu$m, sclera: 700 $\mu$m.

The investigation in human subjects were performed with their informed consent, according to the

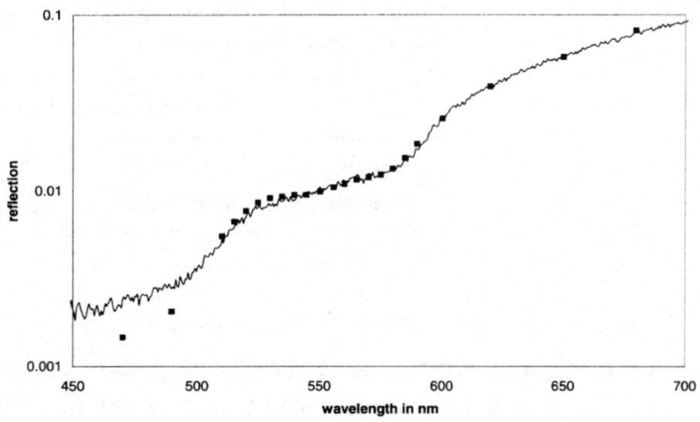


*Figure 1.* Macular fundus reflection spectrum (solid line) and its representation by fitted the adding-doubling model (squares).

guidelines of the Declaration of Helsinki, and under the approval of a local ethic committee. The subjects pupil was dilated to a diameter of at least 6 mm by Tropicamide before the investigation.

### B. Results

A typical macular reflectance spectrum and the reflections calculated by the model described in section 3 fitted to the spectrum at 23 wavelengths is shown in Figure 1. Generally, we found a good agreement between measurement and model. The deviation at the extremely weak reflections below 500 nm is attributed to stray light in the camera which can not be suppressed completely.

Table 1 gives the mean values and standard deviations of the concentrations of xanthophyll, melanin, and hemoglobin in the macula and about two degrees temporal to the fovea in 12 normal subjects. The mean macular concentrations of xanthophyl and melanin in the RPE were found to be in complete agreement with the data published by Wyszecki and Stiles and by Gabel et al. [21] whereas the standard deviations show large inter-individual variations. Toward the para-macular region, the xanthophyll concentration reduces drastically. The melanin concentration in the pigment epithelium is reduced too but this decrease is statistically (Student's t-test, $p = 0.05$) not significant, a result already found *in vitro* by Gabel et al. [21]. In contrast, we found a non-significant increase of the choroidal melanin concentration two degrees temporal to the fovea compared to the macula. This, also, is in agreement with observations by Gabel et al. [28]. Assuming a choroidal thickness of 250 $\mu$m, the total choroidal melanin content in the fovea is as half as high as in the RPE whereas it is 1.5 times higher parafoveal. There was no difference between the choroidal

*Table 1.* Mean values and standard deviations of chromophore concentrations at and besides the macula in healthy subjects

| | Xanthophylls | Melanin (RPE) | Melanin (choroid) | Hemoglobin |
|---|---|---|---|---|
| Macula | $1.01 \pm 0.23$ | $0.99 \pm 0.71$ | $0.020 \pm 0.036$ | $1.71 \pm 1.49$ |
| Para-makular | $0.03 \pm 0.14$ | $0.88 \pm 0.74$ | $0.054 \pm 0.070$ | $1.73 \pm 1.80$ |

hemoglobin concentration at the investigated fundus sites. We found 1.7 mmol/l which is equivalent to a Blood content of the choroid of 20%, however, the inter-individual differences are extremely high.

After the good agreement of the chromophore concentrations derived from *in vivo* measurements at healthy subjects with published data has been shown, the capability of the technique and its limitations was investigated in a few single pathologic cases. Examining a group of four patients suffering from full thickness macular holes, we found no xanthophyll in the fovea (what is obvious from the absence of the retina in this location) but a para-macular xathophyll concentration of $0.13 \times 0.06$ which is significantly higher as that in the normal control group. This finding is explained by the parafoveal deposit of primarily macular retinal tissue as it can be observed also by OCT showing a thickening of the retina besides the hole. The macular measurement of a 72 year old female suffering from non-exsudative age related macula degeneration (AMD) showed a xanthophyll concentration of 0.38 and a melanin concentration of 0.23. All other values were normal. The reduction of the xanthophyll, which is under discussion as a protective agent against oxidative stress in the photoreceptors [30–34], seems to be typical for AMD [29]. The method described here enables the objective measurement of xanthophyll concentration and can, therefore, be beneficial for the early detection of subjects with the risk to develop AMD. The de-pigmentation we found is one of the clinical features of non-exsudative AMD and can be quantified now. Noteworthy are, furthermore, the findings in a 33 year old male patient suffering from juvenile macula degeneration on both eyes: Reduced xanthophylls concentrations (left eye: 0.41, right eye: 0.63) were found in this case too whereas the RPE melanin concentration remained normal in the eye with the higher xantophyll concentration (right eye: 0.94) but was reduced in the other eye (left eye: 0.59). All other values were normal. This also may be a hint on the neuroprotective effect of xanthophyll. That the model introduced in section 3 is able to distinguish between melanin in the RPE and in the choroid despite their absorption spectra are the same was demonstrated in the case of a 60 year old female showing a hyper-pigmentation of the posterior pole. It is known that such hyper-pigmentations are located in the choroids [35]. From our macular measurement we calculated a normal melanin concentration of 0.93 in the RPE but in the choroid a concentration which was 11 times higher as in the normal control group (0.23). The para-macular choroidal melanin concentration was found to be five times higher than the normal one.

However, there were also examples in which our model did not give correct results: In an eye with a myopia of $-13$ diopters we found choroidal melanin concentrations of $-0.033$ (macula) and $-0.028$ (para-macular region) as well as hemoglobin concentration of 0.29 mmol/l and 0.24 mmol/l respectively. The reason of this obviously pathologic values is, however, rather an effect of the reduced thickness of the choroid of the myopic eye than a reduction of the concentrations. Thus, the model needs the input of the correct thickness for all layers. The thickness of the choroid could be measured *in vivo* by high resolution ultrasound sonography which, unfortunately, was not available. From the reflectance spectra of the drusen and hard exsudates in the patient with AMD we found negative values for the pigmentation of the RPE (druse: $-0.3$, exsudate: $-0.23$). All other values were normal. This values are clearly due to enhanced scattering of deposits of hyalin (druse) and lipids (exsudate) and not to a lack in pigmentation. This shows that the scattering coefficients are critical constants of the model which had to be determined separately [26] in pathologic cases.

## 5. Discussion

If its conditions are met, the model given in section 3 yields in reasonable values for the concentration of melanin, hemoglobin, and xanthophyll at the ocular

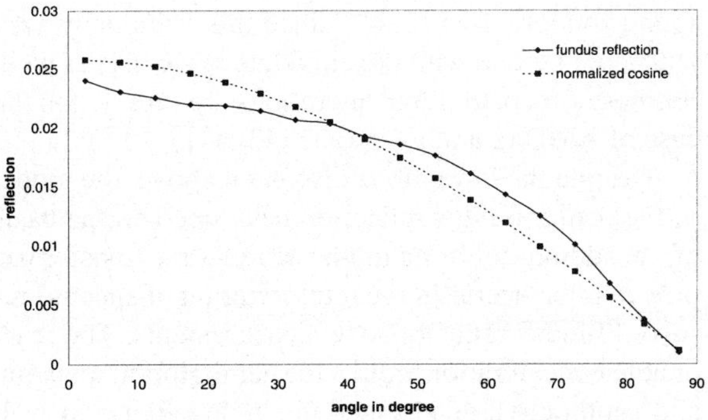


*Figure 2.* Comparison of the ocular fundus reflection modeled by angle resolved Monte Carlo simulation with the suggestions of the lambertian reflection law (cosine).

fundus as is shown by the results presented in section 4.B. Nevertheless, from a theoretical point of view some items need further discussion.

Since the adding-doubling method uses a discretization of the elevation angle it is generally able to represent the illumination and observation geometry of the used fundus camera. In order to average over at least three radiance values within the cone determined by the observation aperture of the camera, 200 quadrature angles were necessary. Inversion of reflection and transmission matrices (Eqs. (1) and (2)), however, needs a computing time proportional to the cube of the number of quadrature angles. This makes the handling of matrices of 200 × 200 values virtually impossible. Therefore, eight quadrature angles were used and collimated incidence of the light was assumed as a reasonable model for fundus camera illumination. In contradiction to the geometry of the fundus camera allowing an observation within 2.15 degrees, we integrated the reflection over all quadrature angles. Since the measured reflection is refered to that of a white reflection target with lambertian characteristics, this integral value represents the fundus reflection into the observation aperture only in the case of a lambertian reflection of the ocular fundus. This has already pointed out by Delori and Pflibsen [2]. Thus, we performed an angle resolved Monte Carlo simulation [8] of the reflection of the fundus with the optical properties given by Hammer et al. [6] at 560 nm. As is shown in Figure 2, the lambertian cosine law of reflection is a good representation for the reflection of the ocular fundus. The mean relative deviation was found to be 0.12 and the deviation within the camera observation pupil, i.e. the error made by the assumption of a lambrertian reflecting fundus, was 0.068.

This error was determined for homogenous layers. The anatomical structure of the retina, however, shows inhomogenities affecting the reflection of light. Especially the waveguide nature of the photoreceptors [36] oriented towards the center of the pupil leads to the so called Stiles-Crawford effect [37] of brighter appearance of light entering the eye through the pupil center. Van de Kraats et al. [38] modelled the receptors as optical antennas not only receiving but also sending out light. This has an anatomic equivalence in the photoreceptor discs which are lipoprotein membranes [19] oriented perpendicular to the direction of the incident light, not only absorbing but also reflecting light. This is called the optical Stiles-Crawford effect: The reflection of light incident through the center of the pupil on axis of the receptors is higher than that of light with oblique incidence. This is used in first clinical attempts for the detection of the misalignment of photoreceptors by De Lint et a1. [39]. However, van Blokiand and van Norren found no optical Stiles-Crawford effect at an eccentricity of the illumination of three millimeter in the pupil plane. Since the fundus camera used in our experiment has an annular illumination aperture with a mean diameter of 2.7 mm, we assumed virtually no light coupled into the receptors and hence the lambertian reflection characteristics is not superimposed by a directed reflex according to the Stiles-Crawford effect. Furthermore, van Blokiand and van Norren showed that the reflection is almost independent from the concentration of the photoreceptor pigments in the case of eccentric illumination. Therefore, these pigments were neglected in the model introduced in section 3.

Since we found the chromopore concentrations determined by our model to be dependent on the thicknesses of the fundus layers and their scattering behavior (see section 4.B), we investigated this point more in detail: Starting from the optical properties of the tissues measured in vitro [6] and from the thickness values given by Rohen [18], we calculated the reflection at the wavelengths 413 nm, 430 nm, 450 nm, 488 nm, 514 nm, 540 nm, 560 nm, 586 nm, 633 nm, and 700 nm. Than, we changed the scattering coefficient of one layer by a factor between 0.125 and 8 or the thickness of one layer by a factor between 0.5 and 2 respectively, keeping all other parameters constant, repeated the calculation and determined the square root of the sum of the quadratic deviations of the reflection at all wavelengths. The results are shown in Figures 3 and 4.

The strongest influences have changes in the RPE. Despite its small thickness and its high absorption

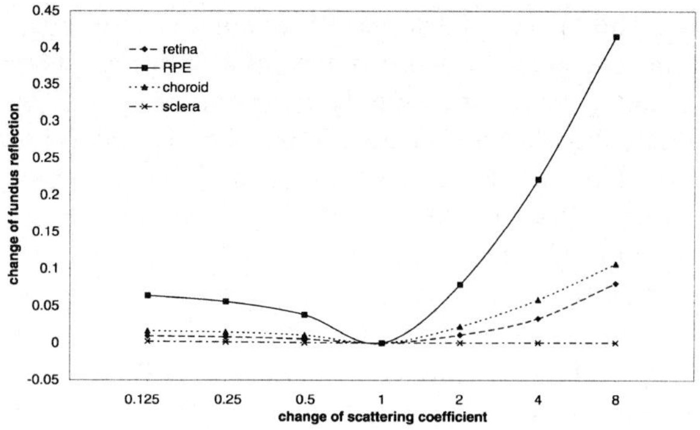


*Figure 3.* Ocular fundus reflection versus variation of the scattering coefficients.

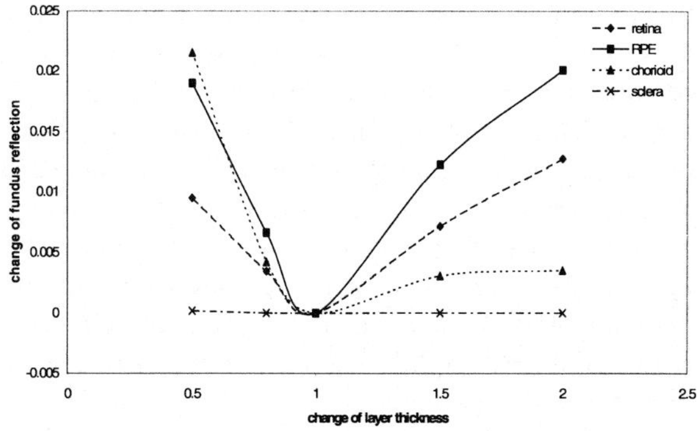


*Figure 4.* Ocular fundus reflection versus variation of the layer thickness.

coefficient, the reflection of this layer is not negligible. This fact, already pointed out in the fundus reflection models by van Norren and Tiemeijer as well as by Delori and Pflibsen, is due to the very high scattering coefficient and the relatively small scattering anisotropy [6] of $g = 0.84$. Virtually no influence have changes in the outermost layer of the eye, the sclera. The parameters of all other layers should be known from separate measurements at best. The scattering properties of the single layers can, in principle, derived from OCT-scans though the signal to noise ratio of these measurements is too poor to determine scattering coefficient and anisotropy with sufficient accuracy at present [26]. OCT is routinely used for the measurement of the thickness of the retina as it was done in this investigation (section 4.A). The depth resolution of the instrument used here was not sufficient to measure the thickness of the RPE. Since, however, the RPE is a monocellular layer, its thickness should be relatively constant within the 9 to 12 microns given by Rohen [18]. Attention should be paid to the measurement of the thickness of the choroid by high resolution ultrasound sonography [42, 41] since this value may vary with age [42] and with the circadian cycle [41] as well. Furthermore, pathologic alterations are reported in the case of AMD42 and glaucoma [43, 44].

Despite the limitations discussed above, the model of the ocular fundus reflection developed on the basis of the adding-doubling model of radiative transfer has proven to be useful in the interpretation of spectral resolved fundus reflectometry measurements. The reasonable concentration values for hemoglobin, melanin, and xanthophyll determined in single patients as well as in a healthy control group gives the reason for the hope that the developed method may be of interest for clinical diagnostics.

## Acknowledgements

The authors are grateful to Dr. Scott A. Prahl for his hints in the implementation of the adding-doubling algorithm.

## References


1. van Norren D, Tiemeijer LF. Spectral reflectance of the human eye. Vision Res 1986; 26: 313–320.
2. Delori FC, Pflibsen KP. Spectral reflectance of the human ocular fundus. Appl Opt 1989; 28: 1061–1077.
3. Schweitzer D, Königsdörffer E, Tröger G, Augsten R, Klein S, Roth H. Möglichkeiten und Grenzen der Fundusreflektometrie zum Nachweis von Veränderungen in einzelnen Schichten des Augenhintergrundes. Folia Ophthalmol 1990; 15: 125–137.
4. Kubelka P, Munk F. Ein Beitrag zur Optik der Farbanstriche. Zeitschr f techn Optik 1931; 11: 593–611.
5. Hammer M, Schweitzer D, Leistritz L, Scibor M, Donnerhacke KH, Strobel J. Imaging Spectroscopy of the human ocular fundus in vivo. J Biomed Opt 1997; 2: 418–425.
6. Hammer M, Roggan A, Schweitzer D, Müller G. Optical properties of ocular fundus tissues – an in vitro study using the double-integrating-spere technique and inverse Monte Carlo Simulation. Phys Med Biol 1995; 40: 963–978.
7. Ishimaru A. Wave propagation and scattering in random media. Academic Press New York, San Francisco, London.
8. Wang L-H, Jacques SL. Monte Carlo modelling of light transport in multi-layered tissues in standard C. The University of Texas, M.D. Anderson Cancer Center, Houston.
9. Star WM, Marijnissen JPA, van Gemert MJC. Light dosimetry in optical phantoms and tissue: L Multiple flux and transport theory. Phys Med Biol 33 437–454.
10. van de Huist HC. A New Look on Multiple Scattering. Unnumbered mimeographed report, NASA Institute for Space Science, New York, 1963.
11. van de Huist HC. Multiple light scattering. Vol 1; Academic, New York, 1980.
12. Prahl SA. Light Transport in Tissue, PhD thesis at the University of Texas at Austin, 1988.

13. Prahl SA. The adding-doubling method. In: Welch AJ and van Gemert MJC (eds), Optical-Thermal Response of Laser-Irradiated Tissue., Plenum Press, New York, 1995.
14. Plass GN, Kattawar GW, Catchings FE. Matrix operator theory of radiative transfer. 1: Rayleigh scattering. Appl Opt 1973; 12: 314–329.
15. Hildebrand FB. Introduction to Numerical Analysis. Dover, New York, 1974.
16. Henyey LG, Greenstein JL. Diffuse radiation in the galaxy. Astrophys J 1941; 93: 70–83.
17. Wiscombe WJ. On initialization, error, and flux conservation in the doubling method. J Quant Spectrosc Radiat Transfer 1976; 16: 637–658.
18. Rohen JW. Anatomie und Embryologie. In: Francois J, Holiwich I (eds) Augenheilkunde in KImik und Praxis, Thieme, Stuttgart, 1977.
19. Naumann GOH. Pathologie des Auges. Vol. 1, Berlin, Heidelberg, New York, 1997.
20. van Assendelft OW. Spectrophotometry of heamoglobin derivatives. Roal Vangorcum, Assen, 1970.
21. Gabel V-P, Birngruber R, Hillenkamp F. Visible and near infrared light absorption in pigment epithelium and choroid. In: Shimizu K, Osterhuis JA (eds) Proc. XXIII consilium ophthalmologicum, Kyoto, (Amsterdam Oxford: Excerpta Medica), pp. 658–662, 1978.
22. Wyszecki G, Stiles WS. Colour science. Wiley & Sons, New York, 1967.
23. van Notren D, Vos JJ. Spectral transmission of the human ocular media. Vision Res 1974; 14: 1237–1244.
24. Pokorny I, Smith VC, Lutze M. Aging of the human lens. Appl Opt 1987; 26: 1437–1440.
25. Brent RP. Algorithms for Minimization without Derivatives. Prentice-Hall, Englewood Cliffs, 1973.
26. Hammer M, Schweitzer D, Thamm E, Koib E. Optical properties of ocular fundus tissues determined by optical coherence tomography. Submitted to Op. Comm. 2000.
27. Hammer M, Schweitzer D, Leistritz L, Scibor M, Donnerhacke KH, Strobel J. Imaging spectroscopy of the human ocular fundus in vivo. J Biomed Opt 1997; 2: 418–425.
28. Gabel V-P, Birngruber R, Hillerikamp F. Die Lichtabsorption am Augenhintergrund. GSF-Bericht A 55, Gesellschaft für Strahien- und Umweltforschung mbH, München, 1976.
29. Schweitzer D, Hupfer U, Hammer M, Scibor M. Discrimination between ARMD-patients and normals by reflectometric data. Invest Ophthalmol Vis Sci 1996; 37: 548.
30. Bone RA, Landrurn JT, Fernandez L, Tarsis SL: Analysis of the macular pigment by HPLC: Retinal distribution and age study. Invest Ophthalmol Vis Sci 1988; 29: 843–849.
31. Gerster H. Antioxidant protection of the ageing macula. Age Ageing 1991; 20: 60–69.
32. Schalch W, Werner P. Vitamins and carotenoids – a promising approach to reducing the rise of coronary heart disease, cancer; and eye disease. Adv Exp Med Biol 1994; 366; 335–350.
33. Sommerburg O, Keunen JEE, Bird AC, van Kuijk FJGM. Fruits and vegetables that are sources for lutein and Zeaxantin: the macular pigment in human eyes. Br J Ophthalmol 1998; 82: 907–910.
34. Young RW. Solar radiation and age related macular degeneration. Surv Ophthalmol 1998; 32: 252–269.
35. Sachsenweger M. Augenheilkunde. Hippokrates Verlag, Stuttgart, 1994.
36. Snyder AW. Photoreceptor optics – theoretical principles. In: Snyder AW, Menzel R (eds) Photoreceptor optics. Springer, Berlin Heidelberg New York, 1975.
37. Snyder AW. Photoreceptor optics – theoretical principles. In: Snyder AW, Menzel R (eds) Photoreceptor optics. Springer, Berlin Heidelberg New York, 1975.
38. van de Kraats J, Berendschot TTJM, van Norren. The pathways of light measured in fundus reflectometry. Vision Res 1996; 36: 2229–2247.
39. DeLint PJ, Berendschot TTJM, van Nonren D. A comparison of the optical stiles-crawford effect and retinal densitometry in a clinical setting. Invest Ophthalmol 1998; 39: 1519–1523.
40. Cristini G, Cennarno O, Daponte P. Choroidal thickness in primary glaucoma. Ophthalmologica 1991; 202: 81–85.
41. Nickla DL, Wildsoet C, Wallman J. Visual influences on diurnal rhythms in ocular length and choroidal thickness in chick eyes. Exp Eye Res 1998; 66: 163–181.
42. Ramrattan RS, van der Schaft TL, Mooy CM, de Bruijn WC, Mulder PO, de Jong PT. Morphometric analysis of Bruch's membrane, the choriocapillaries, and the choroid in aging. Invest Ophthalmol Vis Sci 1994; 35: 2857-64.
43. Kubota T, Jonas JB, Naumann GOH. Decreased chonoidal thickness in eyes with secondary angle closure glaucoma. An aetiological factor for deep retinal changes in glaucoma? Br J Ophthalmol 1993; 77: 430–432.
44. Yin ZQ, Vaegan Millar TJ, Beaumont P, Sarks S. Widespread choroidal insufficiency in primary open-angle glaucoma. J Glaucoma 1997; 6: 23–32.

*International Ophthalmology* **23:** 291–295, 2001.
*J.R. Sampaolesi (ed.), Laser Scanning: Update 1*, 109–113.



# Scattering properties of the retina and the choroids determined from OCT-A-scans

H. Hammer, D. Schweitzer, E. Thamm, A. Kolb & J. Strobel
*University of Jena, Department of Ophthalmology, Jena, Germany*

*Key words:* OCT, tissue scattering, retina, choroid


## Abstract

*Goal*: To determine the coefficient and the anisotropy of scattering as well as the refractive indices in the retina and in the choroid noninvasively *in vivo*. *Methods*: The power of coherent reflected light versus fundus depth is recorded in OCT-A-scans. The ratio of refractive indices is derived from the height of the reflection peaks. Provided that the absorption coefficient is known from fundus reflectometry, the scattering coefficient and anisotropy are calculated from the offset and the slope of the signal behind the reflection peaks on the basis of a single backscattering model. *Results*: We found scattering coefficients of 12/mm (retina) and 27.5/mm (choroid) as well as anisotropy values of 0.97 (retina) and 0.90 (choroid). *Discussion*: The OCT is usually employed for the measurement of intra-ocular distances. The new technique described here gives the unique opportunity to determine further interesting parameters of single ocular layers. The values given above are in good agreement with *in vitro* results.


## Introduction

Recently, optical coherence tomography (OCT) is widely used to observe hidden morphologic structures in the depth of biological tissues. In the ophthalmology, OCT is used to obtain tomographic images of the ocular fundus. This enables the measurement of the thicknesses of retinal structures. However, coherent reflected light carries more information characterising the optical properties of tissue. The goal of this paper is to investigate the feasibility of OCT to determine the refractive index, the extinction coefficient, and the scattering anisotropy of tissues using a single backscattering model. This will be demonstrated at the example of an OCT-scan from a human ocular fundus *in vivo*.

## Methods

OCT is an interferometric technique. Let us consider, e.g., a Michelson interferometer 7with a scattering sample in one arm and a mirror in the other one (reference arm). In order to avoid the superposition of many interfering waves backscattered from the sample, a light source with a high spatial but a short temporal coherence is used. This results in interference only if the optical path difference in both arms of the interferometer is not larger than the coherence length of the source [1]. If $L + \Delta L$ is the distance of the reference mirror and $L$ that of the surface of the sample ($z = 0$) from the beam splitter (50/50), and if there is a reflection $R(z)$ in the depth $z$ in the sample, at the detector the power [2, 3].

$$\begin{aligned} P(z) &= P_0 \times [(1 + R(z))/4 \\ &\quad +1/2\sqrt{R(z)} \times |\gamma(2(\Delta L - nz)/c)| \\ &= \times \cos(4\pi(\Delta L - nz)/\lambda)] \end{aligned} \tag{1}$$

is measured, where $P_0$ is the power of the light source, $n$ is the refractive index of the sample, $c$ is the velocity of the light, $\lambda$ is the wavelength, and $\gamma$ is the temporal coherence function.

As involuntary eye movements make it difficult to observe the interference between waves reflected from the eye and from a stationary mirror, Fercher et al. [4, 5, 6] used a dual beam interferometer: The light of the source is divided by an interferometer so that there are two wavefronts with a path difference equal to the optical length of the eye incident on the eye. Thus, interference between the wave reflected at the cornea and that reflected at the ocular fundus is observed. In order to avoid losses of the signal resulting from the interference of a plane wave reflected at the fundus with a curved wave reflected at the cornea, a diffractive lens focusing 40% of the incident light at the cornea and 60% at the fundus is used. For this optical arrangement, Equation (1) has to be modified:

$$\begin{aligned} P(z)/P_0 &= 0.4R_c + 0.6R(z) + 2\sqrt{0.4R_c}\sqrt{0.6R(z)} \\ &\quad \times |\gamma(2(\Delta L - nz)/c)| \\ &\quad \times \cos(4\pi(\Delta L - nz)/\lambda). \end{aligned} \tag{2}$$

Here, $R_c$ is the reflection at the cornea and $R(z)$ is the reflection at the fundus in the depth $z$. The interference term (third summand of Equation (2)) is separated from the incoherent power by shifting the reference mirror of the interferometer with a constant speed $v = \mathrm{d}L/\mathrm{d}t$ and filtering the detected signal with the Doppler frequency $f_D = v/\lambda$.

If $R(z)$ results from the scattering at cells and cell compartments for which the distance is shorter than the coherence length of the light source, the measured power results from a coherent summation over all scattered waves. To calculate this, the local distribution of the scatterers as well as the amplitude and the phase of their scattering function have to be known. Since this is difficult in biological tissues, a simple single scattering approach [3, 7] is used instead, making the following assumptions: The scatterers are randomly distributed in the sample and their distance is large compared to the wavelength. Only those photons contribute to the coherent signal, which reach the aperture of the detector after a single scattering event. The Stokes-vektor remained unchanged in this backscattering. In compliance, with these requirements we get for the mean power with the frequency $f_D$

$$\langle P(z)\rangle/P_0 = 2\sqrt{0.4 \times 0.6 \times R_c}$$
$$\times\sqrt{\int_{-\infty}^{\infty} \tilde{R} \times (\exp(-2\mu_t z')/CF) \times |\gamma(2n(z-z')/c)|^2\, \mathrm{d}z'}, \tag{3}$$

where $\tilde{R}$ is that fraction of light, which is scattered into the aperture of the detector [8]. The term CF is a correction of the power recorded from a scatterer in the depth $z$ out of the focal plane in a confocal arrangement and is given by Yadlowsky [8] with

$$CF = (1 - (z' - f)/f)^2 + ((z' - f)\pi r^2/\lambda f^2)^2, \tag{4}$$

where $r$ is the radius of the laser beam in the pupil and $f$ is the focal length of the eye. Thus, $\tilde{R}$ and the extinction coefficient $\mu_t$ were calculated solving the non-linear optimization problem

$$\sum_z |\tilde{P}(z) - 2\sqrt{0.4 \times 0.6 \times R_c}$$
$$\times\sqrt{\int_{-\infty}^{\infty} \tilde{R} \times \exp(-2\mu_t z')/CF \times |\gamma(2n(z-z')/c)|^2\, \mathrm{d}z'}|^2$$
$$\rightarrow \text{Min} \tag{5}$$

by the use of Powell's conjugate direction set method [9]. Here, $\tilde{P}(z)$ is the measured coherent power versus depth normalized to the power of the incident beam. $\tilde{R}$ can be determined by the integration of the scattering phase function over the aperture angle of the observation. Assuming the Henyey–Greenstein phase function [10], we get

$$\tilde{R} = \frac{1}{4}\pi \times \int_{\pi}^{\pi+\varphi} (1-g^2)/(1+g^2-2g\cos\theta)^{2/3}\, \mathrm{d}\theta, \tag{6}$$

from which equation the anisotropy coefficient $g$ of scattering can be calculated if the aperture angle $\varphi$ = arctan(pupil radius/axial length of the eye) is assumed.

In OCT-scans of the ocular fundus (Figure 1), peaks resulting from specular reflection at the inner limiting membrane (ILM, border between vitreous and retina) and the retinal pigment epithelium (RPE) are observed. From the height of these peaks, the refractive indices of single layers may be determined using Equation (2) and Fresnel's reflection formula. For small angles of incidence of the laser beam onto cornea and retina, this results in

$$\frac{P}{P_0} = 2\sqrt{0.4 \times 0.6} \times |(n_{\text{air}} - n_{\text{cornea}})/(n_{\text{air}} + n_{\text{cornea}})|$$
$$\times |(n_{\text{vitreous}} - n_{\text{retina}})/(n_{\text{vitreous}} + n_{\text{retina}})|. \tag{7}$$

## Results

The coherent reflection at a human ocular fundus *in vivo* is given in Figure 1. This scan was recorded by

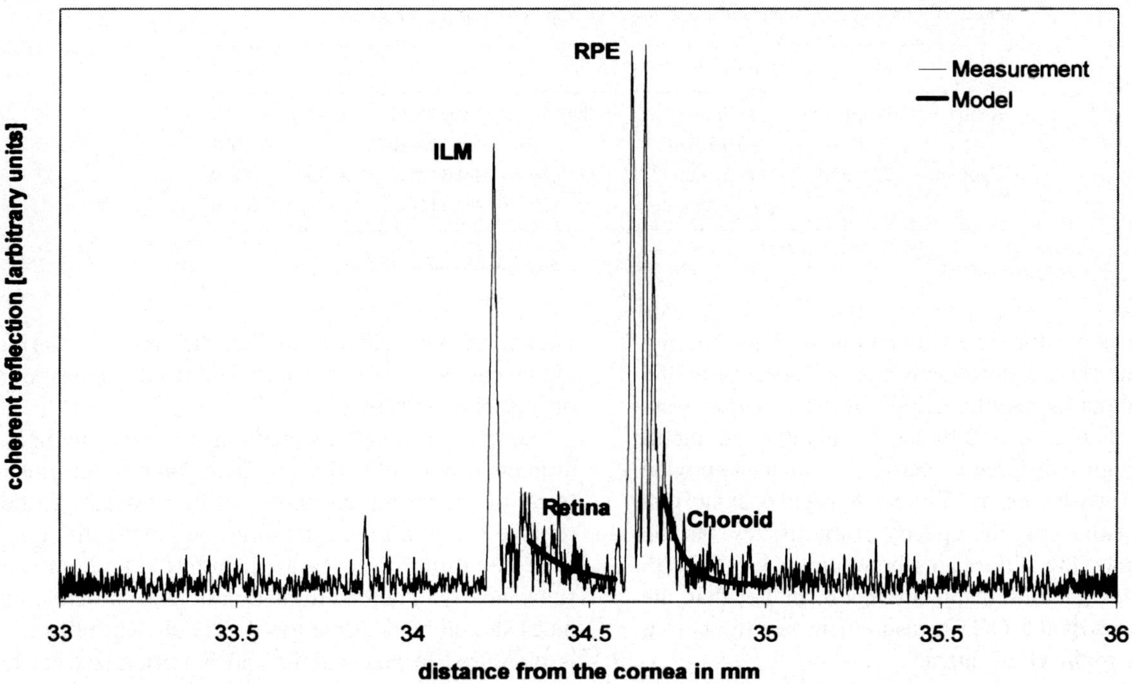


*Figure 1.* Measured OCT-scan through retina, RPE, and choroid as well as the calculated coherent power according to Equation (3) using the parameters given in Table 1.

Baumgartner et al. [11] using a dual beam interferometer. The light source was a superluminescent diode (C86142E, EG&G Optoelectronics, Canada) emitting at $\lambda$ = 855 nm with a coherence length in air of 13 $\mu$m. The power of the beam in the pupillary plane was about 200 $\mu$W at a cross section of 0.6 × 0.8 mm$^2$. The coherence function $\gamma$ was determined from the reflex of a glass plate.

Unfortunately, the data of Baumgartner et al. [11] were not given in power units. Therefore, it was impossible to calculate refractive indices. Equation (7) was employed for scaling the reflectance data using the refractive indices of Gullstrand's eye model [12]: $n_{\text{cornea}} = 1.376$, $n_{\text{vitreous}} = 1.336$, and $n_{\text{retina}}=1.35$.

$\tilde{R}$ and the extinction coefficient $\mu_t$ were determined solving the optimization problem Equation (5) and the scattering anisotropy $g$ was obtained from Equation (6) for the retina (34.33–34.57 mm from the cornea) and for the choroids (34.71–35.01 mm from the cornea). In the calculation for the choroid, the incident power was corrected for the extinction $e^{-\mu_t z}$ of the retina. The reflected coherent power according to the single backscattering model Equation (3) is shown in Figure 1 and all data are given in Table 1.

To obtain an error measure (Table 1), we calculated the variation of one parameter ($\tilde{R}$ or $\mu_t$) needed to obtain a value of the RMS deviation between measurement and model twice as high as the minimum, while the other parameter was kept constant at it's optimal value.

## Discussion

The results in Figure 1 and Table 1 show that the single backscattering model, first employed in the interpretation of OCT-scans by Schmitt et al. [3], is applicable to scans from the ocular fundus too. However, there is one complication: Schmitt et al. [3] assumed the coherence length of the light source to be small compared to the mean free pathlength of the photons in the sample $1/\mu_t$. This allowed them to replace the coherence function $\gamma$ by a Dirac delta function and, therefore, to avoid the solution of the integral in Equation (3). As can be seen from the values of $\mu_t$ in Table 1 and from the coherence length (13 $\mu$m), this is not possible in our case. Therefore, we had to solve the integral numerically.

Comparing the data in Table 1 with *in vitro* measurements from bovine retina and porcine choroids [13], we found complete agreement of the scattering anisotropy. However, the value of the extinction coef-

*Table 1.* Extinction coefficient and scattering anisotropy of retina and choroids

| | $\mu_t$ | $\Delta\mu_t$ | $\tilde{R}$ | $\Delta\tilde{R}$ | $g$ | $\Delta g$ |
|---|---|---|---|---|---|---|
| Retina | 12 mm$^{-1}$ | +120 mm$^{-1}$<br>−10.8 mm$^{-1}$ | 0.0000407 | +0.000061<br>−0.0000407 | 0.97 | −0.037<br>+0.031 |
| Choroids | 27.5 mm$^{-1}$ | +154 mm$^{-1}$<br>−22 mm$^{-1}$ | 0.000146 | +0.000292<br>−0.000146 | 0.90 | −0.159<br>+0.104 |

ficient, which should be comparable to the scattering coefficient since absorption is nearly negligible at 855 nm, is about as double as high in the *in vitro* measurement than measured by OCT. This may be due to morphological differences between human and bovine or porcine tissues or, more likely, to rapid post mortem changes enhancing the opacity of the tissues. On the other hand, the differences between *in vivo* (OCT) and *in vitro* measurements are much smaller than the huge errors of the OCT-measurement resulting from the noise of the OCT signal.

The signal to noise ratio is principally limited by the photon shot noise. Additionally, the thermal noise of the detector and the electronic noise of the amplifier have to be considered. These broadband noise may be reduced by filtering the signal with the Doppler frequency $f_D$, whereas the bandwidth of the filter has to be at least $2v/\Delta$ if $v$ is the speed of the reference mirror and a depth resolution $\Delta$ shall be achieved [14]. Furthermore, the coherent signal is affected by speckle. This could be reduced by an incoherent averaging over measurements at different wavelengths or by a local averaging which, however, reduces the resolution. Therefore, Schmitt et al. [15] used a detector array in order to reduce the speckle by averaging over measurements under slightly different angles. A further enhancement of the signal to noise ratio is possible if the observed depth plane $z$ is always confocal to the detector. This requires a synchronization of the shift of the reference mirror with that of the focus. A solution of this problem is published by Lexer et al. [16].

The theory of single backscattering assumes complete loss of coherence after the second scattering event. The influence of multiple scattered light was estimated by Schmitt et al. [7] calculating the effective detector cross sections for single and multiple scattered light. For a specimen with optical properties similar to that reported here, they found a remarkable contribution of multiple scattered light for depths greater than 0.3 mm. Since the thicknesses of the retina and the choroid are smaller, the single scattering model can be regarded as sufficient for the description of the coherent signal.

Despite all problems mentioned above, the measurement of the extinction coefficient and of the anisotropy of scattering has shown to be possible. To get reliable values for the extinction coefficient, the signal to noise ratio of the measurement has to be improved substantially. The determination of the refractive indices should be possible too if the coherent reflection is measured in power units and is normalized to the incident power.

## Acknowledgement

The authors would like to thank Prof. Fercher and his team for providing the data of the OCT-scan shown in Figure 1.

## References

1. Huang D, Swanson EA, Lin CP, Schuman JS, Stinson WG, Chang W, Hee MR, Flott T, Gregory CA, Pulafito CA, Fujimoto JG. Optical coherence tomography. Science 1991: 254: 1178–1181.
2. Danielson BL, Whittenburg CD. Guided wave reflectometry with micrometer resolution. Appl Opt 1987; 26: 2836–2842.
3. Schmitt JM, Knüttel A, Bonner RF. Measurement of optical properties of biological tissues by low-coherence reflectometry. Appl Opt 1993; 32: 6032–6042.
4. Fercher A F, Mengedoht K, and Werner W (1988) Eye length measurement by interferometry with partially coherent light. Optics Letters 13: 186-189
5. Hitzenberger CK. Optical measurement of the axial eye length by laser Doppler interferometry. Invest Ophthalmol Vis Sci 1991; 32: 616–624.
6. Fercher AF. Optical coherence tomography. J Biomed Opt 1996; 1: 157–173.
7. Schmitt JM, Knüttel A, Grandjbakhche A, Bonner RF Optical characterization of dense tissue using low-coherence interferometry. Proc SPIE 1993; 1889: 197–211.
8. Yadlowsky MJ, Schmitt JM, Bonner RF. Multiple scattering in optical coherence microscopy. Appl Opt 1995; 34: 5699–5707.

9. Brent RP. Algorithms for Minimization without Derivatives. Prentice-Hall, Englewood Cliffs, 1973.
10. Henyey LG, Greenstein JL. Diffuse radiation in the galaxy. Astrophys J 1941; 93: 70–83.
11. Baumgartner A, Hitzenberger CK, Sattmann H, Drexler, W, Fercher F. Signal and resolution enhancement in dual beam optical coherence tomography of the human eye. J Biomed Opt 1998; 3: 45–54.
12. Naumann GOH. Pathologie des Auges, Bd. 1, Springer-Verlag, Berlin, Heidelberg, New York, 1997.
13. Hammer M, Roggan A, Schweitzer D, Müller G. Optical properties of ocular fundus tissues – an in vitro study using the double-integrating-spere technique and inverse Monte Carlo simulation. Phys Med Biol 1995; 40: 963–978.
14. Yadlowsky MJ, Schmitt JM, Bonner RF. Contrast and resolution in the optical-coherence microscopy of dense biological tissue. SPIE Proc 1995; 2387: 193–203.
15. Schmitt JM. Array detection for speckle reduction in optical coherence microscopy. Phys Med Biol 1997; 42: 1427–1439.
16. Lexer F, Fercher AF, Sattmann H, Drexler W, Molebny S. Dynamic coherent focus for transversal resolution enhancement of OCT. Proc SPIE 1998; 3251: 85–90.

*International Ophthalmology* **23:** 297–307, 2001.
*J.R. Sampaolesi (ed.), Laser Scanning: Update 1*, 115–125.




# Retinal pigment epithelium translocation and central visual function in age related macular degeneration: preliminary results

Paulo E. Stanga[1,2], Andres Kychenthal[2], Frederick W. Fitzke[1], Anthony S. Halfyard[1], Roger Chan[1], Alan C. Bird[1,2] & George W. Aylward[2]
[1]*The Institute of Ophthalmology, University College London, London EC1V 2PD, UK;* [2]*Moorfields Eye Hospital, London, UK*

*Key words:* aging, choroidal neovascularization, cLSO, OCT, retinal pigment epithelium translocation, submacular surgery, visual function

## Abstract

*Purpose*: To test the feasibility of a new surgical technique, and to assess visual function over the translocated retinal pigment epithelium (RPE) cells in patients operated upon for subfoveal choroidal neovascularization (CNV) secondary to age-related macular degeneration (AMD). *Materials and methods*: Six patients presenting previously untreated exudative AMD underwent surgical excision of the subfoveal CNV with RPE translocation and were followed from 1 to 10.5 months. The surgery consisted of a standard three port pars plana vitrectomy (TPPPV), excision of the CNV and RPE translocation. Pre and post-operative ocular examination included best-corrected visual acuity measurement, fundus color stereo photography and fundus fluorescein angiography. Optical coherence tomography (OCT) and confocal laser scanning ophthalmoscopy (cLSO) were performed post-operatively. A cross fixation target and a single-point flashing light were projected on different areas of the posterior pole using a cLSO. Photopic 10-2 perimetry, photopic fine matrix mapping, cLSO microperimetry were also performed pre and post-operatively in four patients. OCT cross-sectional scans and cLSO RPE autofluorescence were recorded to detect the presence of viable translocated RPE. Visual acuity, fixation, photopic 10-2 perimetry, photopic fine matrix mapping and cLSO microperimetry were tested for the presence of central visual function. *Results*: RPE could be effectively translocated at the time of CNV removal from the edge of the RPE defect to a subfoveal location. OCT showed the translocated RPE as an area of increased optical reflectivity with optical shadowing external to it. cLSO showed autofluorescence of the translocated RPE. The cross fixation target was seen when projected on the translocated RPE. During eccentric fixation, the patients could see a flashing point-target projected on the translocated RPE. Photopic 10-2 perimetry, photopic fine matrix mapping and cLSO microperimetry showed presence of central visual function. *Conclusions*: The authors propose that translocation of RPE at the time of CNV removal, from the edge of the RPE defect to a subfoveal location, may have a role in the surgical management of AMD.

## Introduction

Age-related macular degeneration (AMD) accounts for about 50% of legal blindness in Europe and North America, and in the majority of patients visual loss is consequent upon choroidal neovascularization (CNV) [1, 2]. Only a few patients with CNV as part of AMD have clinical and angiographic characteristics that meet the eligibility criteria for photocoagulation therapy or photodynamic therapy following the guidelines of the study protocols, which are the only treatments that have been shown to improve the visual prognosis in AMD [3–5].

Surgical excision of subfoveal CNV in the young may result in good vision but has been disappointing in the context of AMD because of loss of retinal pigment epithelium (RPE) at the subfoveal site [6–8]. Attempts have been made to overcome this problem. RPE transplantation has not resulted in significant rescue of foveal function to date [9–11]. Surgical translocation of the fovea to a site with healthy RPE has met with spectacular success in some cases, but torsion has resulted that may be a temporary problem, and there is attendant risk of retinal detachment from proliferative vitreo-retinopathy [12–21]. The American Academy of Ophthalmology's Committee on Ophthalmic Procedures Assessment Retina Panel concluded based on the published literature that there is still no strong evidence that about the effectiveness of this technique or any of its variations. It also concluded that randomized clinical trial are needed to determine whether macular translocation is a safe and effective technique for the treatment of AMD [22].

An alternative approach is to translocate subretinal structures to a subfoveal site. The purpose of the current investigation is to test the feasibility of this surgical technique, and to assess visual function over the translocated RPE cells in patients operated upon for subfoveal choroidal neovascular membrane (CNV) secondary to age-related macular degeneration (AMD).

## Materials and methods

Six patients diagnosed as having a previously untreated classical subfoveal CNV secondary to AMD underwent standard three port pars plana vitrectomy (TPPPV), excision of the CNV and RPE translocation. Full ophthalmologic and systemic pre and postoperative examinations were performed. Preoperative informed consent was obtained.

Preoperative ocular examination included best corrected visual acuity measurement, anterior and posterior biomicroscopy, 35° fundus color stereo photography and fundus fluorescein angiography.

The patients underwent a standard TPPPV as previously described [8]. A posterior vitreous detachment (PVD) was induced with active suction using the cutter (Storz Premiere, Storz, USA). A small retinotomy was performed using a 130° 33 gauge subretinal pick and a posterior retinal detachment overlying the CNV was created by injecting balanced salt solution (BSS). Subretinal horizontal forceps (Synergetics 09.01, Synergetics, USA) were utilized to grasp and remove the neovascular – RPE complex leaving the central macula devoid of RPE. Subretinal vertical scissors (Synergetics 24.01, Synergetics, USA) were introduced through the retinotomy and a 500–1500$\mu$ a pedicled full thickness RPE/choroid autologus graft in the first patient while a free one in the remaining five patients was obtained from the edge of the RPE defect. Subretinal vertical forceps (Synergetics, USA) were used to position the graft subfoveally. Air-fluid exchange was carried out for internal tamponade and the sclerostomies closed. Twenty-four hours facedown positioning was indicated.

The post operative evaluation also included confocal laser scanning ophthalmoscopy (cLSO) (a Zeiss prototype donated by Carl Zeiss, Oberkochen, Germany) and optical coherence tomography (OCT) (Humphrey Optical Coherence Tomography Scanner, Humphrey Instruments, a Division of Carl Zeiss, Inc. California, U.S.A.). Autofluorescence was recorded using previously published techniques in which 40° argon and helium neon reflectance and argon autofluorescence images were obtained using the cLSO [23–25]. Single line scans were obtained using the OCT [26–28]. To test fixation on the graft, a cross pattern fixation target and a single point flashing light were projected on different areas of the posterior pole using the cLSO.

All but one had severe visual loss for at least 3 months in their second eye as a consequence of sub-foveal choroidal neovascularization as part of age-related macular disease (Table 1). The choroidal neovascularization was confirmed by fluorescein angiography, and in each the complex was larger that 1 disc area.

### *Patient no. 1*

A 74 years old white female patient presented with diminished vision in the left eye for 5 months. Best-corrected visual acuity was counting fingers in the right eye and 20/200 in the left eye. The right eye showed an end-stage disciform scar secondary to AMD and the left had signs of a CNV type I, according to D. Gass' classification [29]. Fluorescein angiography confirmed the biomicroscopic findings (Figure 1) of a subfoveal classic CNV type I. A pedicled full thickness RPE/choroid graft was translocated from the superotemporal edge of the RPE defect and positioned subfoveally.

### *Patient no. 2*

An 87 years old white male patient presented with diminished vision in the left eye for 7 months. Best-corrected visual acuity was 20/25 in the right eye and 20/200 in the left eye. The right eye showed a few scattered hard drusen in the posterior pole while

*Table 1.*

| Patient | Sex | Age | Eye | Pre-op va | Post-op va | Complications | Months review |
|---|---|---|---|---|---|---|---|
| 1 | F | 74 | L | 20/200 | 20/200 | None | 10.5 |
| 2 | M | 87 | L | 20/200 | 20/200 | None | 5 |
| 3 | M | 79 | L | 20/200 | 20/200 | Subret bleeding | 4 |
| 4 | M | 71 | L | 20/400 | 20/400 | Inverted graft | 4 |
| 5 | M | 74 | L | CF | HM | PVR – total rd | 4 |
| 6 | M | 78 | L | 20/70 | 20/70 | None | 1 |

Abbreviations: Pre-op va: preoperative visual acuity; Post-op va: postoperative visual acuity; Subret: subretinal; CF: counting fingers; HM: hand motions; PVR: proliferative vitreoretinopathy; RD: retinal detachment.

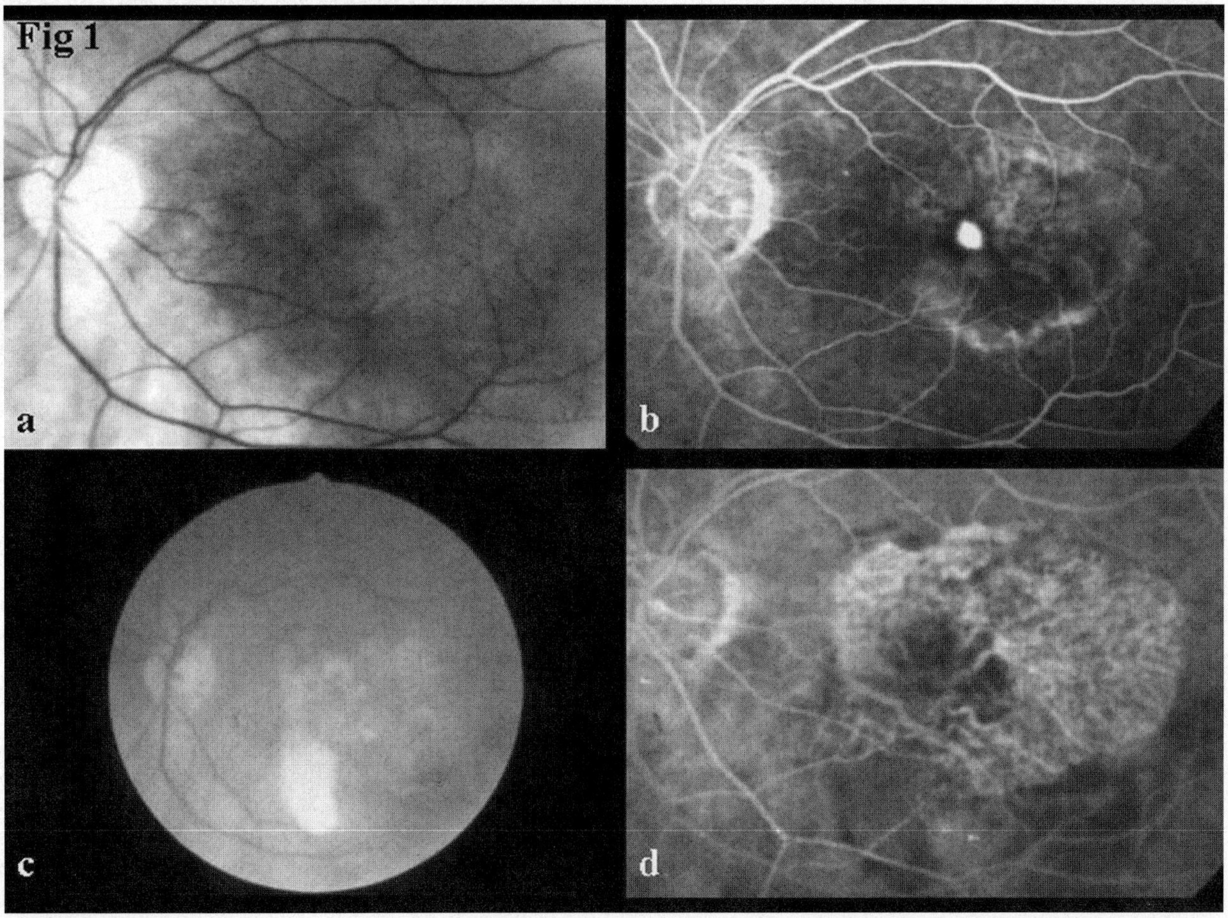


*Figure 1.* Patient no. 2: An 87 years old white male patient presented with diminished vision in the left eye for 7 months. Best-corrected visual acuity was 20/25 in the right eye and 20/200 in the left eye. The right eye showed a few scattered hard drusen in the posterior pole while the left showed signs of a subfoveal CNV type I (a, red-free image). Fluorescein angiography confirmed the presence of a subfoveal classic CNV (b). A free RPE graft was translocated from the inferonasal edge of the RPE defect and positioned subfoveally (c). Fluorescein angiography showed no leakage and/or pooling of dye (d).

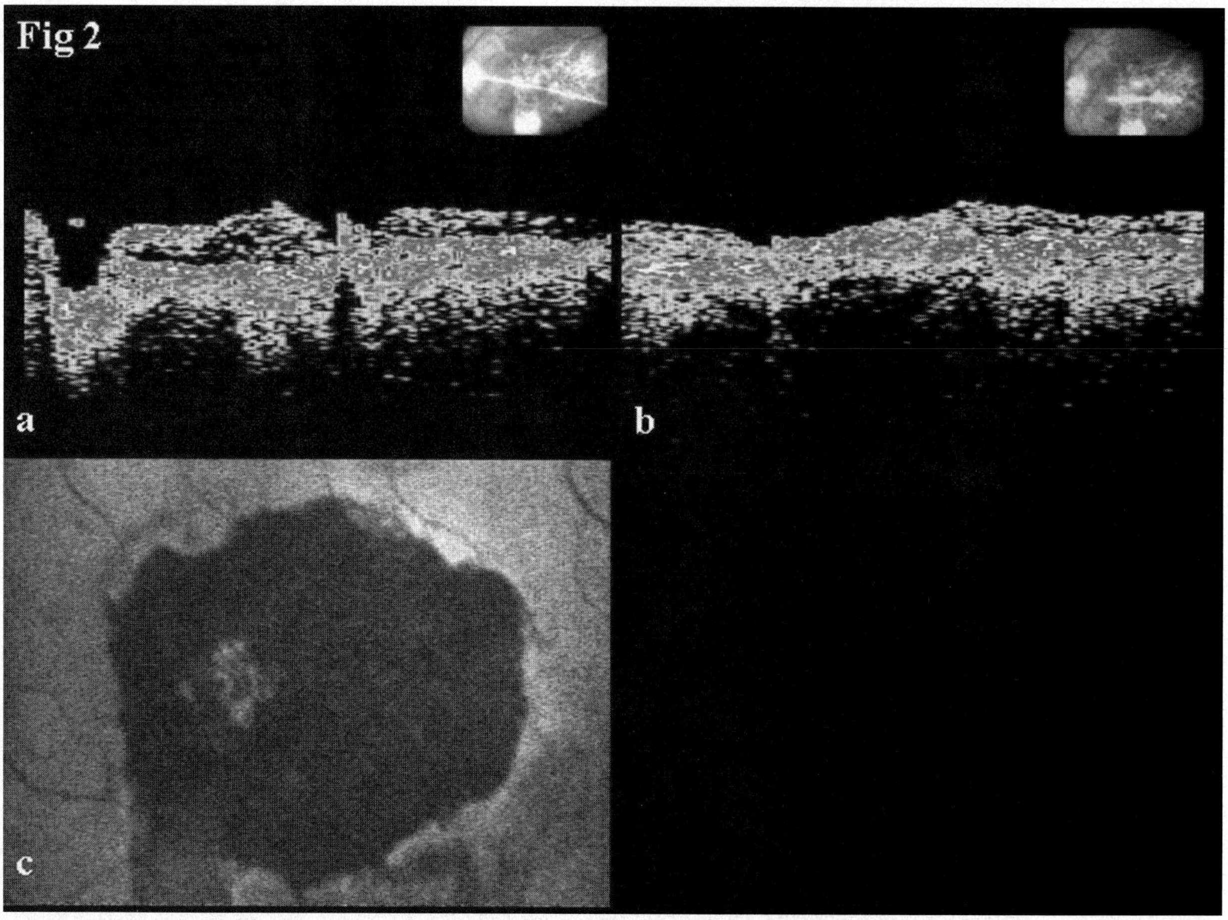


*Figure 2.* Patient no. 2: Five months post-operatively best corrected visual acuity in the operated eye was still 20/200 but the patient claimed a subjective improvement in vision. An 11.31 mm single line post-operative OCT scan across the translocated RPE and extending from the disc towards temporal retina showed a well delimited pattern of increased higher relative reflectivity in the choroidal tissues corresponding to the area devoid of RPE and a well delimited higher relative reflectivity in the layers assigned to the RPE-choriocapillaris complex similar to that observed external to the defect (a). An increasing shadowing effect was present external to it. The area corresponding to the papillo-macular bundle was within normal limits of reflectivity, but the neuroretinal tissue located anterior to the area devoid of RPE lacked the characteristic high relative reflectivity band assigned to the nerve fiber layer and presented a generalized reduction of reflectivity. This area of macular neuroretina appeared thickened nasal to the subfoveally translocated RPE. The area corresponding to the foveal pit appeared wider than normal. An area of relative high reflectivity corresponded to the translocated RPE. Interestingly enough this relative high reflectivity band extended into the neuroretina. A second single line OCT scan of increased resolution showed similar findings (b). cLSO showed, as expected, a lack of autofluorescence in the area devoid of native RPE. The translocated RPE fluoresced as well as the native RPE that surrounded the defect (c).

the left showed signs of a subfoveal CNV type I. Fluorescein angiography confirmed the presence of a subfoveal classic CNV (Figure 2). The same surgical technique was followed except for that in this case a free RPE graft was translocated from the inferonasal edge of the RPE defect and positioned subfoveally.

*Patient no. 3*

A 79 years old white male patient presented with diminished vision in the left eye for 3 months. Best-corrected visual acuity was 20/200 with each eye. The right eye showed an end-stage disciform scar secondary to AMD. Fundus examination of the left eye showed signs of a CNV type I. Fluorescein angiography confirmed the biomicroscopic findings (Figure 3) of a subfoveal classic CNV type I. The same surgical technique was followed except for that in this case a free RPE graft was translocated from the superior edge of the RPE defect and positioned subfoveally.

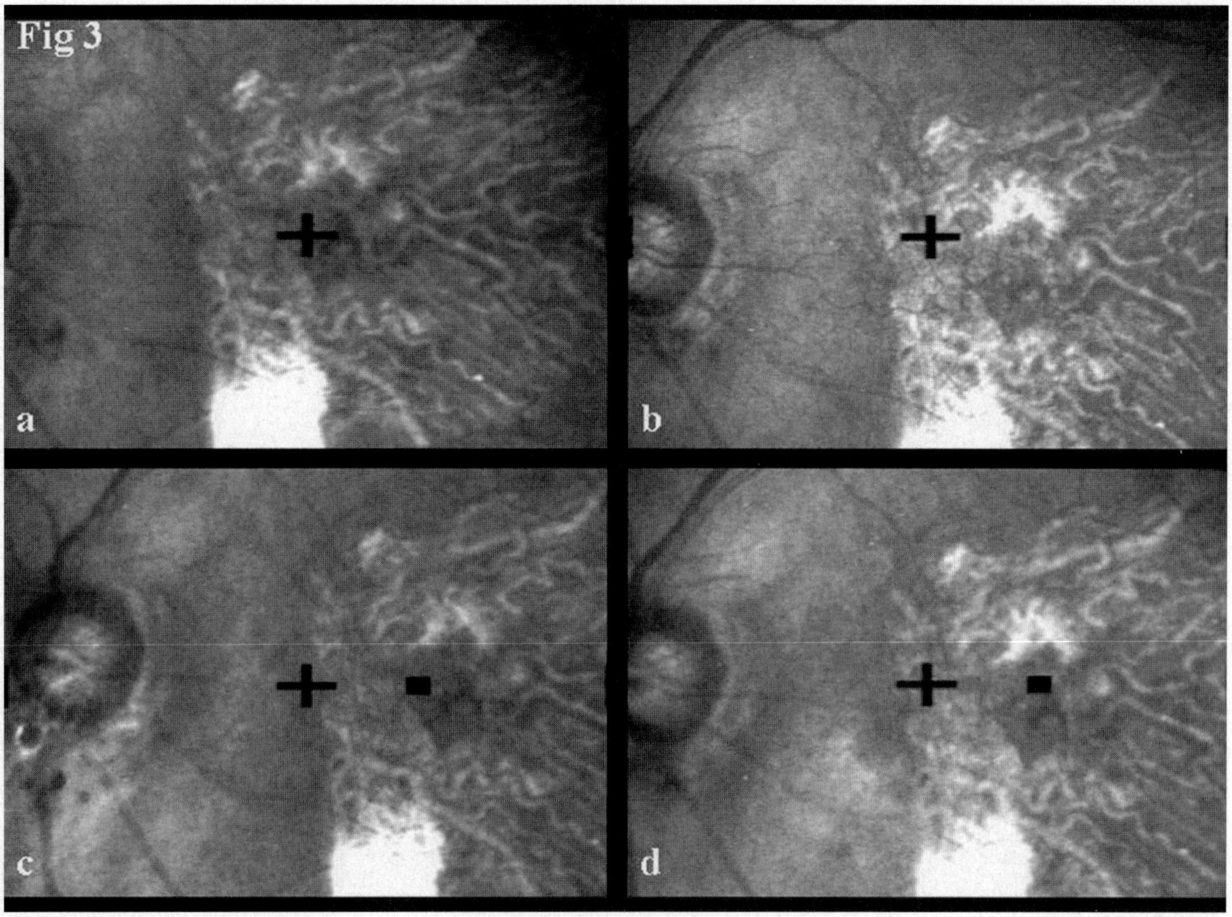


*Figure 3.* Patient no. 2: The cross-pattern fixation target projected on the translocated RPE using the cLSO was only seen by the patient when it was projected either on the translocated RPE (a) or nasal to the RPE defect on the papillomacular bundle. It was not seen when projected on the RPE defect (b). While projecting a flashing point-target on the translocated RPE the cross-pattern target was projected on the papillomacular bundle. At this time the cross-pattern target was seen by the patient (c) while it was not seen when projected over the RPE defect (d).

*Patient no. 4*

A 71 years old white male patient who presented with diminished vision in the left eye or 3 months. Best-corrected visual acuity was counting fingers in the right eye and 20/400 in the left eye. Fundus examination of the right eye showed an end-stage disciform scar secondary to AMD. Fundus examination of the left eye showed signs of a CNV type I. Fluorescein angiography confirmed the biomicroscopic findings (Figure 4) of a subfoveal classic CNV type I. The same surgical technique was followed except for that in this case a free RPE graft was translocated from the superotemporal edge of the RPE defect and positioned subfoveally, though it was post-operatively realized that the graft was upside-down.

*Patient no. 5*

A 74 years old white male patient who presented with diminished vision in the left eye for 3 months. Best-corrected visual acuity was counting fingers in both eyes. Fundus examination of the right eye showed an end-stage disciform scar secondary to AMD. Fundus examination of the left eye showed signs of a CNV type I. Fluorescein angiography confirmed the biomicroscopic findings (Figure 5) of a subfoveal classic CNV type I. The same surgical technique was followed.

*Patient no. 6*

A 78 years old white male patient presented with diminished vision in the left eye for 4 months. Best-corrected visual acuity was counting fingers with the

Fig 4

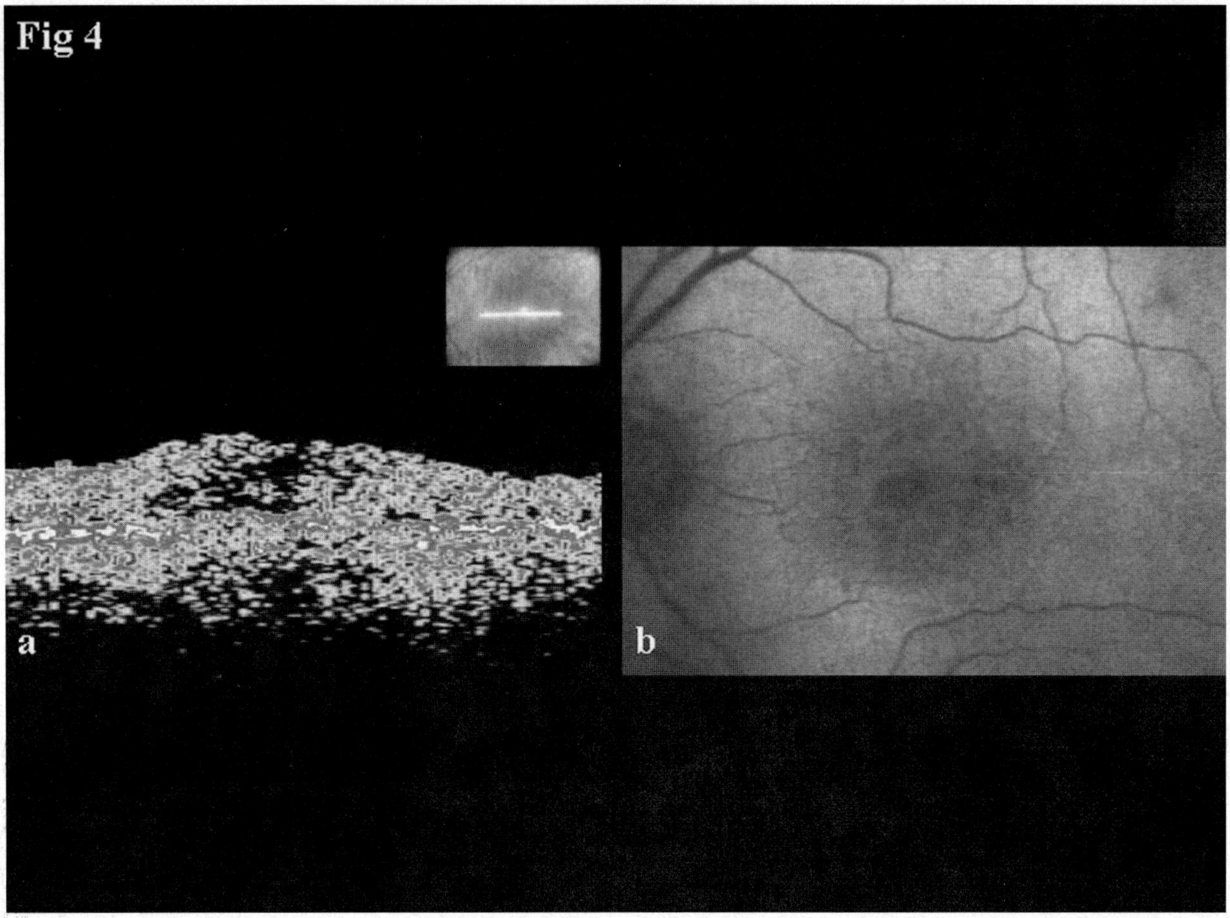


*Figure 4.* Patient no. 6: A 78 years old white male patient presented with diminished vision in the left eye for 4 months. Best-corrected visual acuity was counting fingers with the right eye and 20/70 in the left eye. Fundus examination of the right eye showed an end-stage disciform scar secondary to AMD. Fluorescein angiography of the left eye confirmed the presence of a subfoveal classic CNV type I. A free RPE graft was translocated from the superior edge of the RPE defect and positioned subfoveally. One month post-operatively best-corrected visual acuity in the operated eye was 20/70 and the patient claimed a reduction of metamorphopsia. Pre-operative 4.65 mm OCT single-scan showed localized disruption in the area angiographically assigned to the type I CNV and intraretinal areas of relative low reflectivity corresponding to angiographic intraretinal edema (a). Pre-operative cLSO imaging of the posterior pole showed central autofluorescence slightly diminished in the areas of retinal thickening (b).

right eye and 20/70 in the left eye. Anterior segment biomicroscopy of both eyes showed no lens opacities. Fundus examination of the right eye showed an end-stage disciform scar secondary to AMD. Fundus examination of the left eye showed signs of a CNV type I. Fluorescein angiography confirmed the biomicroscopic findings (Figure 6) of a subfoveal classic CNV type I. The same surgical technique was followed except for that in this case a free RPE graft was translocated from the superotemporal edge of the RPE defect and positioned subfoveally.

## Results

In four out of six patients it was possible to translocate subretinal structures to a subfoveal site and it remained at that site. This was demonstrated by the presence of cLSO autofluorescence at the fovea, and high reflectance on OCT (Figures 6–11).

cLSO showed a lack of autofluorescence in the area devoid of native RPE. The translocated RPE fluoresced as well as the native RPE that surrounded the defect.

Post operative single line OCT scans of the posterior pole showed a well delimited pattern of increased higher relative reflectivity in the choroidal

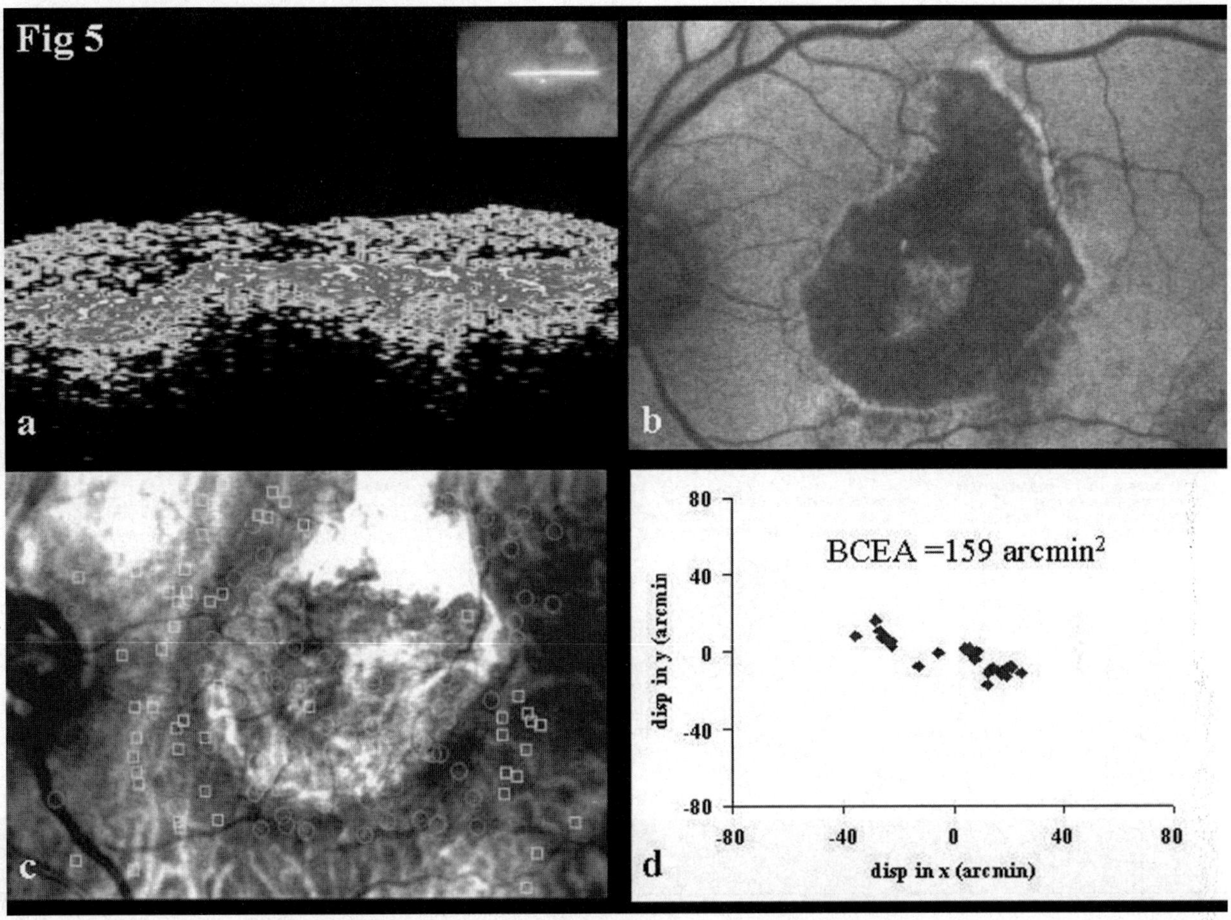


*Figure 5.* Patient no. 6: Post-operative OCT scans and cLSO imaging showed similar findings to those observed in patients no. 2 (a & b). Qualitative (YES or NO) cLSO microperimetry shows as green squares the SEEN-stimuli and as red circles the NOT-seen ones (c) and the fixational stability of this test (d).

tissues corresponding to the area devoid of RPE. A single line scan across the translocated RPE showed a well delimited higher relative reflectivity in the layers assigned to the RPE-choriocapillaris complex similar to that observed external to the defect. An increasing shadowing effect was present external to it.

Fluorescein angiography showed that fluorescein was excluded by the retina centrally.

All four had preferred fixation at the fovea, and could detect a flashing light at the fovea. Loss of RPE and lack of visual function could be demonstrated at the site from that the tissue had been moved (Figure 1). The cross-pattern fixation target projected on the translocated RPE using the cLSO was only seen by the patient when it was projected either on the translocated RPE or nasal to the RPE defect on the papillomacular bundle. It was not seen when projected on the RPE defect. While projecting the cross-pattern target on the papillomacular bundle a flashing point-target was projected on the translocated RPE. At this time the flashing point-target was seen by the patient (Figure 6).

No significant change in pre-operative visual acuity was observed in five out of six patients (see Table 1). The remaining patient's visual acuity changed from counting fingers to hand-motions due to the development of a total retinal detachment as a consequence of PVR.

In patient no. 3 cLSO showed a lack of autofluorescence surrounding the area devoid Of native RPE due to blocking of the native RPE autofluorescence by subretinal blood. Photopic 10-2 perimetry and photopic FMM tests showed presence of central visual function. There was loss of foveal sensitivity on the 10-2 record. Although it appeared that sensitivities had not

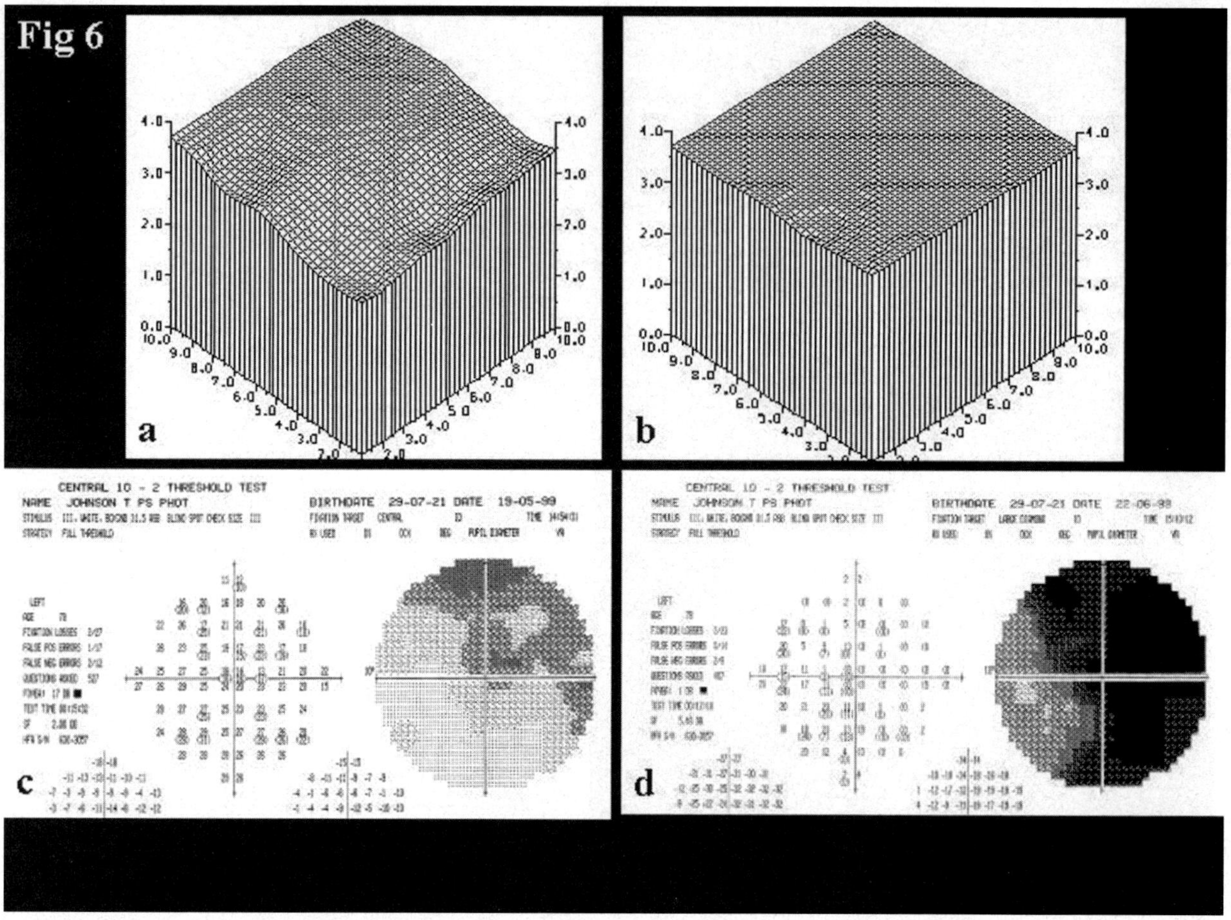

*Figure 6.* Patient no. 6: Photopic 10-2 perimetry and photopic FMM tests show presence of central visual function. Although it appears that sensitivities have not altered according to photopic blue FMM, this is mainly because the functional loss is too great to measure (a: pre-operative and b: post-operative). There is loss of foveal sensitivity on the 10-2 record (c: pre-operative and d: post-operative).

altered according to FMM, this was mainly because the functional loss was too great to measure.

In patient no. 4 the tissue was thought to have flipped-over intraoperatively and this was later confirmed post-operatively. This was correlated with the absence of autofluorescence at the fovea but with high reflectance on OCT. The cross-pattern fixation target projected on the translocated RPE using the cLSO could not be recognized. Photopic 10-2 perimetry and photopic FMM tests still showed presence of central visual function as in the previous patient.

Patients no. 5 developed proliferative vitreoretinopathy (PVR) 7 days post-operatively and a subsequent total retinal detachment. No further post-operative tests were carried out. The retina was subsequently reattached with further surgery and fluid-gas exchange.

## Discussion

In 1991 Thomas and Kaplan published the first report on the surgical excision of subfoveal neovascular membranes [6]. It then became evident that in AMD this was associated with loss of the central RPE and choriocapillaris with consequent poor visual outcome [9, 30]. The observations of Gass in which he defined type 1 and type 2 lesions showed that this is inevitable. He addressed the need to find the means for resurfacing the bed of the surgically excised CNV with RPE [29]. Animal studies showed that the transplanted RPE cells attach to Bruch's membrane and become integrated with the host's RPE and photoreceptors [31].

Age-related changes in Bruch's membrane may interfere in RPE reattachment [32]. Human clinical works to replace diseased RPE with normal human fetal cultured RPE have been carried out in patients presenting exudative and atrophic AMD [10, 11]. Macular edema, fluorescein leakage and gradual reduction in visual acuity considered to be secondary to host-graft rejection were found in disciform lesions over 1–6 months. In the non-exudative group, the RPE cells suspensions transplants showed no evidence of rejection and were associated with the disappearance of drusen. The authors concluded that an intact blood-retinal barrier is likely to protect against rejection. RPE autologus transplantation or RPE translocation might be ways of evading the host-graft rejection limitations. The age of the human cadaver RPE donors may have implications for the success of RPE transplantation in elderly persons and the cells will not grow on a monolayer of damaged or diseased Bruch's membrane [32, 33]. Several animal and human cadaver studies addressed different surgical techniques of harvest, storage and transplantation of pigmented epithelial cells [34–38].

An alternative approach to retinal translocation is to translocate sub-retinal structures to a subfoveal site. We propose a new and feasible surgical technique for AMD patients in which a classic neovascular complex, as shown by fluorescein angiography, is present. Our surgical technique differs from that previously described [39] in that we perform a small posterior pole retinotomy and an RPE-Bruch's membrane-choroid graft is translocated. In our technique, a small retinotomy is used in order not to severe the nerve fibers that account for the nasal peripheral field of view; an important consideration as most of these patients rely on peripheral vision. Because of the small retinotomy there should be less risk of proliferative vitreoretinopathy (PVR) development. No preoperative three-row concentric 360° photocoagulationn procedure is required in our technique, also reducing the chances of visual field damage. Another difference is that we use air as an internal tamponade agent instead of silicone oil. This reduces the surgery time, the possible associated complications of the use of silicone oil and longer surgical time, making the procedure simpler and eliminating the need of a second intraocular procedure to remove the oil.

This approach could eliminate the Creutzfeld–Jacob Disease (CJD) surgical transmission risk from one human being to another [40–42]. Host-graft rejections, as well as the practical, ethical or religious conflicts arising from transplants and the use of human fetal RPE cells, cadaveric donors or RPE cells culture banks. A chronic host-graft rejection process derived from the use of a cadaveric homologous graft could be responsible for the non-improvement of visual acuity and the encapsulation of the graft in the second patient described by others [39]. The cLSO scans using Argon (470 nm) laser beams to illuminate the fundus records the distribution of lipofuscin. This autofluorescent pigment accumulates in RPE as a by-product of metabolic activity reflecting photoreceptor outer segment renewal. As a result of this, lack of autofluorescence is observed either when no functioning photoreceptors are present in a given area or when no RPE cells are present [43]. In the cases we report, the presence of autofluorescence indicates either the presence of functioning photoreceptors that over the translocated RPE, or that residual lipofuscin that accumulated in the cells prior to the translocation is present. We consider this second possibility to be unlikely due to the time span between the surgery and the autofluorescence imaging. The fixation testing results support the presence of viable photoreceptors over the translocated RPE. OCT produces cross-sectional tomographs of the logarithm of reflectivity in the retina are displayed in real time as two-dimensional false-color images [44]. The well-defined increased optical reflectivity observed in the choroidal tissues in the area devoid of RPE was interpreted as an increase in light transmission towards sclera secondary to the absence of RPE. This contrasts with the area of transplanted RPE in which the normal distribution of reflectance.

Thus all the evidence indicates that the transplanted RPE has survived and that it is supporting limited photoreceptor function. There are still many problems to be faced: being a new surgical technique there is a lack of experience, for which ideal surgical instruments are not available and we do not know the most appropriate size of RPE that has to be translocated. Translocation of RPE at the time of the CNV removal from the edge of the RPE defect to a subfoveal location may permit visual function over the translocated RPE cells in patients with AMD, this may be a promising approach to the surgical management of AMD.

Longer follow-up is needed to assess maximum survival time of the translocated RPE cells and function of the overlying macular region. The size of the translocated graft might prove to be crucial for obtaining useful central visual function.

## References


1. Evans J. Causes of blindness and partial sight in England and Wales 1990–1991. Studies on medical and population subjects No. 57. Office of Population Censuses and Surveys, London, UK. Her Majesty's Stationary Office Publications Centre, London, UK 1995.
2. Leibowitz HM, Krueger DE, Mauder LR, et al. The Framingham Eye Study Monograph, 1973–75. Surv Ophthalmol 1988; 24(Suppl): 335–610.
3. Schachat AP. Management of subfoveal choroidal neovascularization. Arch Ophthalmol 1991; 109(9): 1217–1218.
4. Macular Photocoagulation Study Group. Argon laser photocoagulation for senile macular degeneration: results of a randomized clinical trial. Arch Ophthalmol 1982; 100: 912–918.
5. Macular Photocoagulation Study Group. Laser photocoagulation for subfoveal neovascular lesions in age-related macular degeneration. Results of a randomized clinical trial. Arch Ophthalmol 1991; 109: 1220–1231.
6. Thomas MA, Kaplan HJ. Surgical removal of subfoveal choroidal neovascularization in the presumed ocular histoplasmosis syndrome. Am J Ophthalmol 1991; 111: 1–7.
7. Thomas MA, Dickinson JD, Melberg NS et al. Visual results after surgical removal of subfoveal choroidal neovascular membranes. Ophthalmology 1994; 101(8): 1384–1396.
8. Eckstein M, Wells JA, Aylward GW, Gregor Z. Surgical removal of non-age-related subfoveal choroidal neovascular membranes. Eye 1998; 12: 775–780.
9. Hsu JK, Thomas MA, Ibanez H, Green WR. Clinicopathologic studies of an eye after submacular membranectomy for choroidal neovascularization. Retina 1995; 15(1): 43–52.
10. Algvere PV, Berglin L, Gouras P, Sheng Y. Transplantation of fetal retinal pigment epithelium in age-related macular degeneration with subfoveal neovascularization. Graefe's Arch Clin Exp Ophthalmol 1994; 232(12): 707–716.
11. Algvere PV, Berglin L, Gouras P et al. Transplantation of RPE in age-related macular degeneration: observations in disciform lesions and dry RPE atrophy. Graefe's Arch Clin Exp Ophthalmol 1997; 235(3): 149–158.
12. Machemer R, Steinhorst UH. Retinal separation, retinotomy, and macular relocation: I. Experimental studies in the rabbit eye. Graefes Arch Clin Exp Ophthalmol 1993; 231: 629–641.
13. Machemer R, Steinhorst UH. Retinal separation, retinotomy, and macular relocation: II. A surgical approach for age-related macular degeneration? Graefes Arch Clin Exp Ophthalmol 1993; 231: 635–641.
14. Seaber JH, Machemer R. Adaptation to monocular torsion after macular translocation. Graefes Arch Clin Exp Ophthalmol 1997; 235: 76–81.
15. Eckardt C, Eckardt U, Conrad HG. Macular rotation with and without counter-rotation of the globe in patients with age-related macular degeneration. Graefes Arch Clin Exp Ophthalmol 1999; 237(4): 313–325
16. Wolf S, Lappas A, Weinberger AW, Kirchhof B. Macular translocation for surgical management of subfoveal choroidal neovascularizations in patients with AMD: first results. Graefes Arch Clin Exp Ophthalmol 1999; 237: 51–57.
17. de Juan E Jr., Loewenstein A, Bressler NM, Alexander J. Translocation of the retina for management of subfoveal choroidal neovascularization II: a preliminary report in humans. Am J Ophthalmol 1998; 125: 635–646.
18. Lewis H, Kaiser PK, Lewis S, Estafanous M. Macular translocation for subfoveal choroidal neovascularization in age-related macular degeneration: a prospective study. Am J Ophthalmol 1999; 128(2): 135–146.
19. Akduman L, Karavellas MP, MacDonald JC, Olk RJ, Freeman WR Macular translocation with retinotomy and retinal rotation for exudative age-related macular degeneration. Retina 1999; 19(5): 418–423.
20. Ninomiya Y, Lewis JM, Hasegawa T, Tano Y. Retinotomy and foveal translocation for surgical management of subfoveal choroidal neovascular membranes Am J Ophthalmol 1996; 122: 613–621.
21. FujikadoT, Ohji M, Hayashi A, Kusaka S, Tano Y. Anatomic and functional recovery of the fovea after foveal translocation surgery without large retinotomy and simultaneous excision of a neovascular membrane. Am J Ophthalmol 1998; 126: 839–842.
22. Committee on Ophthalmic Procedures Assessment Retina Panel. Ophthalmic Procedure Preliminary Assessment: Macular Translocation. Ophthalmol 2000; 107:1015-1018.
23. Woon WH, Fitzke FW, Chester GH et al. The scanning laser ophthalmoscope. Basic principles and applications. J Ophthalm Photography 1990; 12(1): 17–23.
24. von Rückmann A, Fitzke FW, Bird AC. Distribution of fundus autofluorescence with a scanning laser ophthalmoscope. Br J Ophthalmol 1995; 79: 407–412.
25. von Rückmann A, Fitzke FW, Bird AC. Fundus autofluorescence in age-related macular disease imaged with a laser scanning ophthalmoscope. Invest Ophthalmol Vis Sci 1997; 38: 478–486.
26. Huang D, Swanson EA, Lin CP et al. Optical coherence tomography. Science 1991; 254:1178-81.
27. Hee MR, Izatt JA, Swanson EA et al. Optical coherence tomography of the human retina. Arch Ophthalmol 1995; 113: 325–332.
28. Puliafito CA, Hee MR, Lin CP et al. Imaging of macular diseases with optical coherence tomography. Ophthalmology 1995; 102: 217–229.
29. Gass JDM. Biomicroscopic and histopathologic considerations regarding the feasibility of surgical excision of subfoveal neovascular membranes. Am J Ophthalmol 1994; 118: 285–298.
30. Nasir MA, Sugino I, Zarbin MA. Decreased choriocapillaris perfusion following surgical excision of choroidal neovascular membranes in age-related macular degeneration. Br J Ophthalmol 1997; 81(6): 481–489.
31. Gouras P, Lopez R, Brittis M, Kjeldbye H. The ultrastructure of transplanted rabbit retinal epithelium. Graefe's Arch Clin Exp Ophthalmol 1992; 230(5): 468–475.
32. Ho TC, Del Priore LV. Reattachment of cultured human retinal pigment epithelium to extracellular matrix and human Bruch's membrane. Invest Ophthalmol Vis Sci 1997; 38(6): 1110–1118.
33. Del Priore LV, Tezel TH. Reattachment rate of human retinal pigment epithelium to layers of human Bruch's membrane. Arch Ophthalmol 1998; 116(3): 335–341.
34. Lane C, Boulton M, Marshall J. Transplantation of retinal pigment epithelium using a pars plana approach. Eye 1989; 3 (Pt 1): 27–32.
35. Wongpichedchai S, Weiter JJ, Weber P, Dorey CK. Comparison of external and internal approaches for transplantation of autologous retinal pigment epithelium. Invest Ophthalmol Vis Sci 1992; 33(12): 3341–3352.
36. Rezai KA, Kohen L, Wiedemann P, Heimann K. Iris pigment epithelium transplantation. Graefe's Arch Clin Exp Ophthalmol 1997; 235(9): 558–562.

37. Tezel TH, Del Priore LV, Kaplan HJ. Harvest and storage of adult human retinal pigment epithelial sheets. Curr Eye Res 1997; 16(8): 802–809.
38. Durlu YK, Tamai M. Transplantation of retinal pigment epithelium using viable cryopreserved cells. Cell Transplant. 1997; 6(2): 149–162.
39. Peyman GA, Blinder KJ, Paris CL et al. A technique for retinal pigment epithelium transplantation for age-related macular degeneration secondary to extensive subfoveal scarring. Ophthalmic Surg 1991; 22(2): 102-108.
40. Herzberg L. Creutzfeld-Jakob disease and corneal grafts. Med J Aust 1979; 1(6): 248.
41. Tarkkanen A, Haltia. Creutzfeld-Jakob disease. Trans Ophthalmol Soc UK 1980; 100(Pt 1): 151–154.
42. Cathala F, Moreau-Dubois MC, Brown P. Creutzfeld-Jakob disease: recent advances; biology of unconventional viruses. Pathol Biol (Paris) 1980; 28(8): 545–553.
43. von Ruckmann A, Fitzke FW, Bird AC. Distribution of pigment epithelium autofluorescence in retinal disease state recorded in vivo and its change over time. Graefes Arch Clin Exp Ophthalmol 1999; 237(1): 1–9.
44. Hee MR, Baumal CR, Puliafito CA et al. Optical coherence tomography of age-related macular degeneration and choroidal neovascularization. Ophthalmology 1996; 103:1260–1270.

*International Ophthalmology* **23:** 309–316, 2001.
*J.R. Sampaolesi (ed.), Laser Scanning: Update 1*, 127–134.



# Functional assessment of the native retinal pigment epithelium after the surgical excision of subfoveal choroidal neovascular membranes type II: preliminary results

P.E. Stanga, A. Kychenthal, F.W. Fitzke, A.S. Halfyard, R. Chan, R.H.Y. Asaria, A.C. Bird & G.W. Aylward
*The Institute of Ophthalmology, University College London and Moorfields Eye Hospital, London, UK*

## Introduction

Subretinal neovascular membranes (CNV) can occur in a variety of conditions. SRNVM most commonly occur in age-related macular degeneration (AMD), punctate inner choroidopathy (PIC)/multifocal inner choroidopathy (MIC) and myopia. AMD and CNV most common cause of blindness in western world.

De Juan and Machemer described a technique for removing SRNVM in AMD: visual results were very modest, but the surgical excision was possible. In 1991 Thomas and Kaplan removed 2 membranes from patients with presumed ocular histoplasmosis syndrome (POHS). Visual results were excellent recorded at 20/20 and 20/40.

Gass described two types of membranes: type I; with growth of vessels beneath the RPE and type II; with growth of vessels in the subsensory space. The clinical distinction is not always easy, type II CNV generally occur in patients less than 50 years old, a pigment halo can be present, the borders are sharply defined and the plaque-like elevation of the retinal pigment epithelium (RPE) beyond the edge of the membrane is absent.

Due to the cleavage plane between the native RPE and the reactive RPE that surrounds the CNV and the absence of cellular attachments, type 2 CNV can be surgically excised. Surgical removal of type II CNVs may restore good visual acuity.

## Purpose

The purpose of this work was to assess central visual function and the status of the native RPE after the surgical excision of type II subfoveal CNV in ten patients

## Methods

Ten patients underwent the surgical excision of a subfoveal type II CNV.

Pre and post-operative ocular examination included: best-corrected visual acuity measurement, 35° colour photography and fluorescein angiography.

Post-operative ocular examination included: optical coherence tomography (OCT), confocal laser scanning ophthalmoscopy (cLSO), photopic 10-2 Humphrey® perimetry, photopic blue fine matrix mapping (FMM) and cLSO microperimetry.

In six patients all tests were carried out pre and post-operatively.

The surgical technique consisted of a standard 3 port par-plana vitrectomy. A posterior hyaloid detachment was induced. A 130° angled 36G subretinal was pick used to create a retinotomy A macular detachment was induced by injecting balanced salt solution (BSS®). The CNV was grasped and delivered through the retinotomy. The eye was tamponaded with 30% sulphur hexafluoride (SF6) gas or air. Twenty-four hours of post-operative face down posturing was indicated.

The OCT produces cross-sectional tomographs of the logarithm of reflectivity in the retina and displays them in real time as two-dimensional false-color images.

The cLSO utilizes Argon (470 nm) laser beams to illuminate the fundus and capture lipofuscin's autofluorescence. Autofluorescence due to lipofuscin normally is present due to the phagocytosis of photoreceptor outer segments suggesting photoreceptor/RPE interaction. Loss of autofluorescence is visible in areas lacking RPE and in regions of the retina lacking photoreceptors.

Photopic 10-2 Humphrey® perimetry tests the central 10° of the visual field.

Photopic blue FMM concentrates on the central 4.5° testing 100 locations. In the FMM images the elevation above base line indicates loss of sensitivity.

cLSO microperimetry tests an 18 × 14 degrees area. The stimulus size is 15 arc minutes (approximately Goldmann size II). The ntensity of the stimulus is 0.09 nanoWatts. The fixation cross position results from the mean of 32 images (1.25 seconds of fixation video recording). The individual blue points specify the 32 individual fixation positions.

In the images, seen stimuli are represented as green squares and not seen stimuli in red. An evenly distribution of green squares indicates good fixation.

## Results

Pre-operative OCT showed a well-defined "fusiform-shaped" area of increased optical reflectivity in the layer assigned to the RPE-choriocapillaris complex.

Post-operative OCT showed well-defined relative high optical reflectivity in this area.

cLSO showed autofluorescence in the area corresponding to the CNV both pre and post-operatively.

Histopathology demonstrated features characteristic of those observed in CNV type II.

Photopic 10-2 perimetry, photopic blue FMM and cLSO microperimetry tests showed good central visual function in all patients post-operatively. An objective improvement was observed in the six patients that underwent all tests pre and post-operatively.

The average improvement of best corrected Snellen acuity for all ten patients was 1.3 lines (range 0–5 lines).

## Conclusions

Pre-operative OCT shows multilayered RPE. Pre-operative cLSO shows autofluorescence internal to the neovascular complex.

Post-operative OCT and cLSO show intact native RPE.

Visual acuity improves in some patients.

Photopic 10-2 Humphrey® perimetry, photopic blue fine matrix mapping (FMM) and cLSO microperimetry show improved central visual function.

Uncertainty exists whether the long-term outcome is better than the natural history of the disease. Usually there is a post-operative reduction in distortion.

More psycho-physiological data are needed. As more data is gathered we need to outline which etiological groups will improve.

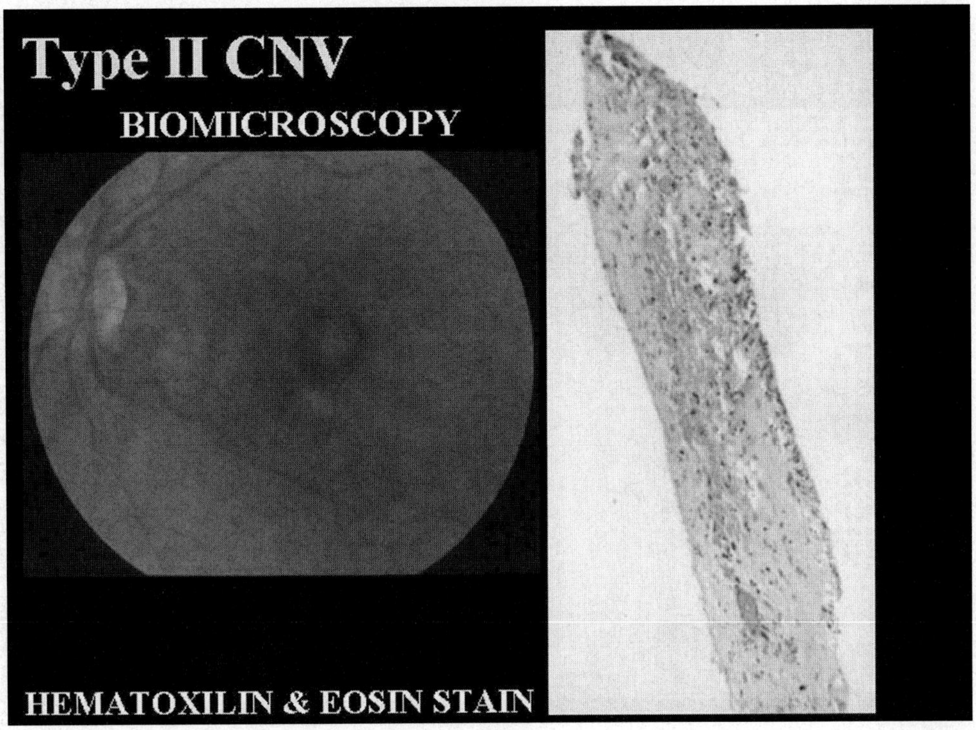


*Figure 1.*

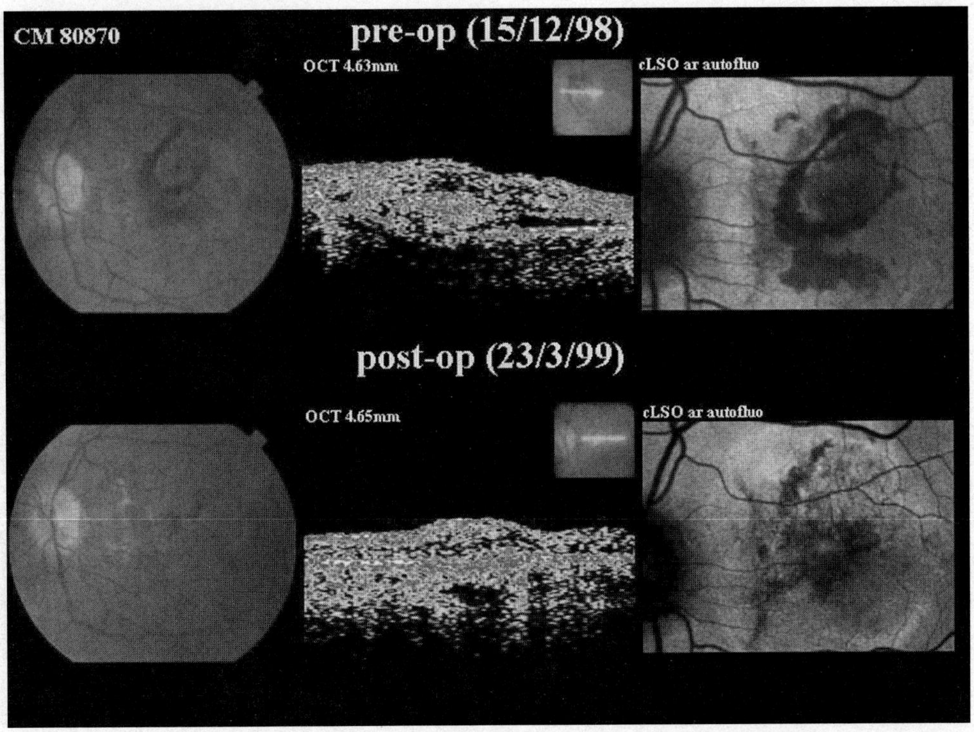


*Figure 2.*

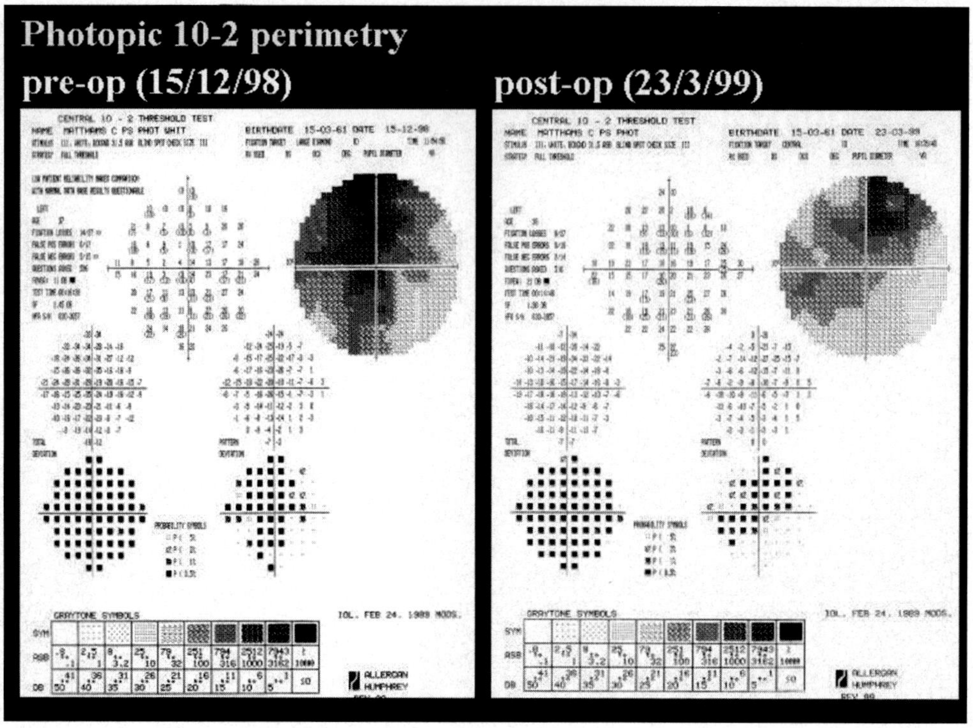


*Figure 3.*

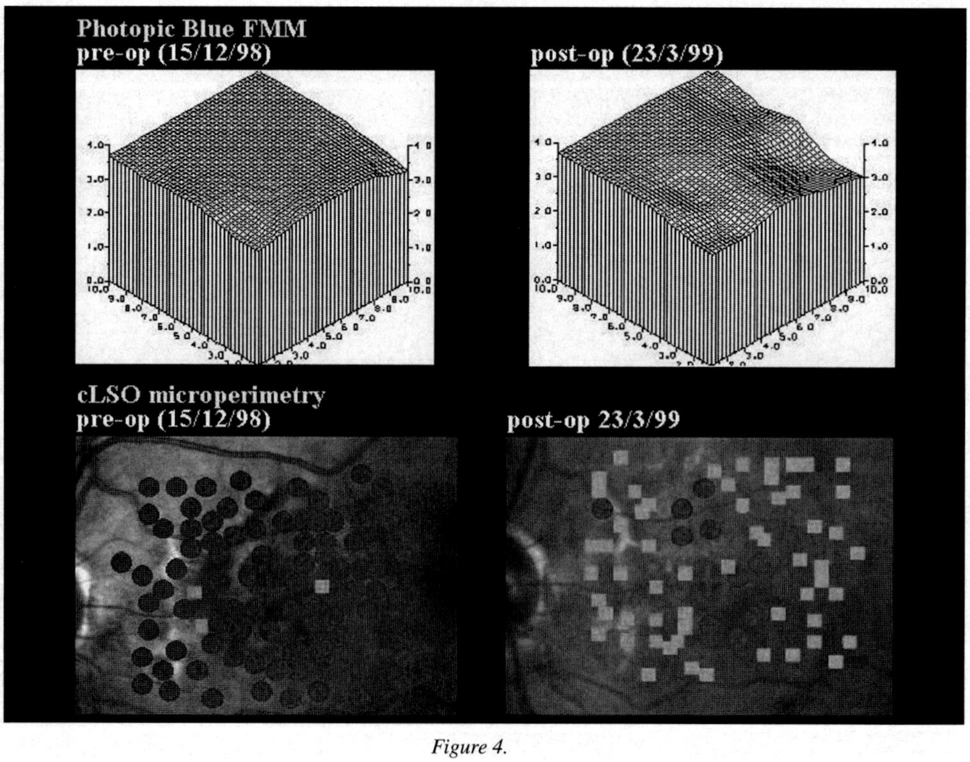


*Figure 4.*

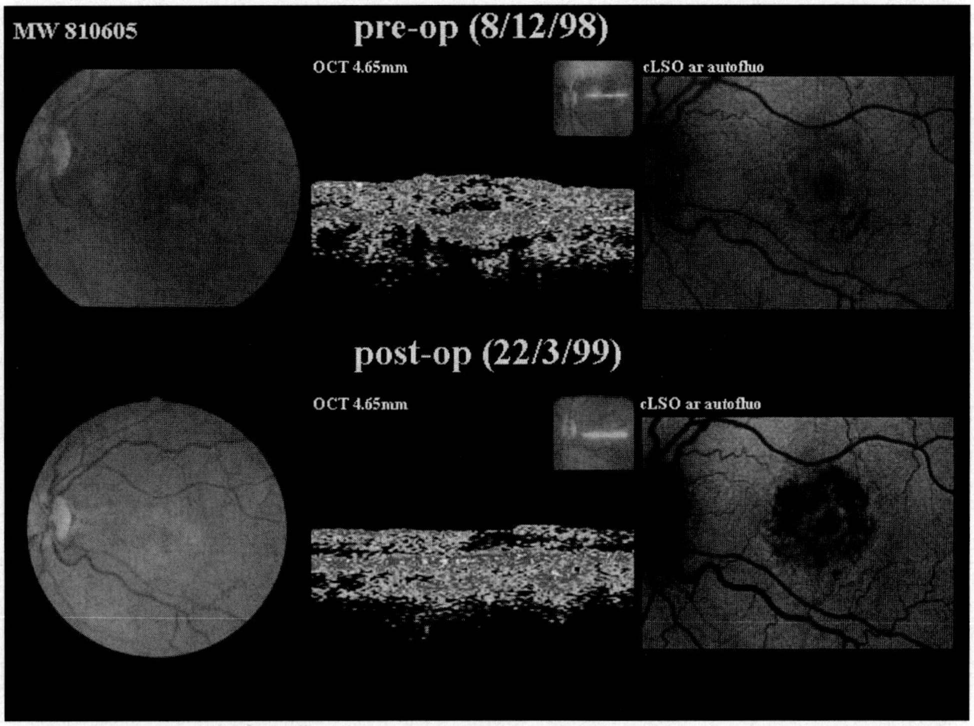


*Figure 5.*

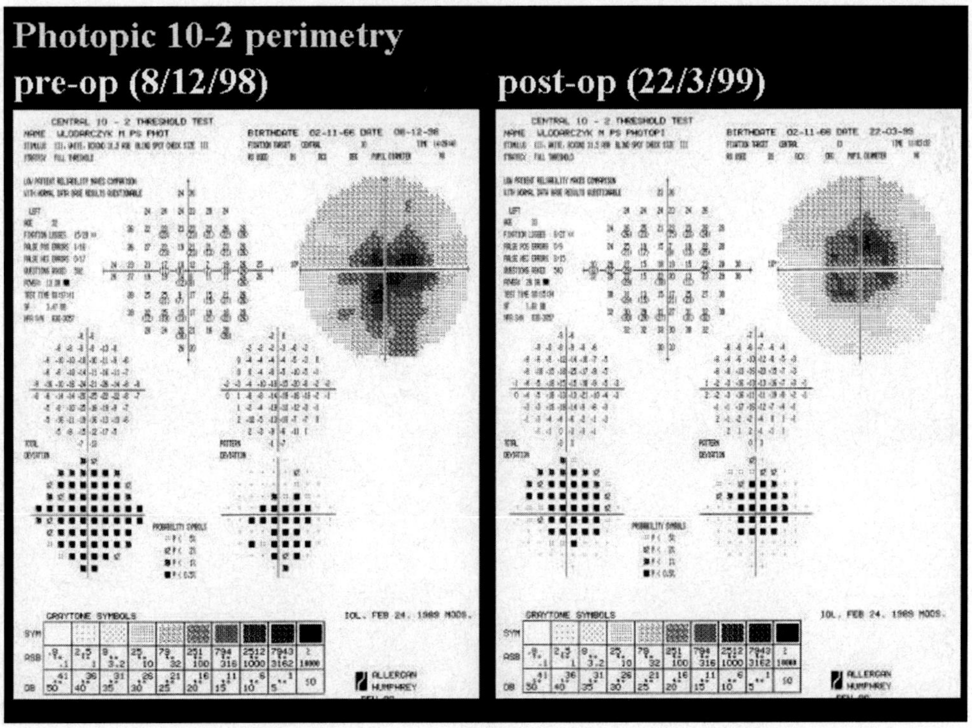


*Figure 6.*

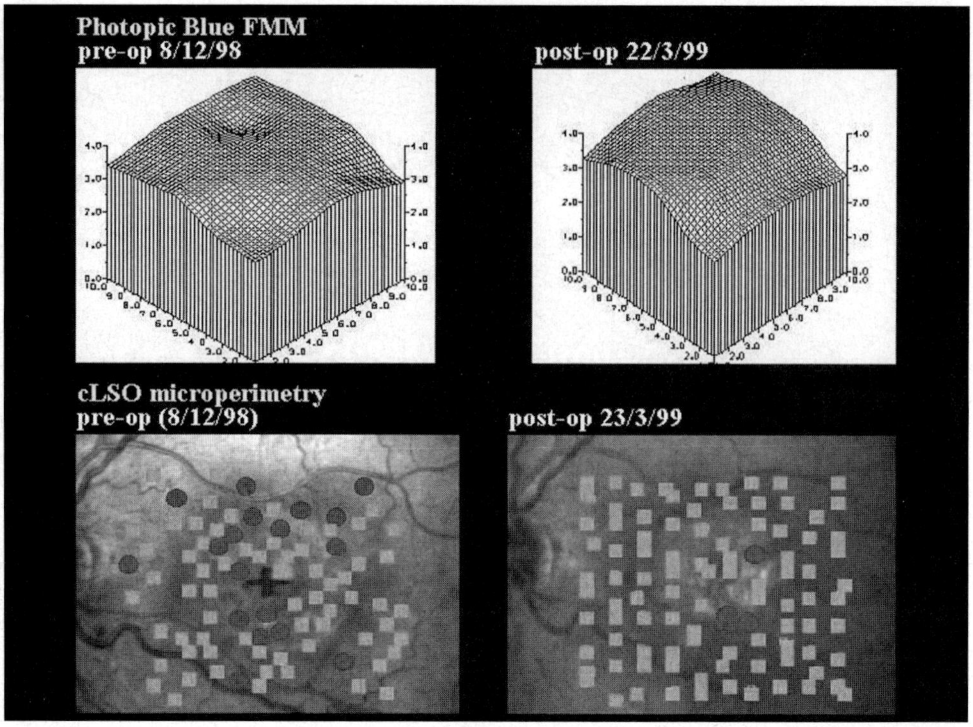


*Figure 7.*

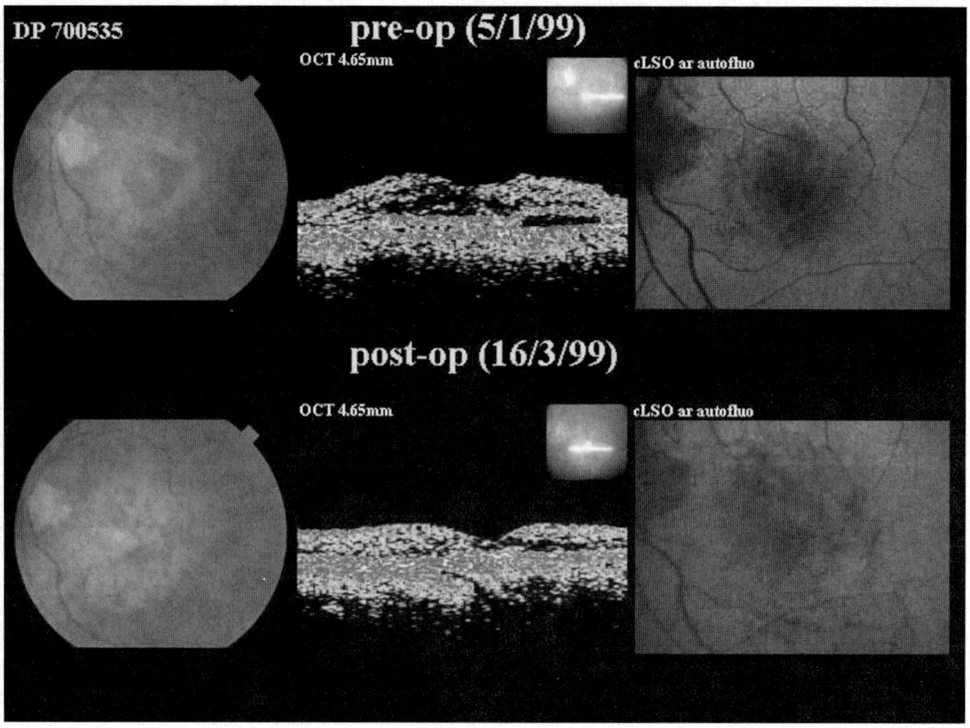


*Figure 8.*

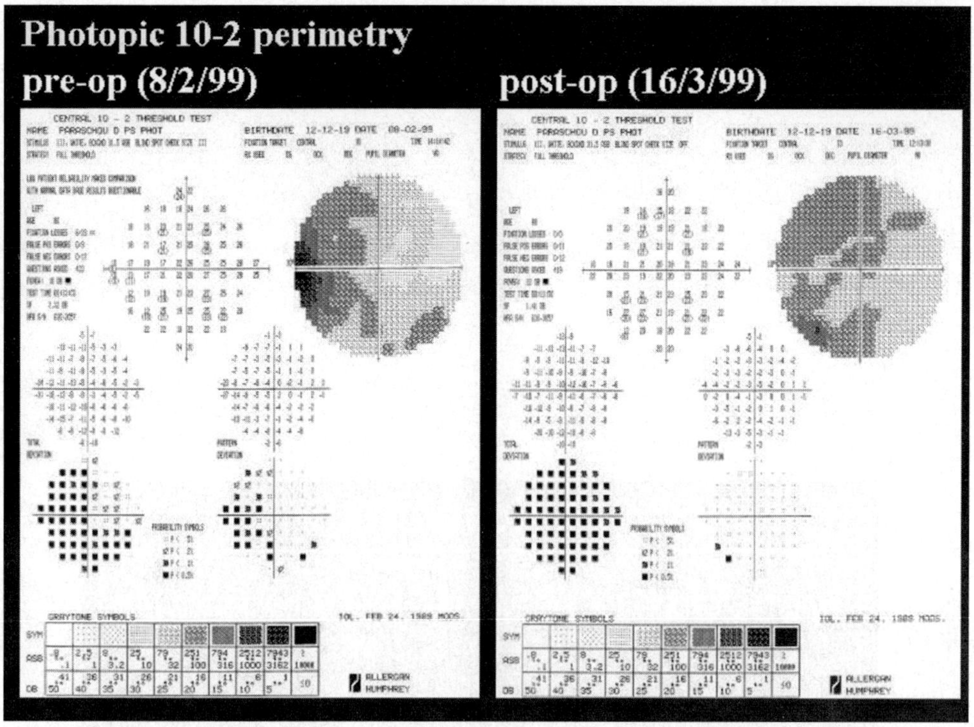


*Figure 9.*

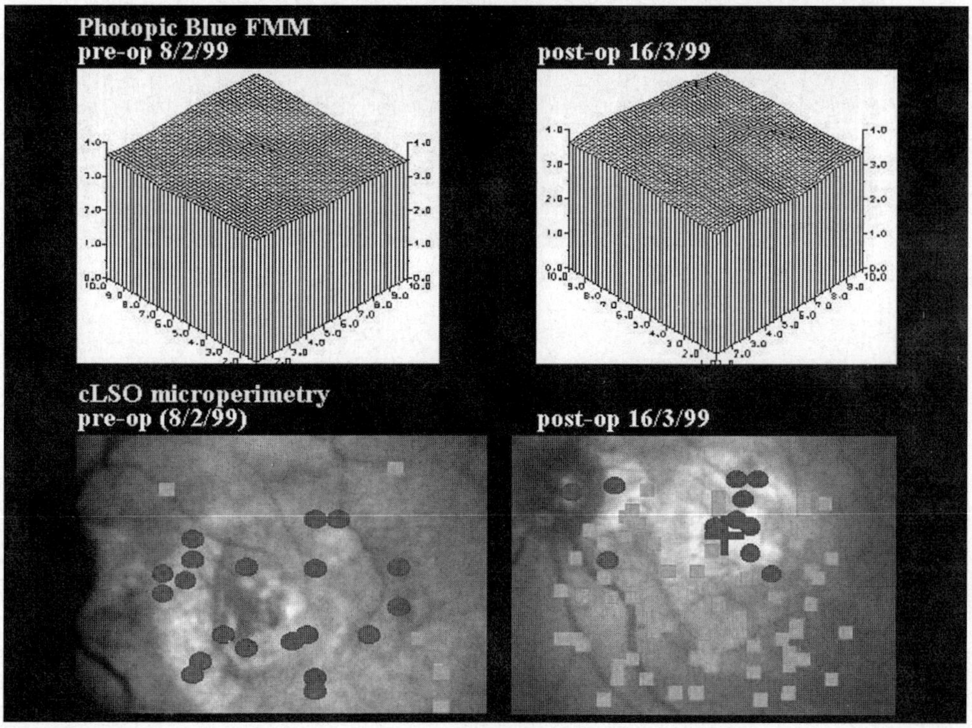


*Figure 10.*

## Summary of Patients' Data

| PATIENT # | GENDER | AGE | EYE | Pre-op VA | Post-op VA | MONTHS FOLLOW-UP |
|---|---|---|---|---|---|---|
| 1 | M | 26 | R | 6/18 | 6/18 | 3 |
| 2 | F | 31 | R | 6/60 | 6/24 | 5 |
| 3 | F | 30 | L | 6/18 | 6/18 | 13 |
| 4 | F | 47 | L | 6/24 | 6/12 | 1 |
| 5 | F | 34 | L | 6/60 | 6/9 | 7 |
| 6 | F | 47 | R | 6/24 | 6/24 | 4 |
| 7 | F | 33 | L | 6/36 | 6/9 | 3 |
| 8 | M | 50 | L | 6/60 | 6/60 | <1 |
| 9 | M | 37 | L | 6/18 | 6/18 | 3 |
| 10 | M | 79 | L | 6/24 | 6/24 | 1 |

*Figure 11.*

International Ophthalmology **23:** 317–320, 2001.
J.R. Sampaolesi (ed.), Laser Scanning: Update 1, 135–138.




# Measurement of eye length and eye shape by optical low coherence reflectometry

Gregor F. Schmid[1], Benno L. Petrig[1], Charles E. Riva[1,2], Eric Logean[1] & Rudolf Wälti[3]
[1]*Institut des Recherches en Ophtalmologie (IRO), Sion, Switzerland;* [2]*Université de Lausanne, Faculté de Médecine, Lausanne, Switzerland;* [3]*Haag-Streit AG, Köniz, Switzerland*

*Key words:* eye length, eye shape, eye growth, myopia, optical low coherence reflectometry


## Abstract

*Background:* The precise and rapid measurement of eye length and eye shape is essential for investigating eye growth regulation and myopia. For this purpose, we developed an optical low coherence reflectometer (OLCR) and present preliminary measurements. *Methods:* The OLCR includes a super luminescent diode (wavelength: 845 nm, coherence length: ~30 $\mu$m) and rotating glass cube to produce longitudinal scans at a velocity of 0.42 m/s and a repetition rate of ~13 scans/s. Heterodyne detection of light reflected from the anterior cornea and the posterior retina permits to measure axial eye length and eye shape (off-axis eye length). Each measurement consists of five consecutive scans. Reproducibility and precision were determined in one volunteer by measuring axial eye length five consecutive times, each time repositioning the eye. Eye shapes were determined in right eyes of four volunteers by measuring eye length every 3.3° from 10° nasally to 10° temporally. *Results:* Axial eye length measured repeatedly in one volunteer did not differ between or within the measurements (one-factor ANOVA). The average standard deviation was 11 $\mu$m. Eye shapes (a) varied substantially among subjects and (b) differed considerably from the corresponding shapes of spherical model eyes with identical axial eye lengths. *Conclusion:* The newly developed OLCR permits the precise and rapid measurement of eye length and eye shape. Such measurements, especially in children, may provide important information about mechanisms of eye growth regulation and the development of myopia.


Previous studies suggest that peripheral visual experience may play an important role in postnatal eye growth regulation preventing or minimizing refractive errors. For example, the eyes of juveniles that experience in the periphery of the ocular fundus a predominantly hyperopic refractive error (a distant image is focused behind the photoreceptor plane) tend to elongate excessively regardless of the refractive error at the fovea [1].

Furthermore, it was reported that large refractive errors are present in the periphery [2], the amount and type presenting a considerable individual variability [3]. We measured eye length (EL) on- and off-axis with laser Doppler interferometry (LDI) [4, 5] in several adult subjects and found that the resulting eye shape (ES) presents a considerable individual variability as well. Based on these observations, we hypothesize that ES (a) chiefly accounts for the individual variability in peripheral refraction, and (b) is a determinant factor in children for the ocular refractive development during postnatal eye growth. The precise and rapid measurement of EL and ES may therefore be essential for the investigation of postnatal eye growth control and the development of refractive errors such as myopia. For this purpose, we have developed an **O**ptical **L**ow **C**oherence **R**eflectometer (OLCR) [6].

## Methods

The principle of the OLCR is similar to that of ultrasound, except that optical rather than acoustical reflection is used to measure ocular dimensions [7–9]. This offers several advantages: (1) The precision as estimated by the standard deviation of repeated readings in measuring the optical distance $D_{opt}$ from the cornea to the retina is better than that of ultrasound. It is determined by the coherence length of the light source. For the super luminescent diode (wavelength: 845 nm, coherence length: ~30 $\mu$m) employed in our OLCR, the precision in measuring $D_{opt}$ is approximately 20 $\mu$m corresponding to a precision of approximately 15 $\mu$m after the conversion to geometrical distance $D_{geom.}$ ($D_{geom.} = D_{opt}/n$, where n is the composite refractive index of the ocular media). Conventional ultrasound has – at best – a precision of about 100 $\mu$m [10], and likely more in clinical situations. (2) The measuring light beam can be focused on the retina to a spot as small as ~10 $\mu$m in diameter [8], which permits the measurement of ES at a greater lateral resolution than with ultrasound. (3) Unlike with ultrasound, the eye does not need to be touched for a measurement.

A rotating glass cube [11, 12] in the reference arm of the OLCR produces longitudinal scans at a speed of 0.42 m/s and a repetition rate of ~13 scans/s. Heterodyne detection of light reflected from the eye permits the measurement of EL and ES. Rotating the cube consecutively matches the optical path length of the reference light with that of the sample light reflected from the corneal and retinal surfaces, each time generating interference fringes. Therefore, an interference signal generally contains several components. Figure 1 illustrates an interference signal obtained from the eye of an adult volunteer. The x-axis indicates the angular orientation of the glass cube converted to an optical distance, the y-axis indicates the intensity of the interference fringes as measured with a photo detector. The two highest components correspond to light reflections from the anterior surfaces of the cornea (AC) and the posterior surface of the retina (PR). Their separation was defined as EL. An optical path compensator reduces the scanning distance from the cornea to the retina to a few millimeters (patent pending). The amplitudes of the components vary between the individual scans due to eye motion and changes in the tear film. In high-amplitude interference signals additional components can sometimes be observed (Figure 1) corresponding to light reflections from the posterior surface of the cornea (PC) and the anterior surface of the retina (AR). For the determination of ES, the measuring light beam is deflected to various off-axis retinal locations at perpendicular incidence on the cornea using a polygonal mirror and a telescope lens system while the subject fixates on a target. Each measurement consists of five consecutive scans. The power of the incident light beam was 300 $\mu$W at the cornea. This corresponds to a power density of 780 $\mu W/cm^2$ (average over an aperture with a diameter of 7 mm) which is permitted for about five minutes at a wavelength of 845 nm, a duration much longer than the time required for alignment (less than 1 minute) and measurement (about 0.4 seconds) [13].

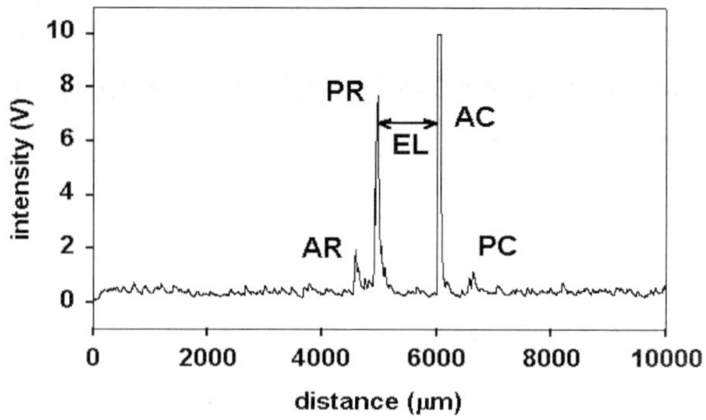


*Figure 1.* Interference signal from the eye of a young volunteer. The components correspond to light reflections from the anterior and posterior surfaces of the cornea (AC, PC) and retina (AR, PR). Eye length (EL) is defined as the distance AC – PR. An optical path compensator reduces the scanning distance to a few millimeters (patent pending).

Reproducibility and precision were determined by measuring EL five consecutive times in an adult volunteer, each time repositioning the eye. ES was determined in the right eye of four adult volunteers by measuring EL in steps of 3.3° from 10° nasally to 10° temporally. All volunteers gave their informed consent to serve as a subject for the study, and the tenets of the declaration of Helsinki were adhered to.

## Results

EL measured repeatedly in one volunteer was not significantly different, neither between nor within the five consecutive measurements (one-factor ANOVA). After the conversion to geometrical eye length, the average standard deviation was 11 $\mu$m. ES varied substantially among subjects and differed considerably from the corresponding shapes of spherical model eyes with identical axial eye lengths as measured by OLCR (Figure 2).

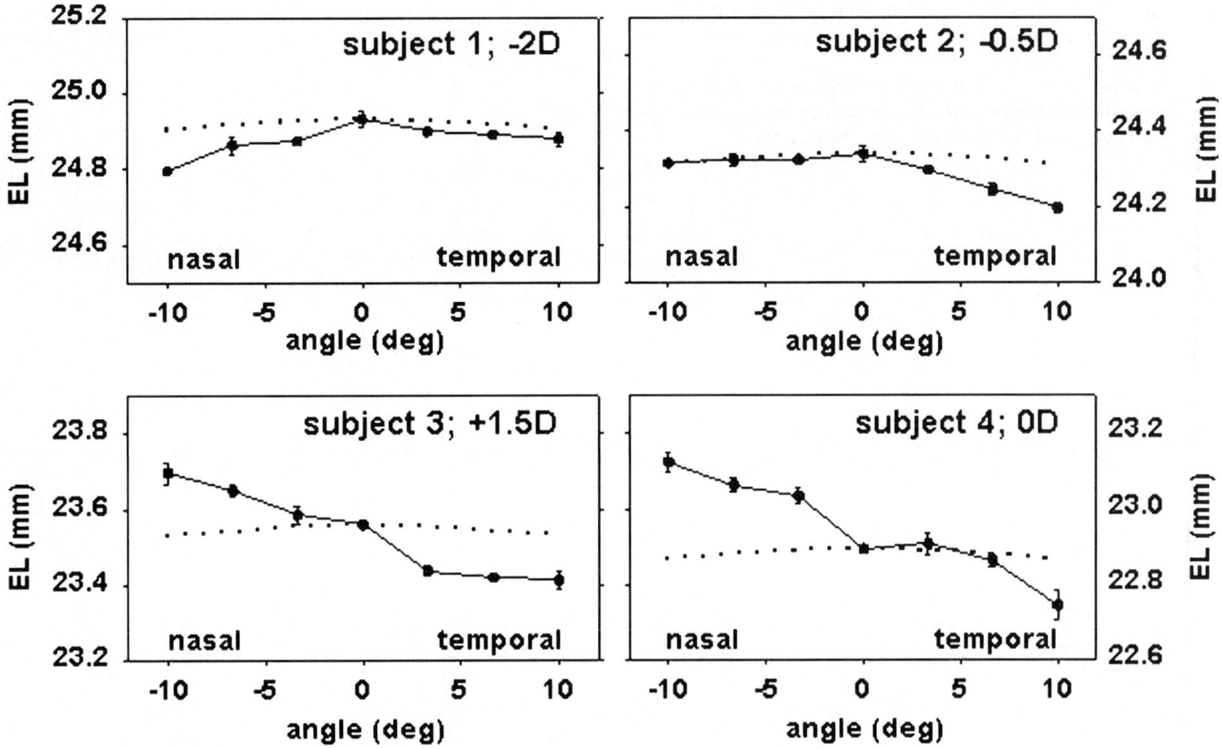


*Figure 2.* Eye shapes (peripheral eye lengths {EL}) measured in the right eye of four young volunteers. Note the considerable differences (a) between volunteers and (b) to the spherical model eye with identical EL (dotted line).

## Conclusions

The newly developed OLCR permits the precise and rapid measurement of EL and ES. ES presents a substantial individual variability and differs considerably from spherical model eyes. Such measurements, especially in children, may provide important information about mechanisms of eye growth regulation and the development of myopia. First measurements in children are under way.

## Acknowledgements

Supported by the Priority Program Optics II (project #492) and the Loterie Romande.

## References

1. Hoogerheide J, Rempt F, Hogenboom, WPH. Acquired myopia in young pilots. Ophthalmologica 1971; 163: 209-215.
2. Millodot M, Lamont A. Refraction of the periphery of the eye. J Opt Soc Am 1974; 64(1): 110-111.
3. Ferree CE, Rand G. The refractive conditions for the peripheral field. The Physical and Optical Societies Report of a Joint Discussion of Vision. 1932: 244-262.
4. Schmid GF, Petrig BL, Riva CE, Laties AM, Mendel MJ, Dunne, MCM. Eye-shape and diurnal eye length variation by laser Doppler interferometry. Vision Science and its Applications. Topical Meeting, Optical Soc Am, Santa Fe, Technical Digest Series 1994; 2: 130-133.
5. Schmid GF. Development of a laser Doppler interferometer to investigate postnatal eye growth regulation. Doctoral thesis; Department of Bioengineering, University of Pennsylvania, Philadelphia, USA, 1996.
6. Schmid GF, Petrig BL, Riva CE, Logean E, Waelti R. Measurement of eye length and eye shape by low coherence reflectometry. Annual meeting of the Swiss Ophthalmological Society (SSO/SOG), Montreux, September, 1999.
7. Fercher AF, Mengedoth K, Werner W. Eye-length measurement by interferometry with partially coherent light. Opt Lett 1988; 13: 1867-1869.
8. Hitzenberger CK. Optical measurement of the axial length by laser Doppler interferometry. Invest Ophthalmol Vis Sci 1991; 32: 616-624.
9. Schmid GF, Petrig BL, Riva CE, Shin KH, Stone RA, Mendel MJ, Laties AM. Measurement of intraocular distances in humans and chicks with a precision of better than $\pm 20$ $\mu$m by laser Doppler interferometry. Applied Optics 1996; 35(19): 3358-3361.
10. Byrne SF, Green RL. Ultrasound of the eye and orbit (Mosby Year Book, St. Louis, 1992), Chap. 6, pp. 215.

11. Clivaz X, Marquis-Weible F, Salathé RP, Nowak RP, Gilgen HH. High-resolution reflectometry in biological tissues. Opt Lett 19921 17(1): 4-6.
12. Chavanne P, Bonvin P, Gianotti R, Salathé RP. High speed, high precision broad band reflectometer. In: Proceedings Applied Optics & Optoelectronics,York, Sept. 1994.
13. American National Standards Institute: Safe use of lasers, ANSI Z 136.1. New York, American National Standards Institute, 1986.

*International Ophthalmology* **23:** 321–325, 2001.
*J.R. Sampaolesi (ed.), Laser Scanning: Update 1*, 139–143.



# Comparison of optical coherence tomography and fluorescein angiography in assessing macular edema in retinal dystrophies: preliminary results

P. E. Stanga, S. M. Downes, R. M. Ahuja, N. H. V. Chong, R. Antcliff, A. C. Reck & A. C. Bird
*The Institute of Ophthalmology, University College London and Moorfields Eye Hospital, London, UK*

## Purpose

The purpose of our work was to compare optical coherent tomography (OCT) and fluorescein angiography (FA) in detecting cystoid macular edema (CME) in retinal dystrophies and to quantify the response to oral acetazolamide.

## Materials and methods

Patients from the Retinal Genetics Clinic at Moorfields Eye Hospital were examined for the presence of macular edema. 17 patients were recruited over a 6-month period. 12 patients were female while 5 were male. The mean age was 43.5 years (range 21–60 years).

The Retinal Dystrophies affecting these group of patients were as follows: 1 cone-rod dystrophy, 4 Usher syndrome, 12 Retinitis Pigmentosa (autosomal recessive = 10, autosomal dominant = 2). A total of 34 eyes were studied.

Pre-treatment visual acuities, biomicroscopy, color photography, fundus fluorescein angiography (FFA), and optical coherence tomography (OCT) were performed. 1 month of acetazolamide (Diamox SR®) 250-mg twice a day by mouth was administered. Post-treatment visual acuities and OCT were performed. Any reported improvement was recorded.

OCT Technique: 3-mm single line scans were obtained in the macula, through the fovea. Horizontal scans were analyzed. Bilaminar scans were obtained with two bands of high intensity signal as previously published. Limitations on achieving this were poor fixation, posterior capsular thickening or cataract.

OCT Analysis: Raw images were aligned and normalized using the software provided by the OCT Scanner (Humphrey Instruments, San Leandro, California, USA). These images were analyzed by counting pixels using the software callipers using Adobe Photoshop® (Adobe Systems Inc, San Jose, California, USA). 500 pixels make up the 2 mm scan depth; so that 1 pixel = 4 $\mu$m.

Seventeen patients underwent FFA. Intravenous sodium fluorescein (5 ml 20%) was administered. Late photographs were taken at 7 minutes. An experienced medical retina specialist masked to the OCT findings assessed the angiograms.

## Results

One patient was unable to tolerate acetazolamide. Cystic spaces observed on OCT in 34 eyes. Edema observed on FFA in 31 eyes. The pre-treatment all eyes average visual acuity was 0.371 (SD = 0.28), while the post-treatment all eyes average visual acuity was 0.396 (SD = 0.30). Statistically not significant: student paired 't' test $p = 0.075$.

In the Acetazolamide treatment group the pre-treatment average retinal thickness was 302.5 $\mu$m (SD = 83.7). Range 196–504. The post-treatment average retinal thickness was 276.5 $\mu$m (SD = 57.3). Range 196–464. Statistically significant: p = 0.000274. Foveal thickness = 150–200 $\mu$m. Retinal thickness = 250–300 $\mu$m.

## Conclusions

OCT is as least as sensitive as FFA for identifying edema. OCT is a useful tool in assessing response to therapy by comparative measurements of retinal thickness. OCT is a safe, non-invasive and relatively fast procedure

Subjective response to therapy correlates well with decreased retinal thickening, this is not so for visual acuity.

Future investigations such as visual fields and dark adaptation may correlate better with subjective and anatomical improvement than visual acuity.

## References

1. Fercher AF, Hitzenberger CK, Drexler W, et al. In vivo optical coherence tomography. Am J Ophthalmol 1993; 116: 113–4.
2. Hee MR, Puliafito CA, Wong C, et al. Quantitative assessment of macular edema with optical coherence tomography. Arch Ophthalmol 1995; 113: 1019
3. Hee MR, Puliafito CA, Wong C, et al. Optical coherence tomography of central serous retinopathy. Am J Ophthalmol 1995; 120: 65–74.
4. Hee MR, Puliafito CA, Duker JS, et al. Topography of diabetic macular edema with optical coherence tomography. Ophthalmology 1998; 105: 360-70.
5. Hirakawa H, Ikjima H, Gohdo T, Tsukahara S. Optical coherence tomography of cystoid macular edema associated with retinitis pigmentosa. Am J Ophthalmol 1999; 128: 185–91.
6. Puliafito C, Hee MR, Lin CP, et al. Imaging of macular diseases with optical coherence tomography. Science 1991; 254: 1178–81.
7. Toth CA, Narayan DG, Boppart SA, et al. A comparison of retinal morphology viewed by optical coherence tomography and by light microscopy. Arch Ophthalmol 1997; 115: 1425–8.
8. Yanuzzi LA, Rohrer KT, Tindel LJ, et al. Fluorescein angiography complication survey. Ophthalmology 1986; 93: 611–7.

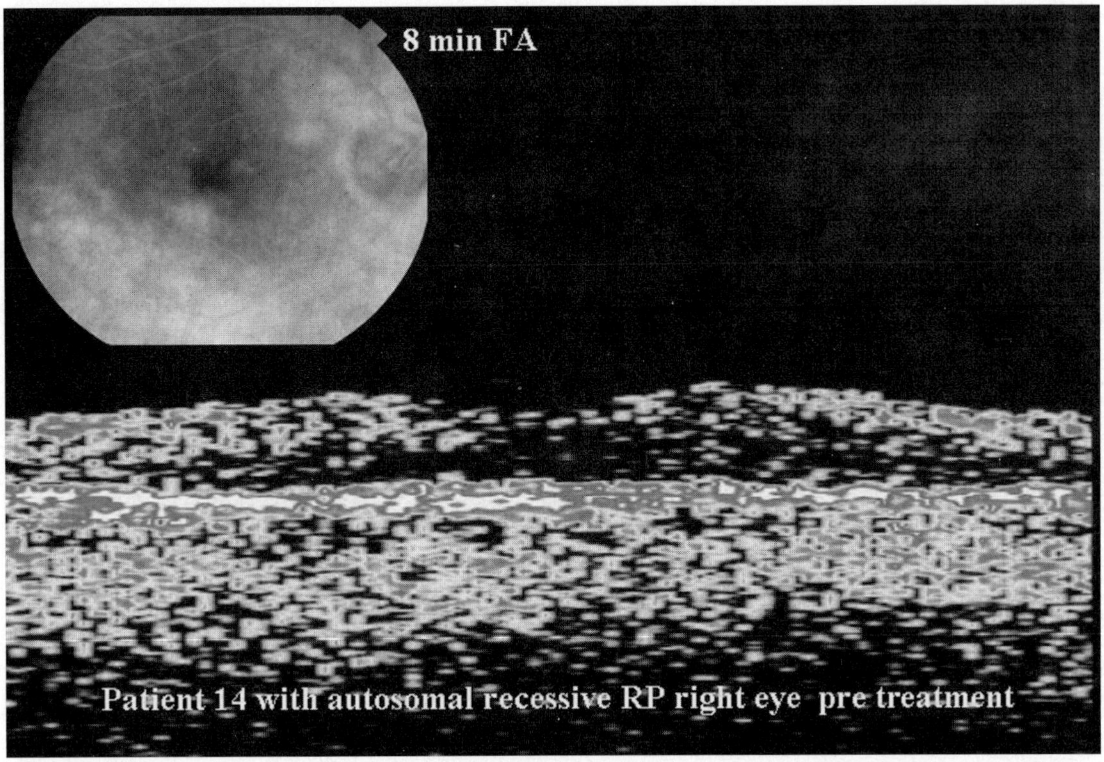


*Figure 1.*

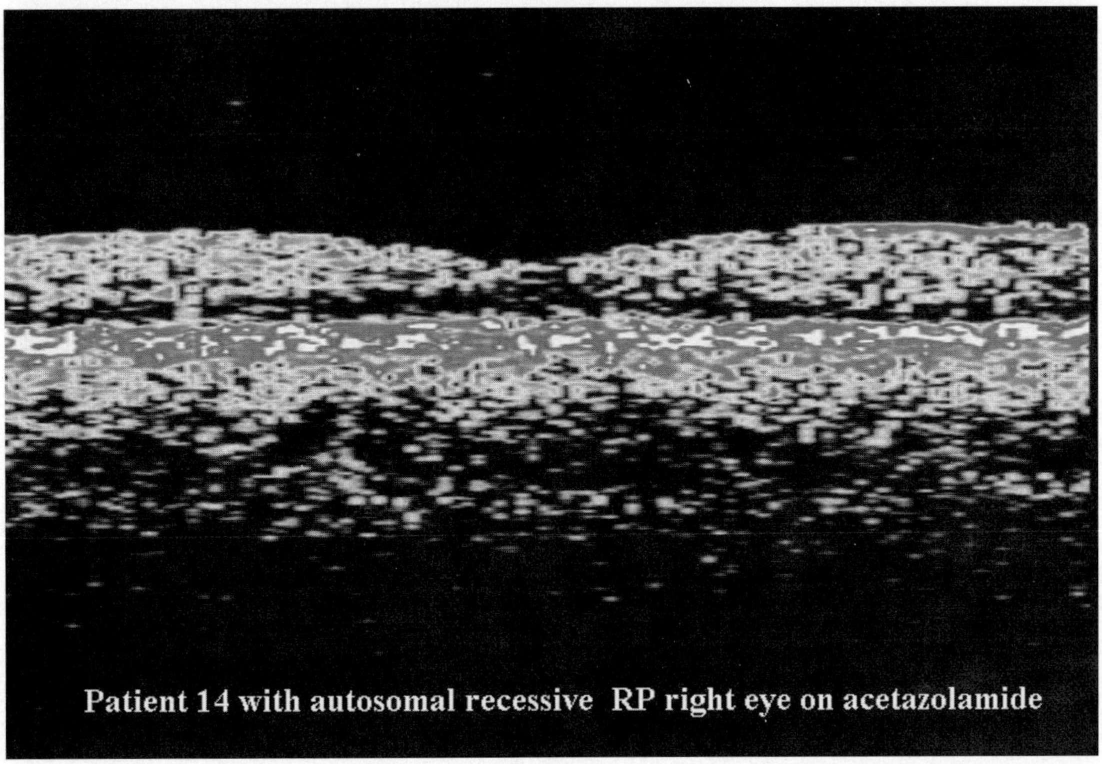


*Figure 2.*

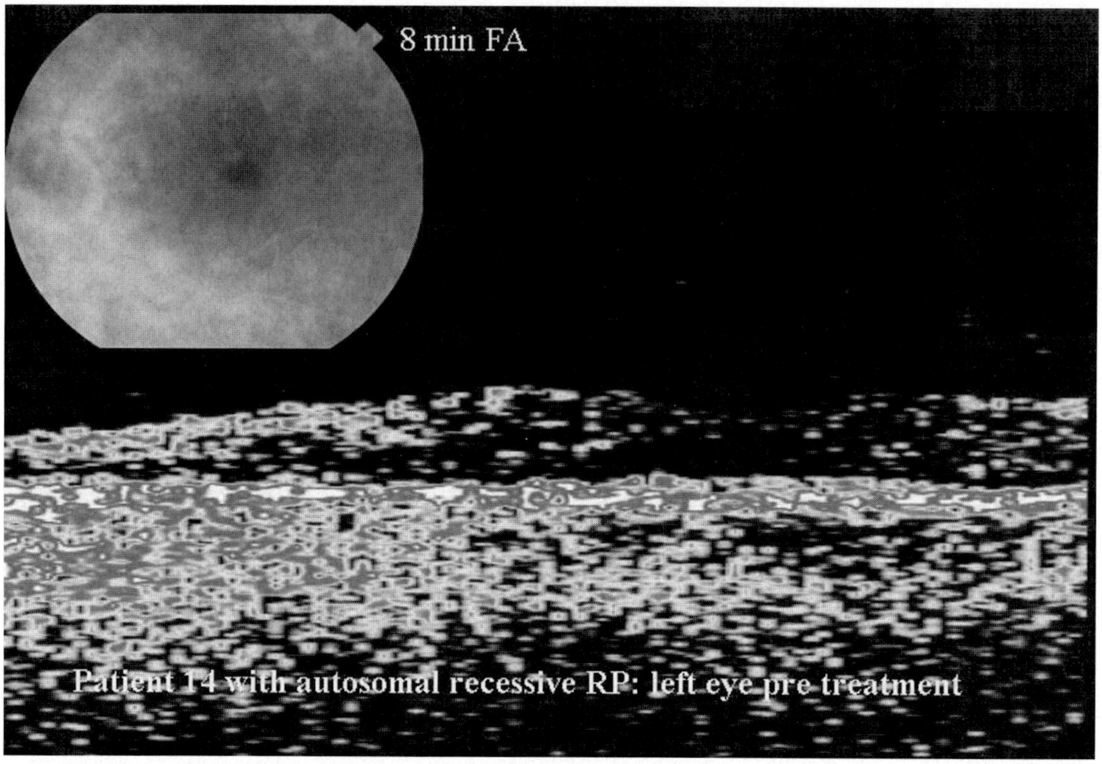


*Figure 3.*

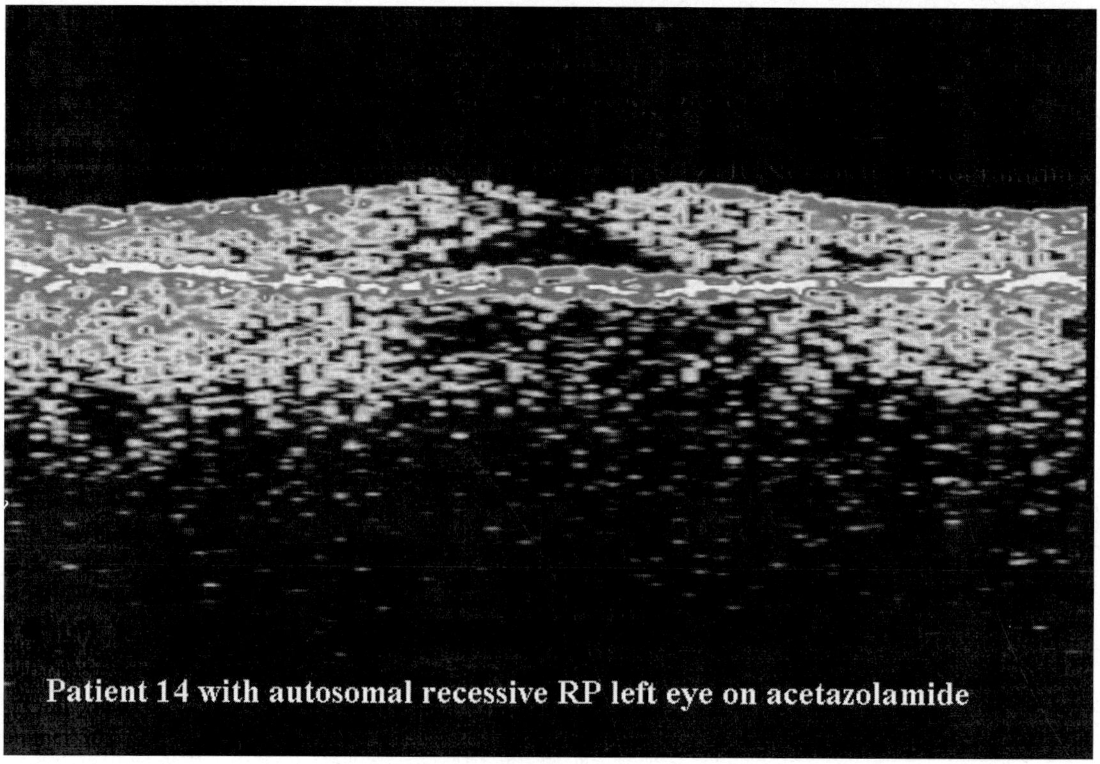


*Figure 4.*

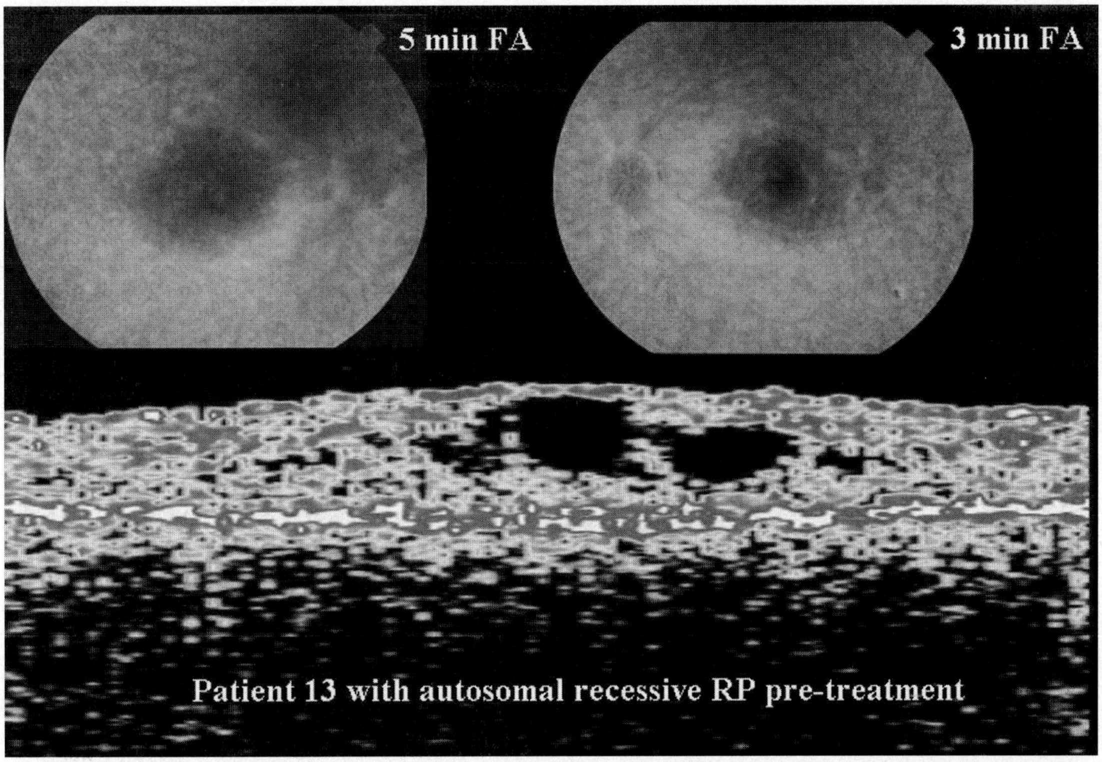


*Figure 5.*

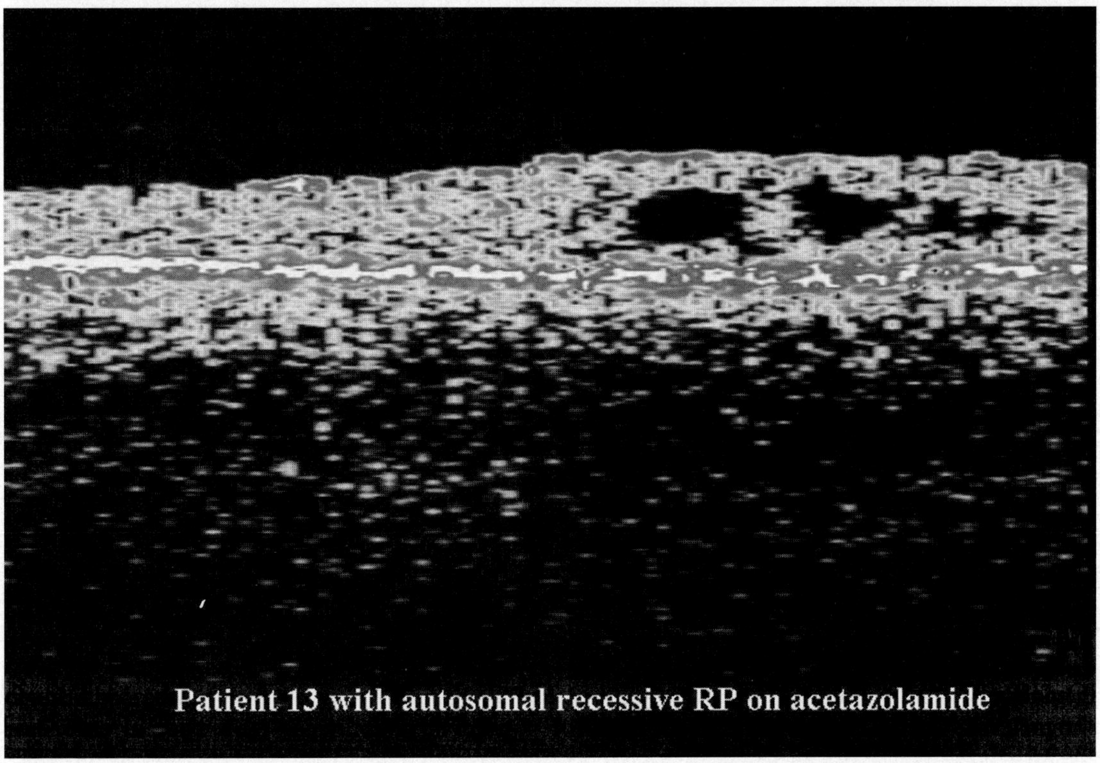


*Figure 6.*

*International Ophthalmology* **23:** 327–335, 2001.
*J.R. Sampaolesi (ed.), Laser Scanning: Update 1*, 145–153.




# Functional imaging of the retinal microvasculature by Scanning Laser Doppler Flowmetry

Georg Michelson, Jürgen Welzenbach, Istvan Pal & Joana Harazny
*Universitäts Augenklinik, Erlangen, Germany*

*Key words:* average distance to the next capillary, functional imaging of the retinal circulation, Scanning Laser Doppler flowmetry – AFFPIA

## Abstract

*Purpose:* to image functionally perfused retinal vessels and to assess quantitatively the intercapillary space of the retinal microvasculature. *Method:* The base of functional imaging and the quantitative assessment of the retinal vasculature is the two-dimensional map of the retina encoded by the laser Doppler frequency shift. By Scanning Laser Doppler Flowmetry (HRF. Heidelberg Engineering) the laser Doppler frequency shift of 16.384 retinal sites (256 pixels × 64 lines, spatial resolution 10 *mu*m) of a retinal area of 2.7 × 0.7 mm was gained. The image processing was performed by a recently described algorithm (AFFPIA). Using the data of the laser Doppler frequency shift of every retinal site, a color-coded retinal image was established showing *perfused* vessels and capillaries. By automatic pattern analysis of this image vessels and capillaries were identified and segmented. Based on this image the distances in [$\mu$m] of every retinal site to the next vessel or capillary were calculated ("distance to next capillary"). The functional imaging of the retinal perfusion was demonstrated in (1) normal retina, (2) retinal arterial occlusion, and (3) proliferative retinopathy. *Intra*observer reliability of the quantitative assessment of the parameter "distance to next capillary" was estimated by measuring 10 eyes of 10 subjects at 5 different days by one observer. *Inter*observer reliability of the quantitative assessment was evaluated by analysing 10 perfusion maps by 5 different operators. In 93 eyes of 71 normal subjects (mean age 40.4 $\mu$ 15 years) the juxtapapillary retina was quantitatively evaluated. *Results: Qualitative evaluation:* The functional images of the retinal perfusion of eyes with normal retina, with retinal arterial occlusion, and with proliferative retinopathy corresponded well with the fluorescein angiography. Perfused vessels and capillaries became visible in a high local resolution. *Quantititative assessment:* The coefficient of reliability of the *intro*observer and *inter*observer reproducibility of the parameter "mean distance to next capillary" was 0.74, and 0.95, respectively. The quantitative assessment of the perfusion showed that the major part of the retinal sites (>700%) had distances to the next capillary lower than 30 $\mu$m 46% of the retinal area had distances to the next capillary from 0–20 $l\mu$m 26% of the retina had distances from 20–30 $\mu$m, 12% of the retina had distances from 30–40 $\mu$m 7% of the retina had distances from 40–50 $\mu$m, 4% of the retina had distances from 50–60 $\mu$m, and 4% of the retinal sites showed distances to the next capillary greater than 60 *mu*m. The mean distance to the next capillary or vessel was calculated with 21 ± 6.5 $\mu$m. *Conclusion:* By non-invasive Scanning Laser Doppler Flowmetry in combination with adequate software it is possible to perform a functional imaging of the retinal vasculature and to measure all index for the functional density of retinal capillaries and vessels.

## Introduction

There is great clinical need for non-invasive functional imaging and quantitative assessment of the retinal microcirculation. For quantitative assessment of the retinal vasculature one important parameter might be the capillary density. The distance between the retinal neuronal cell and next lying capillary resembles the diffusion path length from the nourishing capillary to the oxygen-consuming neuronal cell. The longer the distance between capillary and neuronal cell, the lower the oxygen supply. Many ocular diseases are associated with areas of decreased capillary density or avascularity, leading to a loss of neuronal function in these area due to an impaired transport of oxygen and other substances. Till now only by invasive techniques (fluorescein angiography, etc.) it is possible to perform a functional image of the retinal vasculature in a high resolution and to assess the capillary density of the retinal meskwork [1].

The purpose was to describe a *non-invasive* method for functional imaging and quantitative assessment of the retinal vasculature.

## Patients and methods

All examinations were performed adherent to the "World Medical Association Declaration of Helsinki: Recommendations Guiding Medical Doctors in Biomedical Research Involving Human Subjects".

By Scanning Laser Doppler-technology (HRF. Heidelberg Retina Flowmeter. Heidelberg Engineering, Germany) in combination with a recently described algorithm for the analysis of the perfusion image (AFFPIA) [2] the retinal microcirculation was two-dimensionally imaged. The principles of function of SLDF are described in detail elsewhere [3–6]. Briefly, Scanning Laser Doppler Flowmetry (SLDF) measures the Laser Doppler frequency shift in each of 16,000 retinal sites of a retinal area of 2.7 × 0.7 mm with a spatial resolution of 10 × 10 $\mu$m. The acquisition time is 2 s. By the analysis tool named "automatic full-field perfusion image analyzer" (AFFPIA) the quantification of the laser-Doppler shift in any point of the scanned sector was performed, giving results independently of the reflectivity of the retinal site. AFFPIA calculates the laser-Doppler frequency shift of each pixel according to the theory of Bonner & Nossal [7]. The laser Doppler shift resembles the degree of movement of blood cells in this distinct retinal site. The data of the laser Doppler frequency shift of 64 × 256 retinal sites is the base for the *perfusion image* and for further automatic image processing. The *perfusion image* was coded by the laser Doppler frequency shift of each pixel. For a valid measurement of the laser Doppler frequency shift some assumptions for the intensity image must be fulfilled: adequate reflectivity of each pixel, proper focus, and no artificial movements. To meet these requirements under- and over-exposed pixels, pixels of saccades were excluded automatically by AFFPTA. By the vessel detection-algorithm using the laser-Doppler shift-coded image patterns of pixels with increased laser Doppler shift were identified as perfused capillaries and vessels. In the Appendix, the algorithm is described in detail. Table 1 and Figure 1 showed the steps of the image processing. Briefly, the vessel detection algorithm had four steps. First it is performed an enhancement of structures with an iterative nonlinear contrast transformation on the Gaussian filtered *perfusion image*. In the second step the *perfusion image* was binarized. In step 3 the enhanced image and the binarized image were joined. In step 4 the resulting image was binarized again. Thus perfused capillaries or vessels were identified automatically in the laser Doppler frequency shift image. Based on this topographical map of segmented perfused retinal vessels and capillaries the morphologic parameter "distance to the next perfused capillary or vessel" in [$\mu$m] was calculated automatically. After finding the next lying capillary for each retinal site, the distance of each retinal site to the next lying capillary as calculated. This was performed separately for the temporal, and nasal area.

*Table 1.* Steps of image processing

1. Fast Fourier Transformation of all intensity-time curves
2. Calculation of the laser Doppler frequency shift
3. Interactive marking of saccades
4. Automatic exclusion of
   Pixels with incorrect DC-values (under-, overexposed pixels)
   Pixels of artificial movements (saccades)
5. Automatic characterization of the retinal vessel tree by a vessel detection-algorithm based on the perfusion-image
6. Interactive marking of the rim area
7. Quantitative assessment of the perfusion image: calculation of the distances to the next lying perfused capillary or vessel of the temporal, nasal and rim area
   Distribution of percentages of retinal area [in %] with a given distance to the next capillary
   Mean distance
   Standard deviation

### *Qualitative evaluation of perfusion images*

Sets of perfusion images of three examples were evaluated and compared with corresponding fluorescein angiograms: example 1: normal retina. example 2: arterial retinal occlusion, example 3: proliferative retinopathy after central vein occlusion.

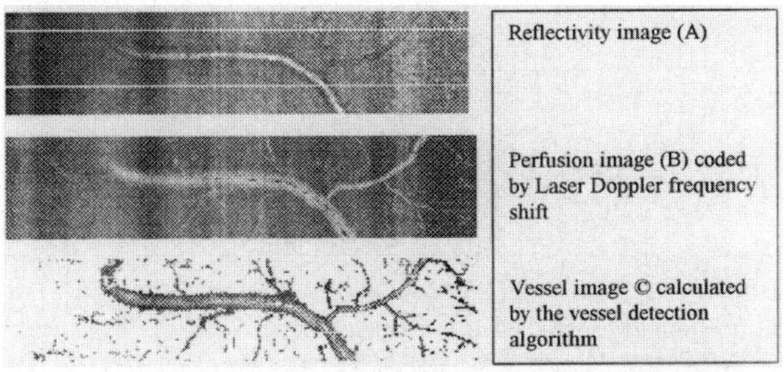


*Figure 1.* Example for the automatic identification of capillaries and vessels of the retinal branch tree by the vessel detection-algorithm using the perfusion-image.

*Quantitative evaluation of perfusion maps*

*Reliability: Intra*observer reliability of the calculated parameter "distance to next capillary or vessel" was estimated by measuring 10 eyes of 10 persons on 5 different days by one observer. *Inter*observer reliability was evaluated by comparing the results of 5 different operators analyzing 10 different perfusion maps. The coefficient of reliability was calculated by the Equations (1)–(3). The reliability coefficient $r_1$ was calculated by the estimation of the variances of the biological component and the error component The measured value is a combination of the true mean value, the biological variability component, and the error component (Equation 2). The variances of the two components are defined in Equation (3). The error component was differentiated from the biological component by the variance component-analysis calculated by VAR COMP-procedure from SAS-STAT (SAS STAT. users guide release 6.03 edition. Cary, NC: SAS-institute, Inc., 1988; pp. 967–978).

$$Y_{ij} = u + A_i + e_{ij}, \tag{1}$$

where $Y_{ij}$ = measured value; $A_i$ = biological component, $e_{ij}$ = measuring error component; $I = 1, \ldots, n$ eyes; $j = 1, \ldots, m$ measurements.

$$\mathrm{Var}(Y_{ij}) = \mathrm{Var}(A_i) + \mathrm{Var}(e_{ij}), \tag{2}$$

where $\mathrm{Var}_A$ = variance of biological component; $\mathrm{Var}_E$ = variance of error component;

The reliability coefficient $r_1$ is defined by the following equation:

$$r_1 = \mathrm{Var}(Y_{ij})/(\mathrm{Var}(Y_{ij}) + \mathrm{Var}(e_{ij}), \tag{3}$$

*Subjects:* The capillary density of the juxtapapillary retina in 93 eyes of 71 normal subjects was evaluated. The mean age of the subjects was 40 ± 15 years.

## Results

*Imaging of perfused structures:* The perfusion image showed perfused vessels and capillaries in a high local resolution. In example 1 three scanned sectors were joined to map the whole retinal vasculature of a healthy eye. In Figure 2a is depicted the reflectivity images of 3 sectors, showing different intensity of illumination. Figure 2b gives the functional images of the retinal vasculature of these 3 sectors, showing similar perfusion in all 3 sectors. Please note, that the "perfusion image" is purely calculated by the laser Doppler frequency shift. Figures 3a–d show the functional images of in eye with an arterial branch occlusion. Figure 3a presents the reflectivity images of 5 sectors, in Figure 3b are depicted the "perfusion images" of the retinal vasculature of these 5 sectors. Figure 3d shows the fluorescein angiography of this eye. In the perfusion image (Figure 3b) becomes visible a triangular area distal from the embolus with a rarefication of perfused structures indicating a nonperfusion in this area. Please notice the well perfused chorio-retinal anastomosis in the occluded area. The fluorescein angiogram confirms these findings. Example 3: In Figure 4a and Figure 4b are depicted 2 sectors of the retina with a proliferative retinopathy after central vein occlusion, given as reflectivity and as perfusion image. The qualitative comparison with the corresponding fluoresein angiogram (Figure 4c) reveals clearly a coincidence between perfused vessels gained by two different types of data acquisition.

*Quantitative assessment*

*Reliability:* The *intra*observer reliability for the parameter "distance to next capillary" had a coefficient of reliability of 0.78 and 0.95, respectively.

*Distance to the next perfused capillary or vessel:* Based on the *perfusion image* the major part of the retinal sites had distances to the next capillary lower than 30 $\mu$m. In the juxtapapillary temporal (nasal) retina 46 ± 15% (47 ± 18%) of the retinal area showed distances to the next capillary from 0–20 $\mu$m, 26 ± 4% (25 ± 5%) of the retinal area showed distances to the next capillary from 20–30 $\mu$m, 12 ± 4% (12 ± 5%) of the retinal area showed distances to the next capillary from 30–40 $\mu$m, 7 ± 3%) (6 ± 4%) of the retinal area showed distances to the next capillary from 40–50 $\mu$m, 4 ± 3% (5 ± 4%) of the retinal area showed distances to the next capillary from 50–60 $\mu$m, and 4 ± 5% (5

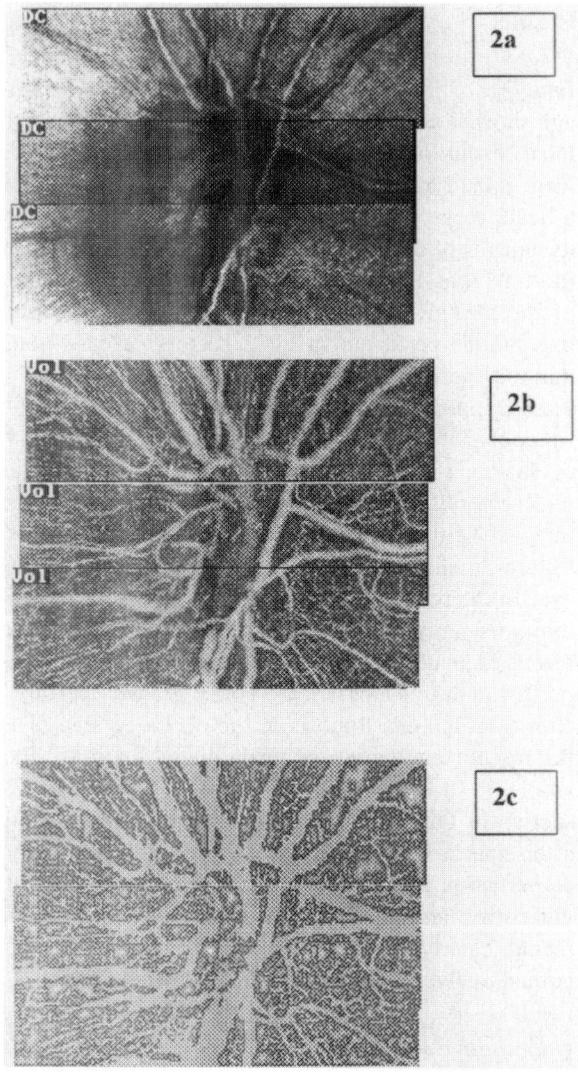


*Figure 2.* Example 1: Functional imaging of an eye with a normal retinal perfusion. (a) Reflectivity image. (b) Perfusion image coded by *laser Doppler frequency shift*. (c) Distance map coded by the *distance to the next capillary*.

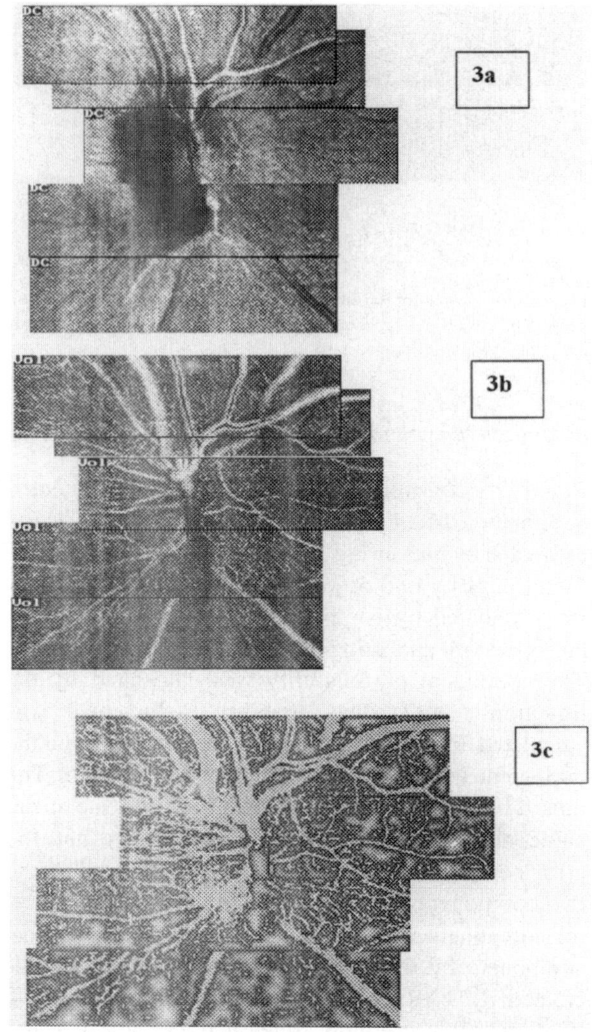


*Figure 3.* Example 2: Functional imaging of an eye with a retinal arterial branch occlusion: (a) Reflectivity image. (b) Perfusion image coded *laser Doppler frequency shift*. The triangular dark area distal from the embolus indicates non-perfusion. Please note, that the darker the pixel the lower the perfusion. (c) Distance map coded by the distance to the next capillary. Please note that the brighter the pixel the more longer is the distance to the next capillary.

± 6%) of the retinal area showed distances to the next capillary showed distances to the next capillary greater than 60 $\mu$m. Figure 5 presents the mean percentages of retinal area with a given distance to the next lying perfused capillary or vessel. The mean distance to the next capillary or vessel was temporally 21 ± 6.5 $\mu$m and nasally 21 ± 6.9 $\mu$m. We found no significant difference between the nasal and temporal area. In Figure 7 is depicted the histogram of the mean distance of all eyes of the temporal area.

### *Dependency of age, blood pressure, and intraocular pressure*

We observed no significant correlation between "mean distance to next capillary" and age, blood pressure, and intraocular pressure.

## Discussion

A. *Functional imaging of the retinal perfusion.* The imaging of perfused capillaries and vessels was per-

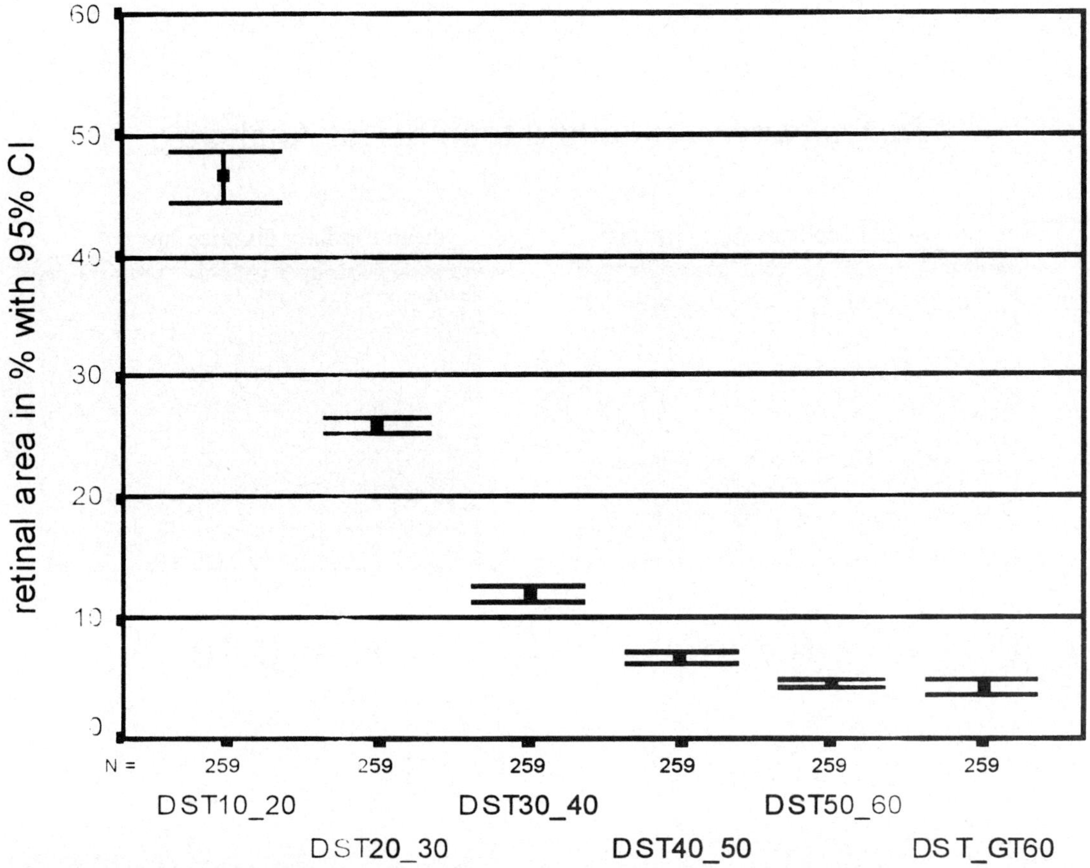


*Figure 5.* Quantitative assessment of normal eyes: percentages of the retinal area [in %] with a given "distance to the next capillaries or vessels" (temporal juxtapapillary retina).

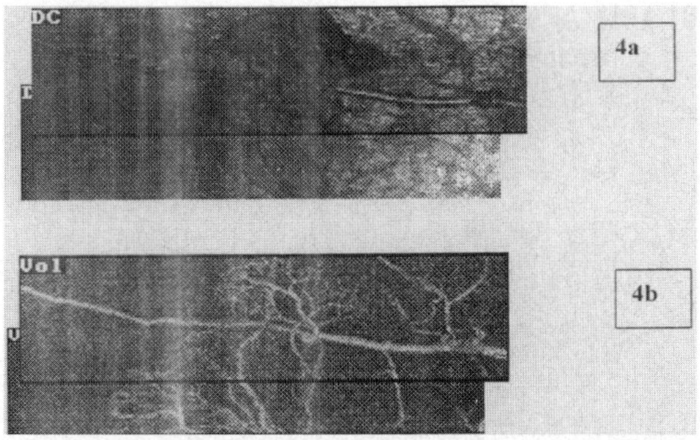


*Figure 4.* Example 3: Functional imaging of an eye with proliferative retinopathy after central vein occlusion. (a) Reflectivity image. (b) Perfusion image coded by *laser Doppler frequency shift.* Please note the perfusion in the proliferative structure.

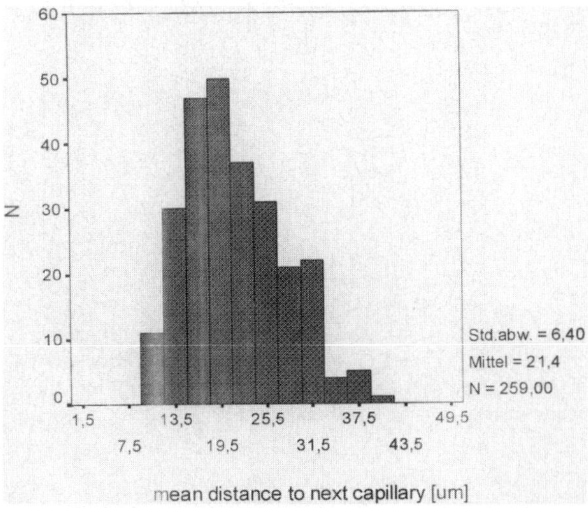


*Figure 6.* Histogram of "mean distance to the next capillary or vessel" of all eyes (temporal area).

formed by measuring topically the laser Doppler frequency shift of retinal area of $2.1 \times 0.7$ mm with a spatial resolution of 10 $\mu$m. Thus, a functional image of the retinal vasculature was established showing

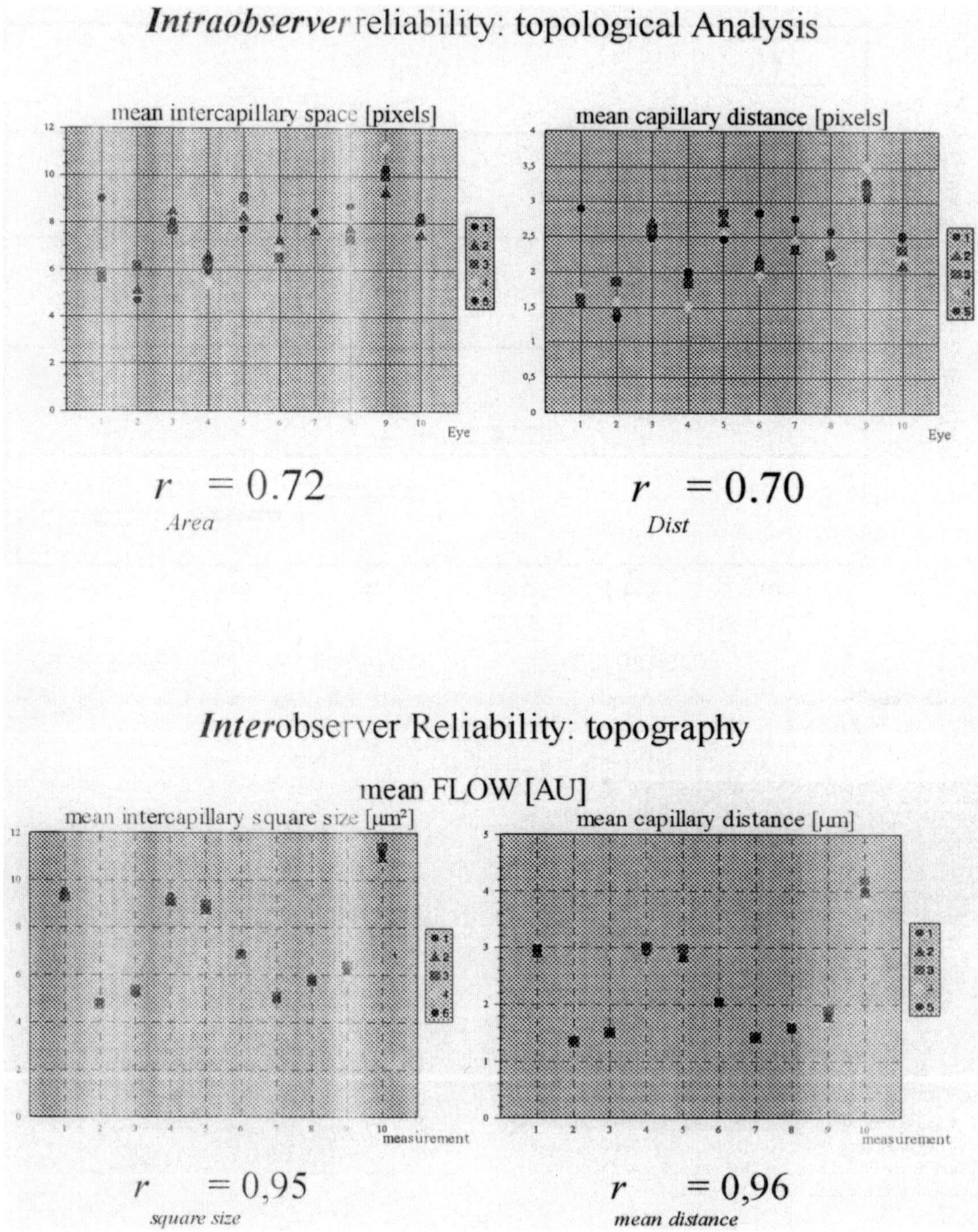
Intraobserver reliability: topological Analysis
mean intercapillary space [pixels]
mean capillary distance [pixels]
Eye
Eye
r = 0.72
Area
r = 0.70
Dist
Interobserver Reliability: topography
mean FLOW [AU]
mean intercapillary square size [µm²]
mean capillary distance [µm]
measurement
measurement
r = 0,95
square size
r = 0,96
mean distance

*Figure 7.*

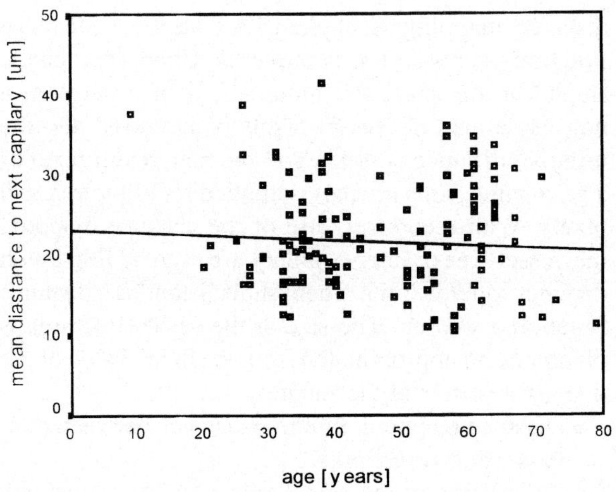


*Figure 8.*

perfused vessels and capillaries in high local resolution. The obvious coincidence between the visible vessels and capillaries in the fluorescein angiogram and the corresponding 'perfusion image" strongly suggests that the presented method allows to image 2-dimensionally perfused vessels and capillaries in a high local resolution.

B. *Calculation of the capillary density.* Based on the *perfusion image* encoded by the laser Doppler frequency shift and the automatic segmentation of the vessel and capillaries the distance between all retinal sites and the next lying capillary was calculated. A differentiation between arteries and veins was not performed. Thus, the data of the distances of all retinal sites to the next capillary or vessel allowed a second type of functional imaging based on the capillary density. Figures 2c and 3c give the functional images in respect to capillary density. Please note, that the pixels of the "distance map" becomes brighter when the distance to the next capillary increases.

*Reliability of quantitative measurements.* The *intra*observer reliability was 0.78 measuring different parts of the retina at different sessions. This allowed reproducible quantitative measurements. The observed reproducibility was in the same range than published by other investigators when measuring blood flow by Scanning Laser Doppler Flowmetry [8–10]. The *inter*observer reliability of AFFPIA analyzing perfusion maps was very high (0.95) due to the largely automatic procedure.

*Quantitative measurement of retinal density.* We observed in the juxtapapillary retina a *mean* capillary distance of 21 $\pm$ 7 $\mu$m. We found no significant difference between the capillary density of the juxtapapillary temporal and nasal area. The capillary density decreased slightly with age, but reached no statistical signifcance.

*Imaging of the capillary density.* In addition to the "perfusion map" which was coded by the laser Doppler frequency shift of each pixel, a "distance map" was created, in which the *distance to the next capillary* of each retinal site was visualized by color coding. The greater the distance to the next capillary is, the brighter the pixel is imaged. Thus, retinal areas with increased distances to the next nourishing capillaries or vessels become visible. This feature might be helpful for marking retinal areas with rarefication of retinal capillaries. Figure 3c depicts the "distance image" of the eye with an arterial embolus in a arterial branch. In this image the triangular area distal from the embolus showed a bright area indicating increased distances to the next lying perfused capillary or vessel. The bright area marks a rarefication of perfused structures indicating low perfusion in this area.

*Limitations.* The major technical limitations of the presented method are the data acquisition time of 2 sec and the proper focussing. The method is only applicable in subjects with the capability to focus a target for 2 sec without eye movements. Focussing with the HRF was performed by sharpening the live image on the computer screen. In case of incorrect focussing of the retina the images were not valid.

Due to the confocal technique the thickness of the measured layer is 300 $\mu$m Correct focussing of the retina (= limitans interna in the middle of the confocal plane) resulted in sharp perfusion images with visible capillaries.

In conclusion functional imaging of vessels and capillaries by "Scanning Laser Doppler Flowmetry" in combination with the "automatic full-field perfusion image analyzer" (AFFPIA) allowed a functional imaging of the retinal vasculature.

## Appendix

*Description of the capillary and vessel enhancement*

The vessel enhancement algorithm is based on an iterative nonlinear contrast transformation of the laser Doppler shift image. The input image $f(N \times M)$ is filtered with a Gaussian filter in order to lessen the noise. The filtered image $f_F(N \times M)$ is binarized with a threshold function. The enhancement of capillaries in this image is performed by this: A mask $h_0$ with size $n \times n$ where $n = 2j + 1$, $j = [1, \ldots, 4]$. In the mask four direction fields are given 11 with the directions 0°, 45°, 90°, 135°. Each Pixel (in one direction field are $n$ pixels) of one direction field is summed up, which is denoted with $s_i$. Let the minimum and maximum of direction sum be $s_{\min} = \min_{\forall i}(s_i)$ and $s_{\max} = \max_{\forall i}(s_i)$. The contrast 12 is defined as

$$C = (s_{\max} - s_{\min} s)/(s_{\max} + s_{\min})$$

and the contrast transformation as

$$f_{ij} = \lambda \cdot f_{ij} \cdot C^{\mu},$$

where $f_{ij}$ ($i \in [(n-1)/2, \ldots N - (n-1)/2]$, $j \in [(n-1)/2, \ldots, M - (n-1)/2]$) denotes the reference point of the middle point of the mask. The $\lambda$ and the $\mu$ are experimental constants image.

The reference point of the mask is shifted from pixel to pixel on the image $f_f(NM)$. To each pixel $f_{ij}$ the direction mask $h_{n=3}$ is applied and the maximal and minimal direction values $s_{\max}$, $s_{\min}$ computed. The reference pixel $f_{ij}$ is changed by the contrast transformation, so it results in the image $f_{Ci=1}$. This contrast transformation is applied on the image $f_{Ci=1}$ again, 3 or 4 times, so that in every iteration the size of the mask to the next odd number changes, and the number of the direction is four and it stays unchanged. The transformation steps is repeated iteratively on the new image $f_{Ci=1}$ with the new mask $h_{n+2}$ until $n = 9$ lid the subresult images $f_{Ci=2}$, $i \in 11 - -13$ are established. This iterative contrast transformation results in an image $f_C = f_{Ci=4}$, which shows the capillaries. The image $f_C$ is added to the image $f_B$ and is binarized with a threshold function Th( ) again. Thus, all vessels and capillaries become visible on the result image $f_R = Th(f_B + f_C)$.

*Distance mapping*

The vessel-free area is an important feature of the circulation on the eye background. First the so called distance mapping is applied. All arbitrary distance function is chosen for example the Chebyshev or the Euclidean distance, the minimal distance between a non-vessel and all vessel pixels is measured and the minimal distance is ordered to the non-vessel pixel as a gray value. This process is applied for all non-vessel pixels. With recursively use of the distance mapping the vessel-free or inter-capillary area can be filled with disjunct squares if the Chebyshev distance as distance function is chosen. The size of the vessel-free area is obtained and approximated from the mean value of the five greatest size of the squares.

- Distance mapping: minimal distance between non-vessel and vessel pixel.
- Calculating of the vessel-free area 11: recursively use of the distance mapping by filling with disjunct squares (Chebyshev distance)

$$d_{CS(px1,y1,px2,y2)} := \max(|x1 - x2|, |y1 - y2|)$$

is the distance function

1 $P_v :=$ {set of vessel points on the image}
2 $P_{nv} :=$ {set of non-vessel image points}
3 $\forall p \in P_{nv} p \leftarrow p^A := \min_{\forall p v}(d_{CS}(p, p_v))$
4 $P_M := \{p \in P_{nv} \mid p = \max_{\forall p}(p^A)\}$
5 $P_R(p_i \in P_M) := \{p_j \in P_{nv} \mid d_{CS}(p_i, p_j) \le p_i^A\}$
6 if $(\exists p_m \in P_M \wedge P_R(p_m) \cap P_v = \emptyset)$ then $P_v := P_v \cup P_R(p_m)$; $P_{nv} := P_{nv} \backslash P_R$; $P_M := P_M\{p_m\}$ goto 6
7 if $(\exists p^A : p^A > 1)$ goto 3

- Mean value of the 5 greatest size of the squares.

## References


1. Arend O, Remky A, Redbrake C, Arend S, Wenzel M, Harris A. Retinal hemodynamics in patients with normal pressure glaucoma. Quantification with digital laser scanning fluorescein angiography. Ophthalmologe 1999; 96(1): 24–29.
2. Michelson G, Welzenbach J, Pal I, Harazny J. Automatic full-field perfusion image analyze. Brit J Ophthalmol 1998; 1294–1300.
3. Michelson G, Zinser G, Schmauß B. Verfahren und Gerät zur Messung der Fließgeschwindigkeit, insbesondere des Blutes. 1.6.94, European Patent No. 94108430.3.
4. Michelson G, Groh M, Langhans M, Schinauss B. Zweidimensionale Kartierung der retinalen und papillären Mikrozirkulation mittels Scanning-Laser-Doppler-Flowmetrie. Klin Monatsbl Augenheilkunde 1995; 207: 180–190.
5. Michelson G, Scmauß B. Two-dimensional mapping of the juxtapapillary perfusion by scanning laser Doppler flowmetry. British J Ophthalmol 1995; 79: 1126–1132
6. Michelson G, Schmauß B, Langhans MJ, Harazny J, Groh MJM. Principle, validity, and reliability of scanning laser Doppler flowmetry. J Glaucoma 1996; 5: 99–105.

7. Bonner RE, Nossal R. Principles of laser-Doppler flowmetry. In Laser-Doppler Blood Flowmetry, Vol. 7. Sheperd Alber P, Öberg P Ake (eds) Boston: Kluwer Academic Publishers, 1990; 17–45.
8. Strenn K. Menapace R, Rainer G, Findl O, Woltzt M, Schmetterer L. Reproducibility and sensitivity of scanning laser Doppler flowmetry during graded changes in Po2. Br J Ophthalmol 1997; 81: 360–364.
9. Chauhan B. Confocal scanning laser Doppler flowmetry of the retina and optic nerve head. In: Anderson DR, Drance SM (eds) Encounters in Glaucoma Research 3. How to Ascertain Progression and Outcome. Amsterdam: Kugler 1996; 263–276.
10. Chauhan B, Smith FM. Confocal scanning laser Doppler flowmetry: experiments in a model flow system. J Glaucoma 1997; 6: 237–245.
11. Pál I. Michelson G, Niemann H, Welzenbach J: Erkennung von Mikrozirkulationsstörungen der Netzhaut mittels "Scanning Laser Doppler Flowmetrie". In Lehmann T, Scholl I, Spitzer K (eds) Bildbverarbeitung für die Medizin: Algorithmen-Systeme-Anwendungen Proceedings des Aachener Workshops, Verlag Der Augustinus Buchh., Aachen, Nov. 1996; 89–94.
12. Kutka R, Stier S. Blood vessel and feature extraction based on direction fields. In: Proceedings of the Workshop on Image Registration and Segmentation, Brussels, April 1994.
13. Zhou RW, Quek C. Ng GS. A novel single-pass thinning algorithm and an effective set of performance criteria. IEEE Trans on Pattern Recognition Letters 1995; 16: 1267–1275.

International Ophthalmology **23:** 337–342, 2001.
J.R. Sampaolesi (ed.), Laser Scanning: Update 1, 155–160.




# Pulsatile ocular blood flow in patients with pseudoexfoliation

Andrea Mistlberger[1], M. Gruchmann[1], W. Hitzl[2] & G. Grabner[1]
[1]*Department of Ophthalmology, County Hospital Salzburg* ; [2]*Office for Applied Statistics, Salzburg*

*Key words:* pseudoexfoliation, glaucoma, nerve fiber analysis, pulsatile ocular blood flow

**Abstract**

*Purpose:* Pseudoexfoliation syndrome is one of the most frequent causes of open-angle glaucoma and is statistically significant associated with a high risk of hypertension, angina, myocardial infarction or stroke and retinal vein thrombosis. The aim of this study was to evaluate the pulsatile ocular blood flow (POBF) in pseudoexfoliation syndrome without (PEX) and with glaucoma (PEG). *Methods:* Seventeen eyes with PEX, 17 with PEG and 11 normal eyes of age-matched patients were enrolled. A complete ophthalmological examination included measuring the POBF with the Langham Pneumotonometer as well as the nerve fiber layer thickness by scanning laser polarimetry (GDx$^{TM}$). *Results:* The blood flow parameters, pulse volume and POBF, were statistically significant different between normals and patients with PEG ($p < 0.003$, t-test). A negative correlation between the intraocular pressure and the POBF was found for all eyes tested. Analysis of GDx? parameters showed a negative correlation for the "number" with the POBF and a positive one for ellipse modulation. *Conclusion:* Although pseudoexfoliation is reported to be a systemic disease measurement of the POBF could not detect any difference between normals and PEX, but was statistically significant different in PEG. Assessments of nerve fiber layer thickness as determined by scanning laser polarimetry also showed a correlation with POBF in some parameters tested.

## Introduction

Pseudoexfoliation syndrome (PEX) is one of the most frequent causes of open-angle glaucoma. In 1917 Lindberg reported cases of glaucoma showing thin, transparent membranes detached from the anterior lense surface [1]. Vogt called this "senile exfoliation", very often found in combination with glaucoma [2]. In order to distinguish this disease from genuine exfoliation material seen in glass blowers the term "pseudoexfoliation" was created by Dvorak–Theobald [3].

PEX has been studied extensively over the last years, aetiology and pathogenesis are still unknown, however. Histopathologic findings and electron microscopy verified the systemic character of the syndrome [4]. Pseudoexfoliation material could be identified in the iris, the ciliary body, the anterior lens capsule and – interestingly – also in the skin, lung, heart, liver, kidneys and meninges.

Vogt also first described the association of PEX and glaucoma [2]. Today the high risk of developing glaucoma in patients with PEX is extensively documented [5, 6] and some typical peculiarities are well known: intraocular pressure in PEG is generally higher than in primary open-angle glaucoma, foudroyant progression of visual field loss is more frequent, topical therapy is often less effective and cataract surgery has a higher rate of complications [7, 8].

The aim of this study was to evaluate the pulsatile ocular blood flow (POBF) in PEG and PEX, since it has been reported as a systemic vascular disease [9, 10]. Nerve fiber layer thickness, a parameter for the early detection and follow-up of glaucoma was – to our knowledge – not previously shown to correlate with POBF [11, 12].

## Methods

Forty-five eyes were enrolled and according to the inclusion and exclusion criteria they were separated into 3 groups: patients with PEX, PEG and age-matched normals.

Visual acuity was better than 0.4, refractive error not exceeding ±5.00 diopter sphere and ±2.00 diopters cylinder. The axial length was limited to 28 mm. Any kind of previous intraocular surgery, coexisting retinal diseases or the use of vasoactive systemic therapy were considered to be exclusion criteria.

All patients underwent a complete ophthalmic examination including slit-lamp biomicroscopy in miosis and mydriasis to identify even minor forms of pseudoexfoliation and for the evaluation the optic nerve head, optic nerve head stereophotography, standard Goldmann applanation tonometry, achromatic automated perimetry, A-scan echography, nerve fiber analysis by scanning laser polarimetry and pulsatile ocular blood flow measurements.

Normal subjects had an intraocular pressure (IOP) of 21 mm Hg or less, a normal disc and a perimetry with a Glaucoma Hemifield Test (GHT) within normal limits. A normal disc was defined as having a cup/disc ratio ≤0.6, an intact neuroretinal rim without notches or peripapillary hemorrhages.

For inclusion of patients in the PEX group, pseudoexfoliation material (any degree) had to be found with biomicroscopy. In all other respects they met all criteria of normal subjects as mentioned above.

Patients with PEG had at least once an IOP of 22 mmHg or more as well as an abnormal optic disc and visual field testing. Antiglaucomatous therapy was acceptable for inclusion.

Achromatic automated perimetry was performed with the Humphrey Field Analyzer (Humphrey Systems, Inc, Dublin, CA; program 30-2). Visual field reliability criteria included fewer than 33% fixation losses and false-positive and false-negative rates.

Nerve fiber layer thickness was measured with the Nerve Fiber Analyzer (GDx$^{TM}$, Laser Diagnostic Technologies, Inc. San Diego, CA) with undilated pupil [13]. This technology measures the thickness of nerve fiber layers which is correlated with the phase shift of the reflected light beam. This is also known as retardation caused by the birefingence of nerve fiber layers [11, 12, 14–16]. For the correlation with POBF only the global indices (number, maximal modulation, ellipse modulation, average thickness, ellipse average) were used for the calculations [12].

POBF was measured using the Ocular Blood Flow System of Langham (O.B.F. Labs Ltd, Wiltshire, UK), a pneumatic tonometer evaluating volume changes due to the pulse rate [16-18]. After topical instillation of a drop of anesthetic (oxybuprocaine) three measurements were recorded. The means of the blood flow parameters were used for all further calculations.

## Statistics

Statistical analysis was computed with Statistica 5.5 [19]. In order to compare the means in all three groups dependent and indepedent (one- and two-sided) t-tests for the two-sample case were applied [19]. Multiple comparisons demand the use of a type I error rate adjustment in order to protect against an increase of the overall type I error rate. Assuming statistical independence, the individual type I error rates based on the Bonferroni adjustment to end up with reasonable overall type I error rates was computed [20]. Correlations between several parameters had a 95% confidence interval for the correlation coefficients.

## Results

Forty-five eyes of 38 patients were enrolled in the study. Seventeen carried the diagnosis of pseudoexfoliation with and 17 without glaucoma, 11 normal eyes were used as a control group. There was no statistical difference between the three groups regarding age or visual field indices (Table1). Intraocular pressure, when measured by applanation and pneumatic tonometry, was equal for both techniques in each group ($p > 0.013$) (Table 2).

As was to be expected, the intraocular pressure measured with the POBF was highest in patients with PEG. Pulse volume and the POBF could also distinguish between normal eyes and those with PEG. There was no statistically significant difference between normals and PEX, and PEX and PEG regarding pulse amplitude, pulse volume, and pulsatile ocular blood flow (Table 3).

A moderate negative correlation between POBF and intraocular pressure was found for all eyes (n = 45), which supports the notion that low POBF is a result of high intraocular pressure and *vice versa* (Table 4).

Nerve fiber analysis demonstrated a positive correlation between "ellipse modulation" and POBF and

*Table 1.* Patient demographics and visual field indices

| | Normal n = 11 | PEX n = 17 | PEG n = 17 | Total n = 45 | p-value 1* | p-value 2* | p-value 3* |
|---|---|---|---|---|---|---|---|
| Gender | w = 4 | w = 15 | w = 10 | w = 29 | | – | – |
| | m = 7 | m = 2 | m = 7 | m = 16 | | | |
| Age | 64.6 ± 7.4 | 69.6 ± 7.5 | 67.6 ± 8.3 | 67.6 ± 7.8 | 0.1 | 0.33 | 0.47 |
| | (56, 79) | (53, 78) | (50, 78) | (50, 79) | | | |
| MD | −2.0 ± 3.3 | −0.94 ± 2.2 | −4.1 ± 8.2 | −2.4 ± 5.5 | 0.3 | 0.42 | 0.13 |
| | (−7.4, 2.15) | (−6.0, 3.2) | (−29.5, 1.6) | (−29.5, 3.2) | | | |
| CPSD | 1.6 ± 0.8 | 2.1 ± 1.2 | 2.5 ± 3.0 | 2.1 ± 2.0 | 0.29 | 0.35 | 0.60 |
| | (0.0, 3.1) | (0.4, 4.6) | (0.0, 12.0) | (0.0, 12.0) | | | |

MD = mean deviation, CPSD = corrected pattern standard deviation.
p-value 1: normal versus PEX-group; p-value 2: normal versus PEG-group; p-value 3: PEX- versus PEG-group.
* A difference is statistically significant, if and only if the corresponding p-value is less than $0.05/9 \simeq 0.006$.

*Table 2.* Applanation tonometry versus pneumatic tonometry

| | Normal n = 11 | PEX n = 17 | PEG n = 17 | Total n = 45 |
|---|---|---|---|---|
| IOP (mmHg) (OBF) | 13.7 ± 2.8 | 15.8 ± 2.6 | 22.5 ± 5.4 | 18.0 ± 5.5 |
| | (9, 17) | (10, 20) | (12, 34.5) | (9, 34.5) |
| AT (mmHg) (Appl.) | 12.6 ± 3.5 | 14.9 ± 3.6 | 21.0 ± 5.9 | 16.7 ± 5.7 |
| | (8.5, 17.6) | (7.4, 20.9) | (12.2, 33.8) | (7.4, 33.8) |
| p-value* | 0.81 | 0.1 | 0.26 | 0.09 |

p-values of corresponding dependent t-tests.
* A difference is statistically significant, if and only if the corresponding p-value is less than $0.05/4 \simeq 0.013$.

*Table 3.* OBF-Parameters

| | Normal n = 11 | PEX n = 17 | PEG n = 17 | Total n = 45 | p-value 1 | p-value 2 | p-value 3 |
|---|---|---|---|---|---|---|---|
| IOP (mmHg) | 12.6 ± 3.5 | 14.9 ± 3.6 | 21.0 ± 5.9 | 16.7 ± 5.7 | 0.1 | 0.0001* | 0.0004* |
| | (8.6, 17.6) | (7.4, 20.9) | (12.2, 33.8) | (7.4, 33.8) | | | |
| PA (mmHg) | 3.1 ± 0.7 | 3.2 ± 1.1 | 2.9 ± 1.0 | 3.1 ± 0.9 | 0.6 | 0.73 | 0.41 |
| | (2.1, 4.4) | (1.9, 5.4) | (1.1, 4.8) | (1.1, 5.4) | | | |
| PV ($\mu$l) | 8.6 ± 2.4 (5.4, 11.3) | 7.1 ± 2.5 (4.1, 11.1) | 4.9 ± 2.0 (1.6, 8.8) | 6.4 ± 2.6 (1.6, 11.1) | 0.2 | 0.0007* | 0.008 |
| | (5.4, 11.3) | (4.1, 11.1) | (1.6, 8.8) | (1.6, 11.1) | | | |
| POBF ($\mu$l/sec) | 17.5 ± | 16.0 ± | 11.8 ± | 14.8 ± | 0.4 | 0.003* | 0.012 |
| | (9.4, 23.8) | (9.1, 25.1) | (4.7, 22.6) | (4.7, 25.1) | | | |

IOP = intraocular pressure measured with the OBF; PA = pulse amplitude: maximum IOP – minimum IOP; PV = pulse volume: volume changes; POBF = pulsatiler ocular blood flow.
p-value 1: normal versus PEX-group; p-value 2: normal versus PEG-group; p-value 3: PEX versus PEG-group.
* A difference is statistically significant, if and only if the corresponding p-value is less than $0.05/12 \simeq 0.004$.

*Table 4.* Correlation between POBF and IOP measured with applanation tonometry (95% confidence interval for the correlation)

| | Normal n = 11 | PEX n = 17 | PEG n = 17 | Total n = 45 |
|---|---|---|---|---|
| POBF ($\mu$l/sec) AT (mmHg) | (−0.9, 0.2) | (−0.7, 0.2) | (−0.4, 0.5) | (−0.6, −0.1)* |

* Indicates a moderate negative correlation.

*Table 5.* Correlation between POBF and GDx$^{TM}$ parameters (95% confidence interval for the correlations)

| | Normal n = 11 | PEX n = 17 | PEG n = 17 | Total n = 45 |
|---|---|---|---|---|
| Number | (−0.7, 0.4) | (−0.6, 0.3) | (−0.6, 0.3) | (−0.6, −0.1)* |
| Max Mod | (−0.6, 0.5) | (−0.2, 0.7) | (−0.2, 0.7) | (−0.1, 0.5) |
| Ellipse Mod | (−0.4, 0.7) | (0.1, 0.8)** | (−0.1, 0.7) | (0.2, 0.7)** |
| Av Thick ($\mu$m) | (−0.6, 0.5) | (−0.1, 0.8) | (−0.5, 0.5) | (−0.1, 0.5) |
| Ellipse Av ($\mu$m) | (−0.7, 0.5) | (−0.7, 0.8) | (−0.5, 0.4) | (−0.1, 0.5) |

* Indicates a moderate negative correlation.
** Indicates a moderate positive correlation.
Max Mod = maximum modulation; Ellipse Mod = ellipse modulation; Av Thick = average thickness; Ellipse Av = ellipse average.

a negative one between "the number" and POBF (Table 5).

## Discussion

Nowadays the definition of glaucoma as an "optic neuropathy" associated with characteristic glaucomatous changes of the optic nerve head is well accepted. The pathogenesis of this disease, however, generates still a lot of discussion and great research efforts. The intraocular pressure is considered a major, but only one of many risk factors. Various investigations have studied neuroprotection and ocular blood flow [21]. Several technologies are evaluated in early clinical studies to measure ocular hemodynamics and the effect of intraocular pressure-reduction by medication or surgery on blood flow to the eye [10, 22, 23].

For this study the POBF was measured with a pneumatic tonometer. Changes in intraocular blood volume, presumably in the choroid, that occur with each pulse are detected [16-18, 24].

Former studies report the influence of IOP on POBF in glaucoma in general, or using antiglaucomatous Medication [24-27]. James et al. and Quaranta et al. were able to prove a reduction of POBF in patients with primary open and low-tension glaucoma [26, 27]. On the other hand POBF increases in patients after trabeculectomy in a supine position [22].

Pseudoexfoliation syndrome is one of the most frequent causes of secondary open- angle glaucoma and glaucoma itself is a risk factor for the occurrence of retinal vein thrombosis [4, 9, 28]. This study was designed to demonstrate possible variations in blood flow in patients with PEX.

Only for pulse volume and POBF a difference between normals and patients with PEG was confirmed, all other OBF-parameters show no statistically significant difference between the groups (Table 3). For the whole cohort POBF was negatively correlated with IOP (Table 4), as an increase of the IOP is associated with a decrease of POBF. In PEX syndrome these patients carry a risk for higher intraocular pressure, but not necessarily for a danger of retinal vein occlusions due to PEX-material altering the ocular vessels, at least as measured by the OBF.

This result is in good accordance with descriptions of a so-called "PEX-vasculopathy". Helbig et al. reported histopathological damage of the wall of vessels by abnormal matrix-components leading to an anterior segment hypoxia [29]. So far such vascular changes have not been described for posterior segment vessels or the systemic circulation [28]. Another explanation could be that the OBF technology itself is not sensitive enough to detect such subtle vascular or hemodynamic alterations [10].

Furthermore, in this study a correlation between nerve fiber layer thickness and POBF could be demonstrated for the first time. Several studies on scanning laser polarimetry analyzed a variety of single parameters and of their combinations to find the one most predictive in early glaucoma diagnosis and follow-up [12, 30]. Out of the selected global indices "ellipse modulation" was positively and the "number" was negatively correlated with POBF. The "ellipse modulation" numerically represents the difference between the thickest and the thinnest part of the nerve fiber layer along the defined ellipse. In normals a large modulation is correlated with a higher blood circulation and a higher POBF. On the other hand, the worse the blood flow the higher "the number" and the likelihood of glaucoma, as reported in this study. Both correlations confirm the interconnection between nerve fiber layer thickness, blood flow and glaucomatous damage.

In summary, using different diagnostic technologies, it was possible to distinguish normals from glaucoma patients. It was not possible, however, to confirm the reported vascular risk factor of patients with pseudoexfoliation only, by using these devices.

## References


1. Lindberg JG. Clinical investigations on depigmentation of the pupillary border and translucency of the iris. In cases of senile cataract and in normal eyes in elderly persons. Academic Dissertation, Helsinki, 1917. English translation by Tarkkanen A, Forsius H. Acta Ophthal Suppl 1989; 190, vol. 66, University Press, Helsinki.
2. Vogt A. Ein neues Spaltlampenbild des Pupillengebietes: hellblauer Pupillensaumfilz mit Häutchenbildung auf der Linsenvorderkapsel. Klin Monatsbl Augenheilkd 1925; 75: 1-12.
3. Dvorak-Theobald G. Pseudoexfoliation of the lens capsule: Relation to "true" exfoliation of the lens capsule as reported in the literature and role in its production of glaucoma capsulocuticulare. Am J Ophthalmol 1954; 37: 1-12.
4. Schlötzer-Schrehardt U, Koca MR, Naumann GOH, Volkholz H. Pseudoexfoliation Syndrome. Ocular manifestation of a systemic disease. Arch Ophthalmol 1992; 110: 1752-1756.
5. Brooks AMV, Gillies WE. The presentation and prognosis of glaucoma in pseudoexfoliation of the lens capsule. Ophthalmology 1988; 95: 271-276.
6. Naumann GOH, Schlötzer-Schrehardt U, Küchle M. Pseudoexfoliation syndrome for the comprehensive ophthalmologist. Ophthalmology 1998; 105: 951-968.
7. Wollensack J, Becker HU, Seiler T. Pseudoexfoliation syndrome and glaucoma. Does glaucoma capsulare exist? German J Ophthalmol 1992; 1: 32-34.
8. Freyler H, Radax U. Pseudoexfoliationssyndrom – ein Risikofaktor der modernen Kataraktchirurgie? Klin Monatsbl Augenheilkd 1994; 205: 275-279.
9. Mitchell P, Wang JJ, Smith W. Association of pseudoexfoliation syndrome with increased vascular risk. Am J Ophthalmol 1997; 124: 685–687.
10. Repo LP, Suhonen MT, Teräsvirta ME, Koivisto KJ. Color Doppler imaging of the ophthalmic artery blood flow spectra of patients who have had a transient ischemic attack. Correlations with generalized iris transluminance and pseudoexfoliation syndrome. Ophthalmology 1995; 102: 1199–1205.
11. Weinreb RN, Shakiba S, Zangwill L. Scanning laser polarimetry to measure the nerve fiber layer of normal and glaucomatous eyes. Am J Ophthalmol 1995; 119: 627–636.
12. Weinreb RN, Zangwill L., Berry CC, Bathija, Sample PA. Detection of glaucoma with scanning laser polarimetry. Arch Ophthalmol 1998; 116: 1583–1589.
13. Hoh ST, Greenfield DS, Liebmann JM, Hillenkamp J, Ishikawa H, Mistlberger A, Lim ASM, Ritch R. Effect of pupillary dilation on retinal nerve fiber layer thickness as measured by scanning laser polarimetry in eyes with and without cataract. J Glaucoma 1999; 8: 159–163.
14. Anton A, Zangwill L, Emdadi A, Weinreb RN. Nerve fiber layer measurements with scanning laser polarimetry in ocular hypertension. Arch Ophthalmol 1997; 115: 331–334.
15. Choplin NT, Lundy DC, Dreher AW. Differentiating patients with glaucoma from glaucoma suspects and normal subjects by nerve fiber layer assessment with scanning laser polarimetry. Ophthalmology 1998; 105: 2068–2076.
16. Krakau CET. Calculation of the pulsatile ocular blood flow. Invest Ophthalmol Vis Sci 1992; 33: 2754–2756.
17. Krakau CET. A model for pulsatile and steady ocular blood flow. Graefe's Arch Clin Exp Ophthalmol 1995; 233: 112–118.
18. Langham M, To'mey K. A clinical procedure for measuring the ocular pulse. Exp Eye Res 1978; 27: 17.
19. StatSoft, Inc. STATISTICA for Windows [Computer program manual]. Tulsa, OK: 1999; StatSoft, Inc.
20. Miller, J. Simultaneous Statistical Inference, 2nd ed., 1981 New York: Springer Verlag.
21. McKinnon SJ. Glaucoma, apoptosis, and neuroprotection. Current Opinion in Ophthalmol 1997, 8; II: 28–37.
22. James CB. Effect of trabeculectomy on pulsatile ocular blood flow. Br J Ophthalmol 1994; 78: 818–822.
23. Willianson TH, Harris A. Ocular blood flow measurement. Br J Ophthalmol 1994; 78: 939–945.
24. Silver DM; Farrell RA, Langham ME, O'Brien V, Schilder P. Estimation of pulsatile ocular blood flow from intraocular pressure. Acta Ophthalmol Suppl 1989; 191: 25–28.
25. Boles Carenini A, Sibour G, Boles Carenini B. Differences in long-term effect of timolol and betaxolol on pulsatile ocular blood flow. Surv Ophthalmol 1994; 38: 118–124.
26. Quaranta L, Manni G, Donato F, Bucci MG. The effect of increased intraocular pressure on pulsatile ocular blood flow in low tension glaucoma. Surv Ophthalmol 1994; 38: 177–182.
27. James CB, Smith SE. Pulsatile ocular blood flow in patients with low tension glaucoma. Br J Ophthalmol 1991; 75: 466–470.
28. Cursiefen C, Händel A, Schönherr U, Naumann GOH. Das Pseudoexfoliationssyndrom bei Patienten mit retinaler Venenast- und Zentralvenenthrombose. Klin Monatsbl Augenheilkd 1997; 211: 17–21.
29. Helbig H, Schlötzer-Schrehardt U, Noske W, Kellner U, Foerster M, Naumann GOH. Anterior-chamber hypoxia and iris

vasculopathy in pseudoexfoliation syndrome. Ger J Ophthalmol 1994; 3: 148–153.
30. Choplin NT, Lundy DC, Dreher AW. Differentiating patients with glaucoma from glaucoma suspects and normal subjects by nerve fiber layer assessment with scanning laser polarimetry. Ophthalmology 1998; 105: 2068–2076.

*International Ophthalmology* **23:** 343–346, 2001.
*J.R. Sampaolesi (ed.), Laser Scanning: Update 1*, 161–164.




# Oximetry with a multiple wavelength SLO

Ross A. Ashman, Fred Reinholz & Robert H. Eikelboom
*Lions Eye Institute, Perth and Centre of Ophthalmology and Visual Science, UWA, 2 Verdun St Nedlands 6009, Western Australia*

*Key words:* multiple wavelength, oximetry

## Abstract

Oximetry may be useful for understanding pathologies of the eye. We use a prototype scanning laser ophthalmoscope capable of simultaneous multiple wavelength imaging to record fundus images. The system is run under non-confocal conditions in using slit apertures with a width of up to 600 $\mu$m. The laser lines launched into the SLO were the 633 nm line of a HeNe-laser, and the 815 nm line from a tunable (700 to 900 nm) cw Ti:Sapphire-laser. These wavelengths were selected because of their availability and absorption characteristics. The difference in absorption at these two wavelengths is used to assess blood oxygen content. Images were averaged to improve the signal to noise ratio.

This simultaneous method of measuring oxygen content may be preferred to other techniques such as sequential SLO imaging at different wavelengths, or spot analysis using a modified fundus camera. The advantages of this technique are that image registration is not required, and a large area of the retina can be assessed concurrently.

## Introduction

It is known that any alteration in blood circulation such as seen in diabetic retinopathy, hypertension, sickle cell disease, and vascular occlusive diseases results in functional impairment and extensive retinal tissue damage [1]. However little is know about retinal oxygen distribution or consumption and how it changes in response to disease. Measuring these changes non-invasively may improve the ability to diagnose, or monitor the progression of, such eye diseases.

Non-invasive optical methods have been reported [1–7]. These techniques depend on quantifying differences between Hb and HbO light absorption. Methods include two, three and four wavelength approach with fundus cameras and SLO's. SLO's have been used in serial acquisition.

## Method

An SLO set-up for recording multiple wavelengths simultaneously was used [8] (Figure1).

The system is run under non-confocal conditions in using slit apertures with a width of up to 60 $\mu$m. The use of large apertures improves the signal to noise ratio and the detection of absorption. The intensities of individual wavelengths can be adjusted using laser intensity stabilisers (noise eaters). Images are recorded with a size of 512 $\times$ 512 pixels and an 8 bit dynamic range in each channel and averaged over 7 frames.

Three sets of recordings were made for each patient. The first set involved breathing room air. A second series of images was then recorded whilst breathing a mixture of 10% $O_2$ in $N_2$ delivered at a rate of $\sim$10 L/min. A final series is recorded using 100% $O_2$ at the same flow rate. The blood $O_2$ saturation levels were monitored with a pulse oximeter for each series of readings.

The laser lines used were the 633 nm line of a HeNe-laser, and the 815 nm line from a tunable (740 to 840 nm) cw Ti:Sapphire-laser. The 815 nm line corresponds with an isobestic point in the shows no change in absorption with blood oxygen concentration [9]. The 633 nm line however does show a change in absorption with blood oxygen content. The differ-

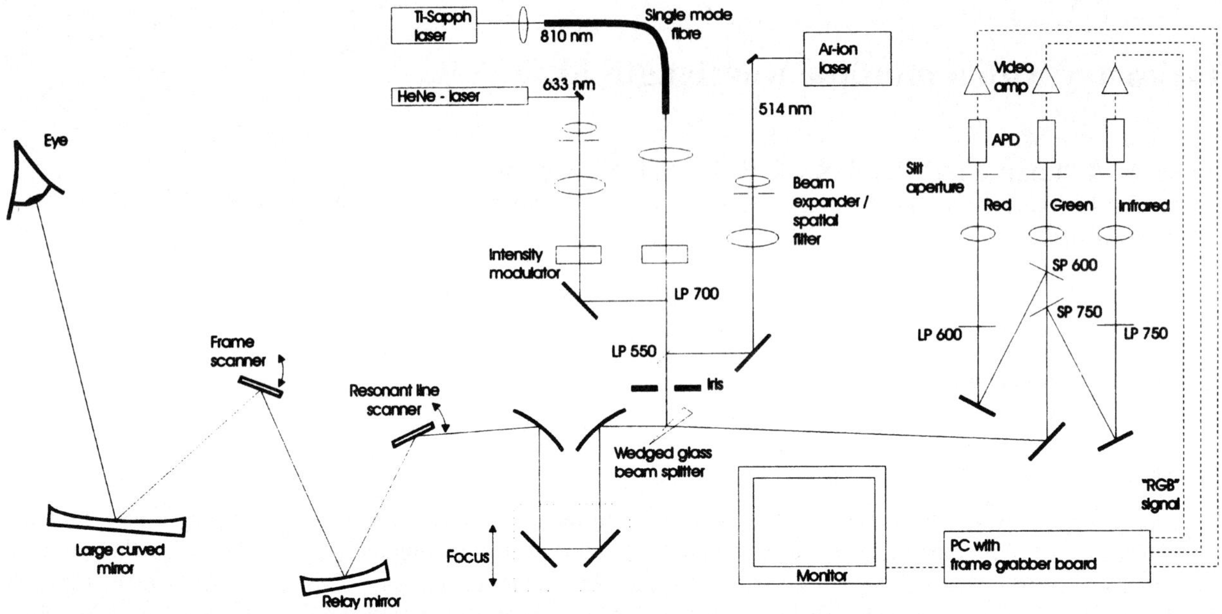


*Figure 1.*

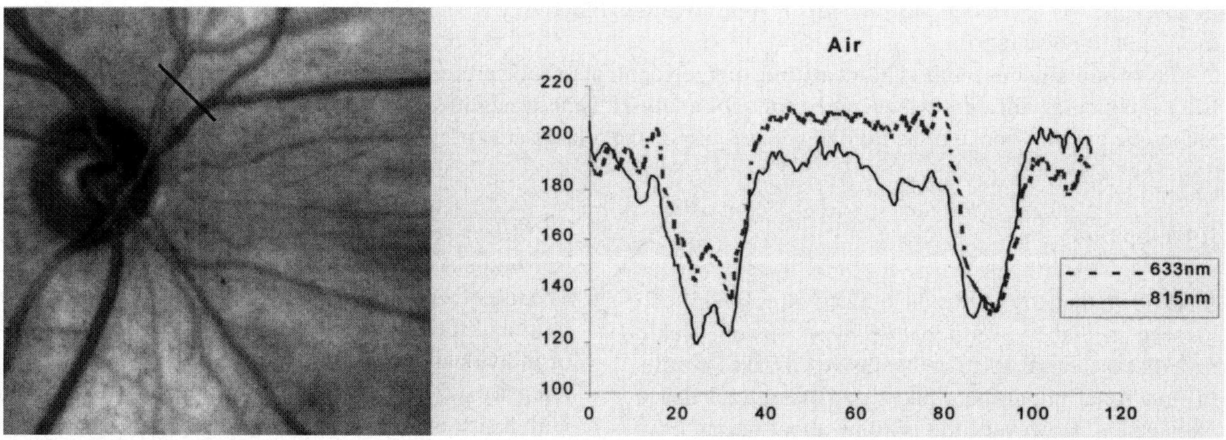


*Figure 2.*

ence in absorption at these two wavelengths is used to assess blood oxygen content.

The line profile of a retinal vein and artery of similar size was taken. Figure 2 and the difference in 633 nm and 815 nm profiles was then determined. This was repeated for each set of gas mixtures.

## Results

The results are shown in Figure 3. From the line profiles for breathing gases with various oxygen concentrations there was no measurable change in blood oxygen content with the SLO. For the pulse oximeter there was also no appreciable difference in oxygen saturation detected with an oxygen concentration of around 98% in all cases.

## Conclusions

Difficulties were encountered in achieving any significant changes in oxygen levels, which could be measured with a pulse oximeter. Qualitatively retinal veins could be distinguished from retinal arteries. However no quantitative assessment of simultaneous dual wavelength oximetry with an SLO could be made

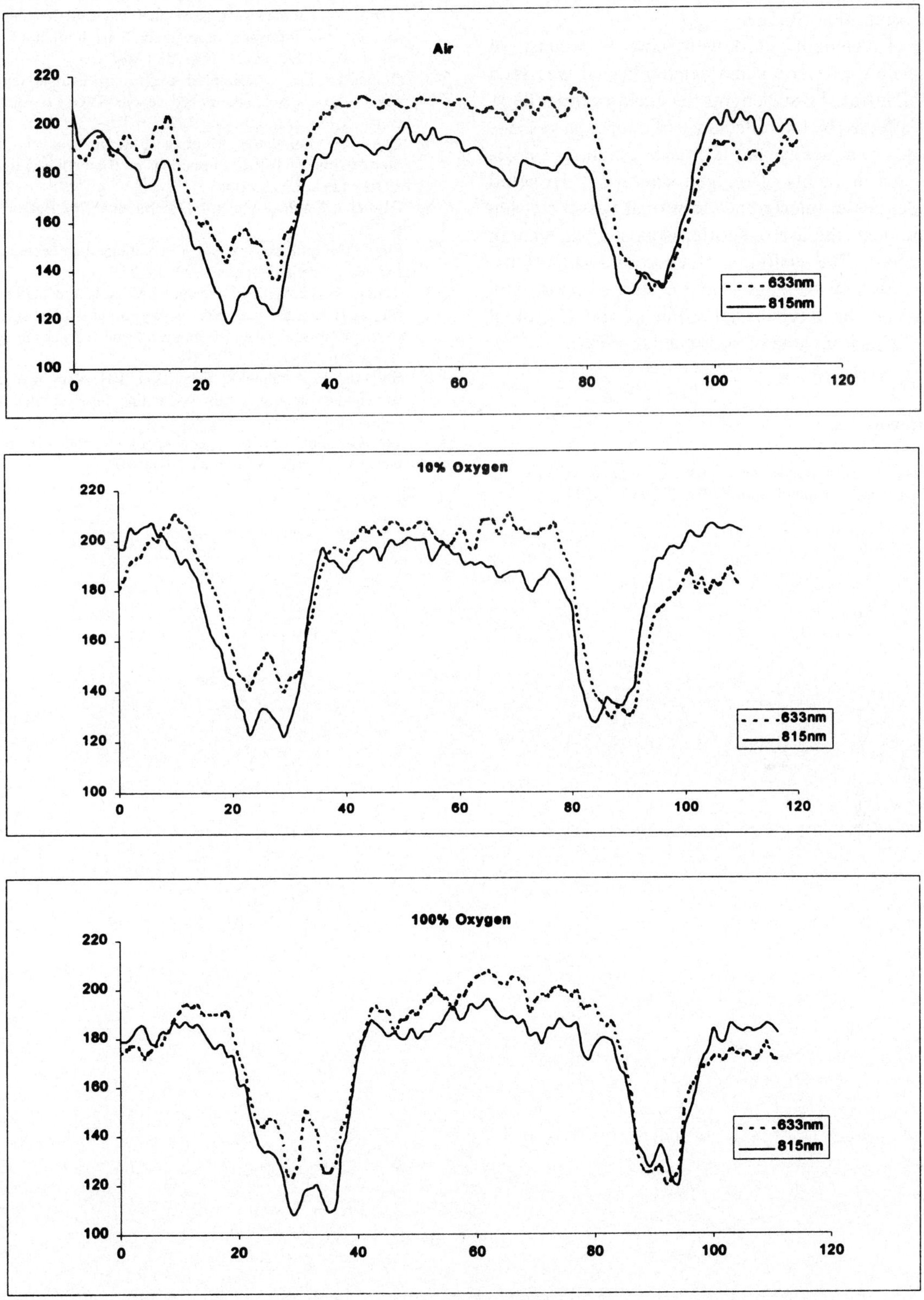
Air
220
200
180
160
140
120
100
0
20
40
60
80
100
120
633nm
815nm
10% Oxygen
220
200
180
160
140
120
100
0
20
40
60
80
100
120
633nm
815nm
100% Oxygen
220
200
180
160
140
120
100
0
20
40
60
80
100
120
633nm
815nm

*Figure 3.*

because of the inability to induce any measurable oxygen saturation change.

The wavelengths used were chosen because of their spectral properties and availability. It was seen that with infrared wavelengths the background reflectance is affected by the reflectance of choroidal vessels. This adds complexity and a greater potential for error in measurements of oxygen saturation. To avoid choroidal vessel interference in retinal vessel oxygen measurements the use of shorter isobestic wavelength is preferred. The visibility of choroidal vessels using the present wavelengths does offer the opportunity to examine the oxygen saturation of the choroidal vessels. This is an area of continued research.

## References

1. Delori FC, Noninvasive technique for oximetry of blood in retinal vessels. Applied Optics 1988; 27: 1113–1125.
2. Beach JM, Schwenzer KJ, Srinivas S, Kim D, Tiedman JS, Oximerty of retinal vessels by dual wavelength imaging: calibration and influence of pigmentation. Journal of Applied Physiology 1999; 86(2): 748–758 (Abstract).
3. Bongard F, Lee TS, Simultaneous *in vivo* comparison of two-versus three-wavelength mixed venous(SVO2) oximetry catheters. J Clinical Monitoring 1995; 5: 329–334.
4. Cohen AJ, Laing RA, Multiple scattering analysis of retinal blood oximetry. IEEE Transactions on Biomedical Engineering 1976; bme-23: 391–400.
5. Gloster J, Fundus oximetry. Experimental Eye Research 1967; 6: 187–212.
6. Smith M, Optimum wavelength combinations for retinal vessel oximetry. Applied Optics 1999; 38: 258.
7. Thamm E, Schweitzer D, Hammer M, A data reduction scheme for improving the accuracy of oxygen saturation calculations from spectroscopic *in vivo* measurements. Physics in Medicine & Biology 1998; 43; 1401–1411.
8. Reinholz F, Ashman R, Eikelboom RH, Simultaneous three wavelength imaging with a scanning laser ophthalmoscope. Cytometry 199; 37: 165–170.
9. van Assendelf OW, In: Spectroscopy of haemoglobin derivatives. CC Thomas, Springfield, IL, 1970.

International Ophthalmology **23:** 347–353, 2001.
J.R. Sampaolesi (ed.), Laser Scanning: Update 1, 165–171.




# A new method for the measurement of oxygen saturation at the human ocular fundus

D. Schweitzer, E. Thamm, M. Hammer & J. Kraft
*Department of Experimental Ophthalmology, University of Jena, Germany*

## Introduction

The investigation of changes in the metabolism as a functional diagnosis is a promising way for the detection of early reversible pathological alterations before they get manifest. Such a functional parameter of the microcirculation is the oxygen saturation of blood. The oxygen saturation in an artery describes in the connection with the blood flow the supply of oxygen in a certain region of the fundus. The consumption of oxygen is proportional to the product of the arteriovenous difference in the oxygen saturation and the blood flow.

According to the definition, the oxygen saturation is the ratio of the content of saturated blood in relation to all blood which is able to transport oxygen within the considered blood layer.

The spectral differences of saturated blood and of reduced blood are the basic information for the calculation of the oxygen saturation. Applying optical methods, the oxygen saturation of the blood in vessels at the living human eye can be determined only in reflection. This limitation makes an exact determination of the oxygen saturation very difficult.

## Components of the measured reflected light

Let us consider a vessel, located in front of a reflecting background (Figure 1). The incident light is influenced by the vessel-wall, the blood, and the background in different manner. As the result, the reflected light consists of light, reflected at the background and transmitted through the vessel. A second part of light is only backscattered at the erythrocytes in the blood column. These parts are influenced by the absorption and the scattering of whole blood. The last important part is the light which is spectacularly reflected at the vessel wall.

For the exact determination of the oxygen saturation, we have to study the following problems:

1. How depend the spectra of the transmitted light and of the internal reflected light on physiological parameters like thickness of the blood layer, hematocrit, and blood velocity?
2. How is the relation between the parts of the reflected light inside a vessel and how does it change in dependence of the vessel diameter and of the background reflectance.

## Spectral properties of whole blood

### *Transmission spectra*

In the following, the transmission and reflectance spectra of whole blood are given. For the measurements of the spectra, blood was filled in quartz-cuvettes, having different calibrated thickness between 25 $\mu$m and 200 $\mu$m. The measured spectra of the transmitted light were normalised on the spectrum, measured from the water-filled cuvettes. As the saturation of blood is a stable stage, the investigations were performed on blood with 100% saturation.

The dependence of transmission spectra on the thickness of the blood layer is demonstrated in Figure 2. When the thickness of the blood layer in the cuvette increases, the transmission decreases at all wavelengths. Considering the transmission at a certain wavelength, a non-linear relation exists between the negative logarithm of the transmission and the thickness of the blood layer (Figure 3). As Lambert–Beer's law would require a linear dependence between both sizes, it is shown that this law does not yield in whole blood.

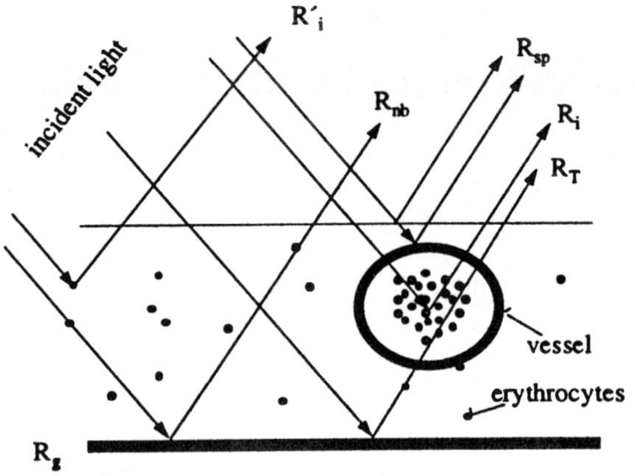


*Figure 1.* Parts of measured light.

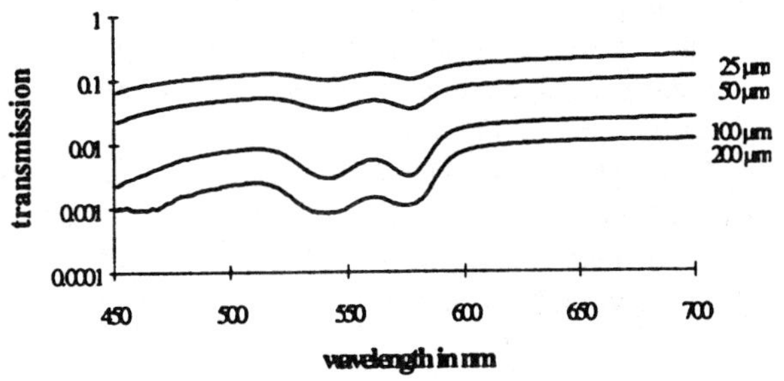


*Figure 2.* Transmission spectra of whole blood in dependence of layer thickness.

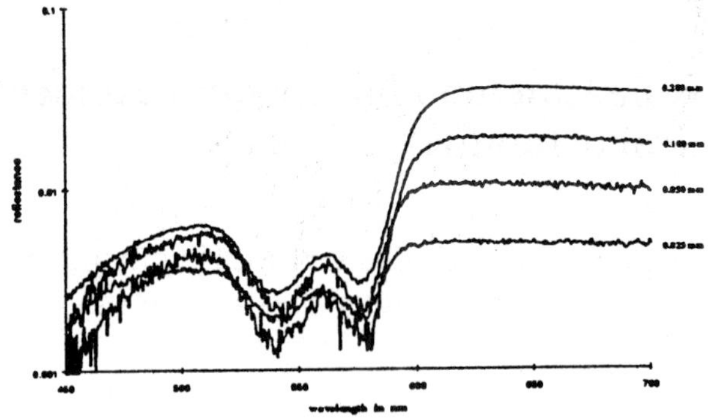

*Figure 4.* Reflectance spectra of whole blood parameter: layer thickness.

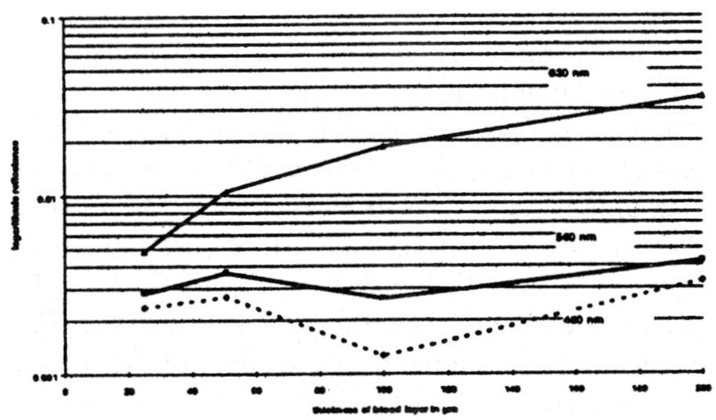

*Figure 5.* Reflectance at 460, 560 and 630 nm as function of the layer thickness.

*Reflectance spectra of whole blood*

The reflectance spectra of whole blood were measured in quartz cuvettes, having a black glass background in it (Figure 4). For the reference measurement, a white standard was used. The light, which was measured under these conditions originated from back-scattering of the incident light at the blood particles (e.g. erythrocytes). No light from the background is measurable under these conditions.

*Figure 3.* Validity check of Lambert–Beer's law: transmission at 460, 560, and 630 nm.

In the reflectance spectra of whole blood two spectral ranges can be distinguished. In the absorption dominated range up to about 586 nm, the alteration of the thickness of the blood layer has only a marginal effect. In the scattering-dominated range for wavelengths higher than about 600 nm the reflectance increases when the thickness of the blood layer increases. This behaviour is completely different in relation to the behaviour of the transmitted light, which decreases with increasing thickness of the blood layer. The logarithmic reflectance at selected wavelengths shows a saturation behaviour reaching a limiting value where the internal reflection is nearly independent of the layer thickness (Figure 5). In this case, the thickness of the layer is greater than the penetrating depth. The measurable light in the absorption-dominated range is much smaller than in the scattering-dominated range and its dependence on the layer thickness is faint. As the internally reflected light is very weak, adjustment errors in the experimental set up have a serious influence. A correction was performed under the assumption that the wavelengths of the isosbestic points both in the transmission spectra and in the reflection spectra of whole blood are equal and independent of the oxygen saturation. The last assumption means, that no differences occurs in the scattering properties of blood when the oxygen saturation will be changed.

The influence of hematocrit is equivalent to changes in the layer thickness. The influence of the velocity on reflectance was also studied. The reflectance of streaming blood increases with increasing velocity and reaches a constant maximal value for velocities higher than 6.3 mm/s.

## Calculation of the relation between transmitted and internally reflected light-influence of the illumination principle

It was shown, that the light from a blood vessel, located in front of a reflecting background, consists of the light, transmitted through a vessel and of the light, reflected at the blood column. It is now of interest, how the relation between these parts changes, when the thickness of the blood layer will be altered, when the pigmentation of the fundus changes or when the principle of the illumination is confocal or when a large field illumination will be applied. For that reason a Monte-Carlo simulation of the radiation propagation in the eye was calculated [1]. In that calculation, the eye was assumed to be a sphere consisting of six layers with different optical properties. The absorption and scattering properties of fundus layers were determined *in vitro* by double sphere measurements [2]. The vessel wall parameters were taken from the literature [3]. In the model, the eye had a pupil of 3 mm and a vessel (diameter 25 to 200 $\mu$m) was located inside the retina centred at the optical axis.

Photon packages were launched in the centre of the pupil and directed into an angle of 2° around the optical axis. The photons underwent absorption and scattering at the fundus. Each photon was traced until it leaves the pupil or was absorbed virtually completely.

Finding out differences between the principles of the large field illumination by a fundus camera and the confocal laser scanning technique, the photons were scored in two different ways. In the modelling of the vessel profile using large field illumination, the photons which went out of the eye through the pupil were scored according to the position which they left at the inner limiting membrane.

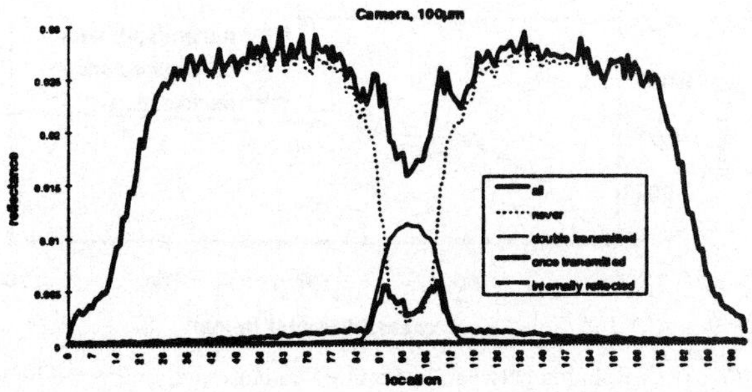


*Figure 6.* Light parts in the simulated vessel profile (camera illumination, diameter 100 $\mu$m).

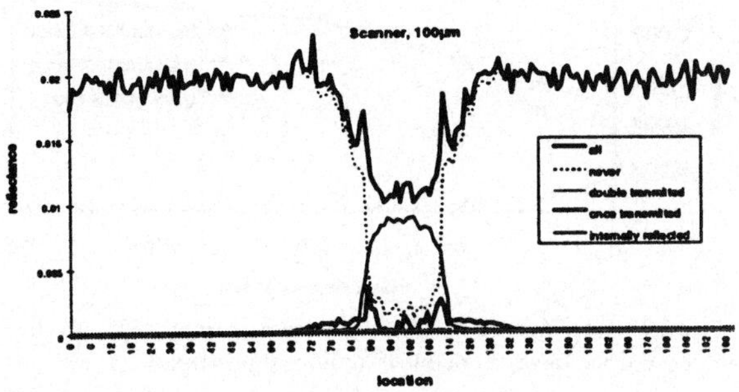


*Figure 7.* Light parts in the simulated vessel profile (scanner illumination, vessel diameter 100 $\mu$m).

In the simulation of the measurement by a laser scanner ophthalmoscope, only photons were scored which passed the inner limiting membrane within a confocal field diaphragm with a diameter of 200 $\mu$m.

To distinguish different pathways of light within the fundus, the photons were registered separately which entered the blood vessel once, twice, or never. To separate the light fraction reflected behind the vessel and transmitted through it from the light which was backscattered from the blood column, the simulation was performed with the eye in which all layers behind the retina were removed. The fraction of the remitted photons was related to the number of photons which would be reflected from an ideal white fundus with Lambertian characteristics.

The different parts of the measured photons in case of camera illumination are demonstrated in Figure 6. The intensity of light, which is transmitted once the vessel and will be reflected at the background, is highest at the borderlines of the vessel but it is detectable also in a certain distance outside from the vessel wall. It decreases inside the vessel. The light which is internally reflected at the blood column is measurable only from the vessel position and is constant in an intersection through the vessel. The light, double transmitted through the vessel, is relat-

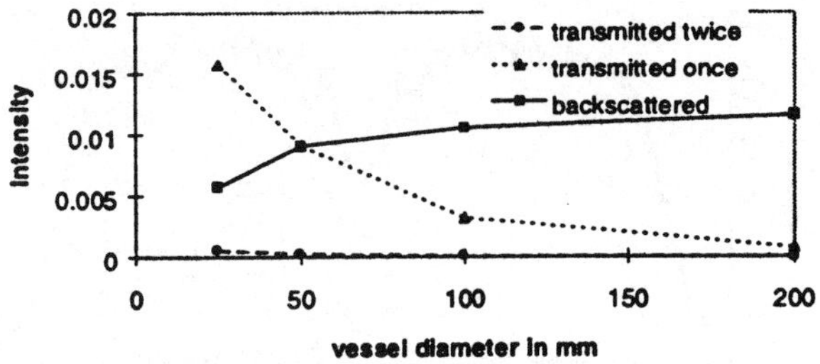


*Figure 8.* Relation between transmitted and internally reflected light in dependence of vessel diameter (camera principle).

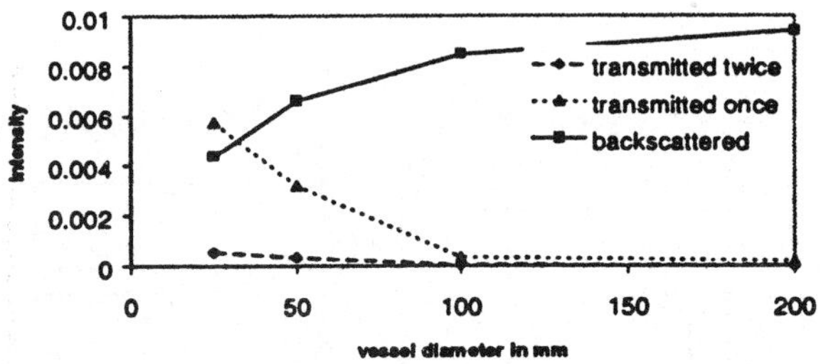


*Figure 9.* Relation between transmitted and internally reflected light in dependence of vessel diameter (confocal principle).

ive small. Though the qualitative behaviour between the parts of light is similar under different detection conditions, the camera principle and the confocal laser scanner principle exhibit serious quantitative differences. The scored intensity of the once transmitted light in case of confocal detection (Figure 7) is only about the half of the corresponding light, detected according to the camera principle. The internally at the blood column reflected light is only a little decreased in comparison with the camera principle.

Let us now consider how these parts of light change, if the vessel diameter changes. For that reason, from Monte-Carlo simulations of the profile at different vessel diameters, mean scored intensity was calculated in an interval of half a vessel diameter centred at the middle of the vessel (Figures 8 and 9).

In case of very small vessels, the single transmitted light dominates under both detection conditions. If the diameter of the vessels increases, the transmitted part of light decreases whereas the internally reflected light is nearly independent of the vessel diameter. Consequently, starting from a certain vessel diameter, the internally reflected light dominates the measurable light. Under the conditions of the large field illumination, in a normal pigmented fundus the transmitted part is about 20% of all the measured light in the centre of vessel (diameter 100 $\mu$m). Near the vessel wall, where the thickness of the blood layer in the vessel is reduced, the transmitted and the internally reflected light are about in the same order. If the fundus is hypopigmented, the relative part of the transmitted light increases.

In case of confocal detection (Figure 9), the critical vessel diameter is considerably smaller than in case of large field illumination and detection. That means, the measured reflected light in vessels of about 30 $\mu$m consists in two parts of light, having already the same order. Under confocal detection, the intensity of the transmitted light at a vessel diameter of 100 $\mu$m is only 10% of the intensity, detected according to the camera principle. As under confocal illumination the internally reflected light considerably dominates the measured light, the oxygen saturation can not exactly be determined, if only the transmission spectra of reduced or of saturated hemoglobin are used in the model. The spectral modulation of the internally reflected light, e.g. in saturated blood is not so marked as of the transmitted light. The detection of weaker differences in the confocally measured spectra would require an improved photometric resolution and an improved signal to noise ratio. The last is determined by the maximal permissible exposure and can not be increased. The consequence is, that the oxygen saturation of whole blood can not be determined exactly enough, if the confocal measuring principle is applied [4].

## Model function

From these investigations it follows that the determination of the oxygen saturation from reflection measurements requires a model function which includes both the transmitted and the internally reflected light. The model function approximates a reflectance spectrum, not a ratio of the background reflection and of the reflection of the vessel. For the determination of the oxygen saturation, the absorption-dominated wavelengths-range between 510 and 586 nm is considered. The model function includes the reflectance spectrum of the background, the transmission spectra of reduced and of saturated whole blood, as well as the spectra of the internal reflection of reduced and of saturated whole blood. A dependence of the vessel diameter is taking into account only for the transmitted part. Any changes in the spectral characteristics of the internally reflected light are assumed to be caused only by changes in the oxygen saturation.

The model function for the determination of the oxygen saturation from normalised reflectance spectra is given in Equation (1).

$$R(\lambda) = w_1 \cdot R_t(\lambda) + w_2 \cdot R_i(\lambda) \qquad (1)$$

with $w_1$, $w_2$ are the weighting factors for the transmitted and for the internally reflected parts of light, $R_t(\lambda)$ is transmitted light, and $R_i(\lambda)$ is the internally reflected light.

In detail, in the transmitted part the influences of the oxygen saturation and of concentration-thickness product on the changed transmission of the whole blood as well as the reflectance of the background are considered:

$$R_t(\lambda) = R_g^{P[1]}(\lambda) \cdot T_{Hb}^{P[2]\cdot(1-OS)}(\lambda) \cdot T_{HbO2}^{P[2]\cdot OS}(\lambda). \qquad (2)$$

In the equation yield: $R_g(\lambda)$ is the reflectance of the background; $T_{Hb}$ is the transmission of the reduced whole blood at a certain layer Thickness, e.g. 100 $\mu$m; $T_{HbO2}$ is the transmission of saturated whole blood (e.g. 100 $\mu$m); P[1] is the exponent of the background reflection; P[2] is the factor considering the concentration-thickness product of whole blood; OS is the oxygen saturation.

Using the experimental determined transmission spectra, $T_{Hb}$ and $T_{HbO2}$ consider both the absorption and the scattering of whole blood.

The formulation of Equation (2) is equivalent to the application of extinction spectra of whole blood, calculated by the logarithm of transmission measurements, and a following raising to a higher power. The part of the measured light, originating from the internally reflected light, will be described in the model function as:

$$R_i(\lambda) = R_{Hb}^{(1-OS)}(\lambda) \cdot R_{HbO2}^{OS}(\lambda). \qquad (3)$$

$R_{Hb}(\lambda)$ and $R_{HbO2}(\lambda)$ are the spectra of the internally reflected light, measured in a cuvette with a black background and containing reduced or saturated blood. This part of light is to be assumed to have only a marginal dependence on the layer thickness.

The background reflection in Equation (2) can be described by different expressions making possible both the determination of the oxygen saturation in the capillarised tissue and in single vessels.

Describing $R_g(\lambda)$ by the expression:

$$R_g^{P[1]}(\lambda) = S \cdot 10^{-P[1]\cdot E_{mel}(\lambda)} \qquad (4)$$

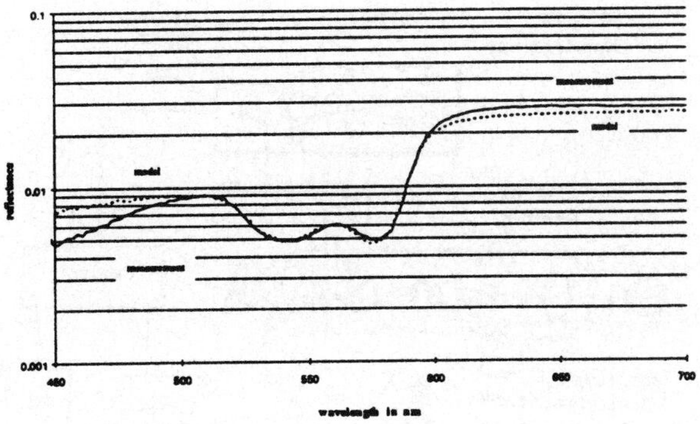


*Figure 10.* Approximation of a blood layer (100 $\mu$m) in front of a coloured background.

allows the determination of the global oxygen saturation in the tissue. In Equation (4) are: S is the reflection of the sclera (e.g. and wavelength-independent), and $E_{mel}(\lambda)$ is the extinction of melanin.

Especially for the application outside the ophthalmology the background reflection can be described by an analytical expression:

$$R_g^{P[1,2]}(\lambda) = S \cdot 10^{-P[1]-P[2]\cdot\log(1/\lambda)}$$

with S = 1.

If the reflection in the neighbourhood of the vessel will be used as $R_g$, the oxygen saturation will be determined inside the vessel. In case of very small vessels, when the internally reflected light can be neglected, the model function corresponds with known models [5].

## Examples

The performance of the described model was demonstrated by the approximation of the reflectance spectrum of saturated whole blood in a 100 $\mu$m cuvette with a coloured background. The reflectance of the coloured background was used as $R_g$ in the model function. The approximation of the reflectance spectrum by the model function as demonstrated in Figure 10 results in oxygen saturation of 101%.

The oxygen saturation was measured in normals and in age-related macular degeneration. The measurements were performed by the imaging ophthalmo spectrometer which is shown in Figure 11. The basic device is a fundus camera RC 250 (Carl Zeiss Jena, Germany), adapted with a spectrograph SPC 200 (Lonjumeau, France) and an intensified ICCD detector system (Princeton Instruments, Trenton NJ, USA). A

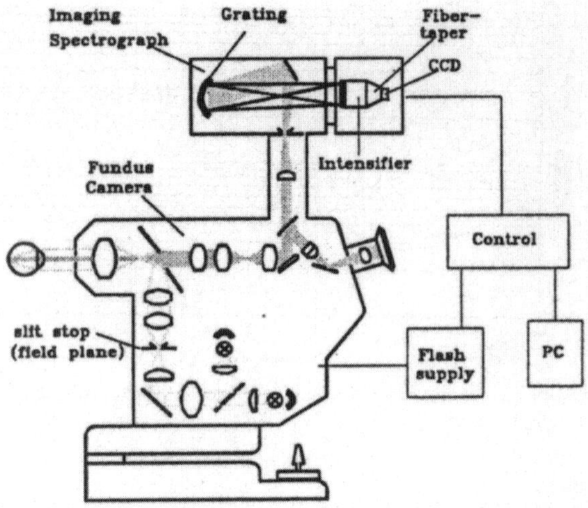


*Figure 11.* Scheme of the applied imaging spectrometer [6].

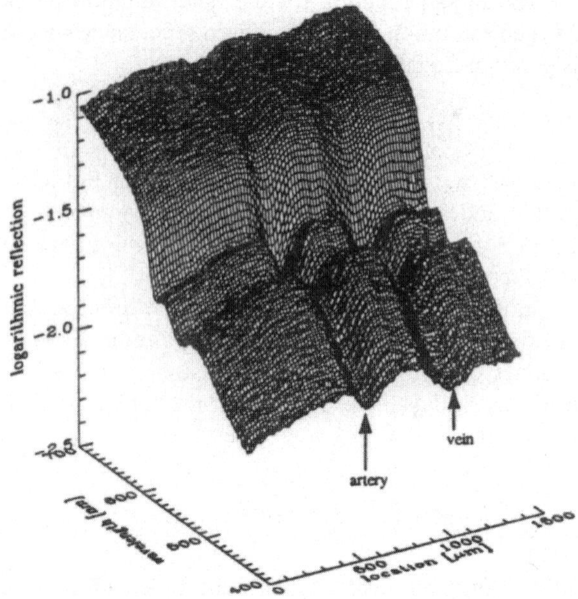


*Figure 12.* Locally resolved imaging spectrum, covering an artery and a vein.

slit-like field of 1.5 mm $\times$ 40 $\mu$m at the fundus is illuminated by a xenon flash. As the field is imaged onto the entrance slit of the spectrograph, the matrix detector system at the exit of the spectrograph registers a reflection spectrum for each illuminated fundus location. The fundus measurement is normalised to the reflection of a white standard in a model eye. An imaging spectrum, covering an artery and a vein, is shown in the Figure 12. The distribution of the oxygen saturation in arteries and veins of normals is demonstrated in Figure 13. The mean arterial oxygen saturation in retinal vessels of normals was $91.8 \pm 9\%$ and the mean venous oxygen saturation was $59.6 \pm 10\%$. An advantage of the imaging spectrometry is the simultaneous measurement of the oxygen saturation in neighbouring vessels under the same conditions. In Figure 14, the arterio-venous difference of the oxygen saturation is presented.

The wide distribution of these values means a large variety in the utilisation of oxygen in the supplied tissue. In normals, the utilisation of oxygen is higher in the temporal than in the nasal quadrants. In a first study, differences were found in the venous oxygen saturation between the temporal quadrants in age-related macular degeneration. In dry AMD the venous oxygen saturation is higher in the superior temporal than in the inferior temporal quadrant. In contrast, in wet AMD the venous oxygen saturation is lower in the superior temporal than in the inferior temporal quadrant. These results demonstrates pathological changes in the retinal microcirculation in AMD which seem to be different in the dry and in the wet form.

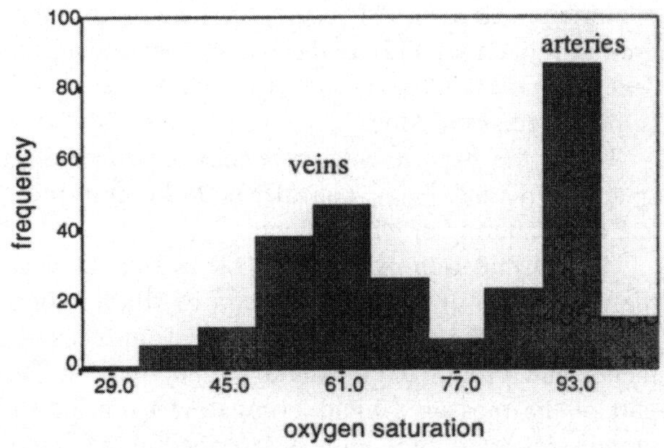


*Figure 13.* Distribution of oxygen saturation in arteries and in veins.

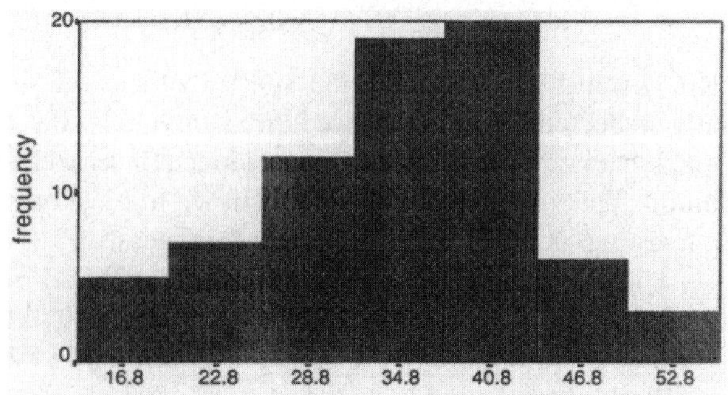


*Figure 14.* Distribution of the arterio-venous difference in the oxygen saturation.

## Summary

Measurements of the transmission and the reflectance spectra of whole blood exhibit a different dependence of the thickness of the blood layer. Whereas the transmission spectra decrease at all wavelengths, in the reflectance two ranges with different behaviour have to distinguish. In the absorption dominated range between about 500 and 600 nm, there is only a marginal influence of the thickness of the blood layer. In contrast, in the scattering dominated range higher than about 600 nm the reflectance increases if the thickness of the blood layer increases. Proofed by Monte Carlo simulation, the reflected light, which will be measured in ophthalmologic devices, consists both of light which penetrates a vessel once and will be reflected at the background and of light which is internally reflected at the blood column. Only in small vessels (diameter $< 50\ \mu$m) the transmitted part dominates. With increasing diameter of the vessels the internally reflected light dominates. Especially under the conditions of a confocal laser scanner, the transmitted part of light is always small.

As the measured light consists of both parts of light, a new model for the approximation of the measured reflected light is given, containing an expression for the transmitted and for the internally reflected light. The model was applied for the approximation of the reflectance of blood in a cuvette with an coloured background, delivering sufficient exact result. The oxygen saturation was determined in 256 measurements of 30 healthy persons using an ophthalmo imaging spectrometer. The mean arterial oxygen saturation was $91 \pm 5\%$, the mean venous oxygen saturation was $59 \pm 9\%$ and the arterio-venous difference in the oxygen saturation was 35%, which is in good agreement with the arterio-venous difference known in the brain. First measurements of the oxygen saturation in age-related macular degeneration point to partly different pathological changes of the retinal microcirculation in dry and in wet AMD.


## References


1. Hammer M, Leistritz S, Leistritz L, Schweitzer D, Thamm E, Donnerhacke HK. Monte-Carlo simulation of retinal vessel profiles for the interpretation of *in vivo* oxymetric measurements by imaging fundus reflectometry. SPIE Series 1997; 3192: 211–218.
2. Hammer M, Roggan A, Schweitzer D, Müller G. Optical properties of ocular tissues – an *in vitro* study using the double-integrating-sphere technique and inverse Monte-Carlo simulation. Phys Med Biol 1995; 40, 963–978.
3. Keijzer M, Jacques SL Prahl SA. Light distribution in artery tissue: Monte Carlo Simulation for finite-diameter laser beams. Laser Surg Med 1989; 9: 148–154.
4. Schweitzer D, Kalve B, Hammer M, Leistritz L. Oxygen saturation is not measurable sufficient exactly by laser scanner technique. Invest. Ophthalmol & Vis Sci 1997; 38: 488.
5. Delori FC. Noninvasive technique for oximetry of blood in retinal vessels. Applied Opt 1988; 27: 1113–1125.
6. Schweitzer D, Hammer M, Scibor M. Imaging spectroscopy in ophthalmology – principle and application in microcirculation and in investigation of pigments. Ophthalmic Res 1996; 28: 37–44.

International Ophthalmology 23: 355–358, 2001.
J.R. Sampaolesi (ed.), Laser Scanning: Update 1, 173–176.




# Sildenafil increases ocular perfusion *

Gianmarco Paris, William E. Sponsel*, Sal S. Sandoval, W. Rowe Elliott, Yolanda Trigo, Donald K. Sanford & Joseph M. Harrison
*South Texas Ocular Imaging Center, Ophthalmology Department, Glaucoma Service, University of Texas Health Science Center at San Antonio, San Antonio, TX, USA; *Author for correspondence*

**Abstract**
*Purpose:* Our goal was to determine whether visual system responses to sildenafil accompany shifts in ocular perfusion. The human choroid, which supports the metabolic function of the outer retina, is an erectile tissue, analogous in many respects to the corpus cavenosum. Methods: Right eyes of 12 normal adults were evaluated before and 2 h after 50 mg oral dose of sildenafil. Pulsatile ocular blood flow (POBF), intraocular pressure (IOP), Heidelberg retinal flowmetry (HRT), 4.26 cpd, 7.5 Hz temporally-modulated contrast sensitivity (CS), full-threshold C-20 frequency doubling technology (FDT) perimetry, and blood pressure (BP) were measured. *Results:* POBF (+29.4%; $p > 0.016$) and CS (+33.6%; $p > 0.014\%$) increased within 110 ($\pm 7.7$) minutes after sildenafil administration. No subject demonstrated decreases in either variable. HRF flow values increase among 7 of the 9 eyes producing stable scans (+8.2%; $p > 0.1$). FDT values did not change significantly, nor did systemic pulse amplitude or mean IOP. *Conclusions:* Since IOP and systemic pulse amplitude both remained stable after sildenafil administration, while POBF values increased to a level nearly one third greater than baseline. It appears sildenafil can induce intrinsic change in the choroidal vasculature, with an apparently positive impact on pericentral contrast sensitivity. This effect may be of clinical utility.

## Purpose

Our goal was to determine whether visual system responses might accompany any shifts in ocular perfusion after administration of oral sildenafil citrate.

## Introduction

Transient visual side-effects have been reported and feature in the product insert for Viagra (Pfizer). Retinal electrophysiologic side-effects associated with orally-administered sildenafil citrate have also been reported [1–5]. It has been suggested that these changes might be secondary to effects of sildenafil on human retinal phosphodiesterase isozyme PDE-6. Sildenafil's action on ocular circulation has received little consideration to date. Given the known actions of this drug systemically, it seemed appropriate to determine whether visual system responses to sildenafil might correspond to any observable shifts in ocular perfusion. The human choroid, which supports the metabolic function of the outer retina, is an erectile tissue, analogous in many respects to the corpus cavernosum. The fenestrated choroidal vasculature is highly responsive to both local and neurogenic stimuli, and the uveal system of which it is part may hold up to 97% of the intraocular blood volume. Choroidal perfusion has recently been reported [6–7] to be decreased in macular degeneration, the leading cause of acquired blindness in North America. Any agent capable of modulating choroidal perfusion might conceivably be of therapeutic value in the treatment of a range of choroidal, retinal and axonal disorders, provided circulatory augmentation could be achieved without metabolic compromise or vascular leakage.

* This study was published as a letter in the New England Journal of Medicine 2000; 342(22): 1680. Presented as a Poster (#204) at the American Academy of Ophthalmology. Dallas, TX, 10-24-2000. Presented as a free paper at 2000 ARVO Meeting. IOVS 2000; 41(4): S98.

## Materials and methods

We studied the right-eye responses among 12 health care professionals (10M, 2F; mean age 36 ± sem 2.7 y) before and 1–2 h after voluntary self-administration of a single 50 mg oral dose of sildenafil (Viagra; Pfizer, Inc., New York, NY). Monitored variables included: (1) pulsatile ocular blood flow (POBF) (OBF Labs, Malmesbury, UK/Paradigm Medical Industries, Salt Lake City, UT), (2) pneumotonometric intraocular pressure (IOP) (OBF Labs (UK) Ltd/Paradigm Medical Industries, Inc, Salt Lake City, UT), (3) Heidelberg scanning laser doppler retinal flowmetry (HRF)(Heidelberg Engineering, Inc. Carlsbad, CA), (4) 4.26 cycles per degree (cpd), 7.5 Hz temporally-modulated (NeuroScientific, Farmingdale, NY) contrast sensitivity (CS), (5) full threshold C-20 frequency doubling technology (FDT) (Welch Allyn, Skaneateles, NY/Humphrey Systems, San Leandro, CA). (6) sphygmomanometer systolic and diastolic blood pressure. Methods 1–3 provide objective numeric data, and methods 4 and 5 are subjective psychophysical tests.

Both POBF and IOP were obtained simultaneously by the OBF pneumotonometer, which produces estimates of POBF when 6 high quality ocular pulse waves are obtained within 25 seconds on the topically anesthetized cornea (Alcaine; Alcon, Fort Worth, TX). This POBF value is primarily a measure of uveal Circulation [8]. The majority of this is choroidal, serving as the primary blood supply for photoreceptors and pigmented epithelium in the outer retina. The HRF projects fixed frequency light via scanning laser, and measures frequency shifts produced by transverse erythrocyte movement within the microcirculation of the retina. Fourier transform analysis of averaged data allows estimates of circulatory volume, and the retinal microcirculatory flow values assessed in this study were derived from HRF velocity and density measurements [9]. Method 5 measures spatio-temporal visual function of the central 8 degrees of visual field (i.e., inside the macula, which subtends the central 10 degrees).

## Results

The 12 subjects demonstrated significant mean increases in POBF (+29.4%; $p = 0.016$) and 4.26 cpd temporally modulated contrast sensitivity (+33.6%; $p = 0.014$; (Table 1) within 110 (±7.7) minutes of sildenafil administration. No subject demonstrated decreases in either variable. HRF flow values increased among 7 of the 9 eyes producing stable scans (+8.2%; 2-tailed paired-$t$ $p = 0.1$). FDT values remained unchanged (mean deviation $-0.7 \pm 0.75$ decibels; $t = 0.18$, $p = 0.86$; pattern standard deviation $+1.16 \pm 0.55$ decibels; $t = -1.4$, $p = 0.18$) as did mean systolic and diastolic blood pressure, systemic pulse amplitude, and mean intraocular pressure.

## Discussion

The POBF receives its predominant contribution from pulsatile choroidal blood flow. A significant increase in choroidal circulatory volume after sildenafil administration might be anticipated given the analogous erectile character of the ocular choroid and penile corpora. The nitric oxide donor hydralazine has been shown to alter choroidal autoregulation in rabbits, with decrease of vasoconstriction that typically accompanies increase in mean arterial pressure [7]. Similarities also apparently exist between the autonomic innervation of the central choroidal and penile vasculature [8–11].

We have previously observed acute, highly significant decreases in temporally modulated 4.26 cpd contrast sensitivity in a comparable population during both hyper and hypocapnia [12]. This visual function decrease was partially modulated but not reversed by topical administration of the carbonic anhydrase inhibitor dorzolamide. We have not encountered a significant learning effect with this psychophysical technique in our prior studies. In the present study, six subjects had extensive prior experiences with all testing methods. Their responses were comparable to those of the other six inexperienced subjects. The observed significant increase in contrast sensitivity among this sildenafil study group may be related to the concomitant increase in ocular perfusion.

These findings invite further investigation of the association of visual function and ocular perfusion in both healthy and diseased eyes under treatment with sildenafil. While it behooves us to be vigilant in observing for any adverse effects with chronic drug administration, we must not overlook the possibility that this widely-used, highly vasoactive medication might also exert potentially desirable effects on eyes with catabolite-accumulative, autoregulatory, or other hypoperfusion-associated degenerative disorders of the eye.

*Table 1.* Changes in pulsatile ocular blood flow (POBF) and contrast sensitivity (CS) associated with ingestion of 50 mg oral sildenafil citrate (see text)

| Test | | Pretreatment | ~1 h s/p sildenafil | |
|---|---|---|---|---|
| POBF | | | $\mu$l/min | $\mu$l/min |
| ($n$ = 12) | Mean | | 915.6 | 1185.0 |
| | Standard error | | 102.8 | 158.3 |
| | Change | +269.4 $\mu$l/min | | |
| | | (+29.4%) | | |
| | Paired-$t$ significance | $p = 0.016$ | | |
| CS 4.26 cpd | | | Units | Units |
| ($n$ = 12) | Mean | | 91.57 | 122.34 |
| | Standard error | | 11.34 | 11.48 |
| | Change | +30.77 units | | |
| | | (□ .13 log units) | | |
| | | (+33.6%) | | |
| | Paired-$t$ significance | $p = 0.014$ | | |

Pharmacologic induction of increased ocular blood flow might be advantageous in appropriately selected and monitored circumstances. We believe that visual function should be carefully monitored among patients receiving sildenafil. While concerns for any retinal PDE-6 affects persist, there is also room for guarded optimism that positive changes might accompany intermittent increases in choroidal perfusion. Appropriate data collection will verify whether sildenafil exerts a positive, neutral, or negative effect among normal individuals and among patients with age-related or metabolic neuroretinal degenerations of the eye.

## Conclusions

IOP and Systemic pulse amplitude both remained stable after sildenafil administration, while POBF values increased to a level nearly one third greater than baseline. It appears sildenafil can induce profound intrinsic change in the choroidal vasculature, with an apparently positive impact on pericentral contrast sensitivity. This effect may be of clinical utility.

## Acknowledgements

Supported in part by an unrestricted grant from Research to Prevent Blindness, New York, NY. Dr. Gianmarco Paris has been supported by the Pan-American Association of Ophthalmology, and Retina Research Foundation. The authors have no extant commercial or other conflict of interest.

## References

1. Vobig MA, Klotz T, Staak M, Bartz-Schmidth KU, Engelmann U, Walter P. Retinal side-effects of sildenafil. Lancet 1999; 353: 375.
2. Zrenner E. No cause for alarm over retinal side-effects of sildenafil. Lancet 1999; 353: 340–341.
3. Vobig MA. Retinal side-effects of sildenafil (letter). Lancet 1999; 353: 1442.
4. Marmor MF. Sildenafil (Viagra) and Ophthalmology. Arch Ophthalmol 1999; 117: 518–519.
5. Marmor MF, Kessler R. Sildenafil (Viagra) and ophthalmology. Surv Ophthalmol 1999; 44: 153–162.
6. Grunwald JE, Hariprasad SM, DuPont J, et al. Foveolar choroidal blood flow in age-related macular degeneration. Invest Ophthalmol Vis Sci 1998; 39: 385–390.
7. Ciulla TA, Harris A, Chung HS, et al. Color Doppler imaging discloses reduced ocular blood flow velocities in nonexudative age-related macular degeneration. Am J Ophthalmol 1999; 128: 75–80.
8. Kiel JW. Choroidal myogenic autoregulation and intraocular pressure. Exp Eye Res 1994; 58: 529–544.
9. Flugel C, Tamm ER, Mayer B, Lutjen-Drecoll E. Species differences in choroidal vasodilative innervation: Evidence for specific intrinsic nitrergic and VIP-positive neurons in the human eye. Invest Ophthalmol Vis Sci 1994; 35: 592–599.

10. Flugel-Koch C, Kauffman P, Lutjen-Drecoll E. Association of a choroidal ganglion cell plexus with the fovea centralis. Invest Ophthalmol Vis Sci 1994; 13: 4268–4272.
11. May CA, Hayreh SS, Furuyoshi N, Ossoining K, Kauffman PL, Lutjen-Drecoll E. Choroidal ganglion cell plexus and retinal vasculature in monkeys with laser-induced glaucoma. Ophthalmologica 1997; 211: 161–171.
12. Sponsel WE, Harrison J, Elliott WR, Trigo Y, Kavanagh J, Harris A. Dorzolamide hydrochloride and visual function in normal eyes. Am J Ophthamol 1997; 123: 759–766.

International Ophthalmology **23:** 359–367, 2001.
J.R. Sampaolesi (ed.), Laser Scanning: Update 1, 177–185.



# Antiglaucomatous drugs effects on optic nerve head flow: design, baseline and preliminary report *

J. Sampaolesi, J. Tosi, V. Darchuk, R.A. Ucha, J. Marengo & R. Sampaolesi
*Department of Ophthalmology, University of Buenos Aires, School of Medicine, Fundación Argentina Oftalmológica Juan Sampaolesi and Instituto de la Visión*

*Key words:* optic nerve head, microcirculation, autoregulation system, glaucoma, drugs

## Abstract

*Purpose:* To determine the effect of the most frequent used antiglaucomatous drugs on the optic nerve head flow. To see the response of the autoregulation system in hypertensive, preperimetric and advanced perimetric glaucomas. Materials: In this preliminary report, optic nerve head parameters and retinal perfusion values were measured in 25 healthy controls as well as in 72 glaucomatous optic nerve heads. The glaucomatous sample was divided into three groups: 24 eyes treated with Betaxolol, 24 with Brinzolamide and 24 treated with Brimonidine (each of these groups was conformed by 8 hypertensive glaucomas, 8 preperimetric glaucomas, and 8 perimetric glaucomas). *Methods:* Patients were examined with scanning laser Doppler flowmetry (wave length 638 nm), and flow, volume and velocity indices were assessed in each report. Perfusion maps were analyzed with the new SLDF software, version 3.2 (automatic full field perfusion image analizer). Examinations were also performed with the Heidelberg Retina Tomograph (wave length 638 nm), using the new standard reference plane. Visual fields were performed with the Octopus 123 or 101 perimeters, programs G1x and G2, with 3 phases completed. Diurnal pressure curve with applanation tonometry and gonioscopy (for classification) were also performed. Results: In this preliminary report, it was found that eyes in Hypertensive or preperimetric stages of glaucoma, seem to use their regulation systems in order to preserve their optic nerve head circulation, while advanced perimetric glaucomas seem not be able to preserve their optic nerve head flow. And that antiglaucomatous drugs were not able to increase optic nerve head flow neither. *Discussion:* Optic nerve head blood supplies is regulated by an autoregulation system. This phenomenon may act in normals, hypertensive and preperimetric glaucomas, but it seems not be able to preserve blood flow in advanced glaucomas. Optic nerve head flow returns to normal values with treatment in glaucoma first stages, while it seems that it could not be restored in advanced glaucomas.

## Purpose

The aim is to determine the effect of the most frequent used antiglaucomatous drugs on the optic nerve head's flow. To find whatever the increase or decrease flow, or if it just remain unchanged. To see if these effects are directly caused by the drug or not. To study the response of the autoregulation system in hypertensive, preperimetric and advanced glaucomas, to the administration of different hypotensive agents.

* Research work supported by Fundación Argentina Oftalmológica Juan Sampaolesi.

## Material

A total of 200 patients were studied. 144 Glaucomatous eyes as the "Glaucoma Group" and 56 normal eyes of 56 subjects as the "Control Group".

The glaucoma group was divided into the following 6 different subgroups according the drug received: Timololo 0.5%, Betaxolol, Brinzolamide, Iso.Unoprostone, Brimonidine, and Latanoprost. Each group was constituted by 24 glaucomatous eyes. Each of these subgroups was divided into 3 subgroups again: 8 eyes were hypertensives glaucomas, 8 eyes were preperimetric glaucomas and 8 eyes were perimetric glaucomas.

Glaucomatous patients were classified into these 3 subgroups according to their clinic history, intraocular pressure, O.N.H. and visual field evaluation.

The meaning of this classification is to study the effect of the different drugs in each evolution period of the disease, based on previous data concerning the behavior of blood flow in different stages of evolution.

Inclusion criteria:

- Only Primary Open Angle Glaucomas were included in this study.
- Clear media of the lens and vitreous.
- No concomitant ocular or systemic diseases.
- Hypertension in range for the selected therapy. (IOP higher than 22 mm Hg and lower than 40 mm Hg in all the cases.)
- Untreated patients. (Patients previously treated for glaucoma were not included in any case.)
- Patients with good fixation and minimal ocular movements.
- Patients eligible for medical therapy and not for surgery at the moment of inclusion.
- Patients with no local or systemic contraindications for antiglaucomatous therapy.
- Minimum age: 22 years.
- Maximum age: 75 years.

## Method

Patients were examined with Anterior and Posterior Segment Biomicroscopy, Gonioscopy , UVA, BCVA, Applanation Tonometry and other records in order to be classified and included in the protocol.

Both normal and glaucomatus eyes undergone the following examinations:

- Scanning Laser Doppler Flowmetry of the optic nerve head (performed with the Heidelberg Retina Flowmeter, wave length 638 nm). Processed and analyzed with the Automatic Full Field Perfussion Image Analizer (AFFPIA, SLDF version 3.2 Software).
- Confocal Tomography of the optic nerve head (performed with the Heidelberg Retina Tomograph, wave length 638 nm).
- Computerized Visual Field examination (performed with Octopus 123 and/or 101, program G1X and G2). Processed and compared with Peridata Software.
- D.P.C.: Daily pressure curve (performed with Applanation Tonometry at 8.00 AM, 11.00 AM, 3.00 PM, and 6.00 PM).

These examinations were performed in 3 different controls called T0 (after patient classification and inclusion and before treatment started), T1 (one month under treatment) and T2 (six months under treatment).

Examinations included in each control:

- T0 and T2 included: DPC, Tomography, Flowmetry and Perimetry
- included: isolated IOP reading, Tomography and Flowmetry.
- The IOP was checked each month between T1 and T2, at 8.00 PM in an isolated reading.

If patients failed to regulate their intraocular pressure at control T1, they were automatically excluded of the protocol, and combined therapy was installed in order to reduce IOP to the normal range in between 72 hours. These patients were controlled as regular patients outside this study.

## Background

In June 1998, we found that normal eyes presented normal flow values, while hypertensive eyes presented increased flow values as long as their intraocular pressure was elevated over the normal range. Advanced glaucomas with visual field defects, presented decreased flow values, regardless of their intraocular pressure level (Figure 1).

This data show that there is a linear correlation between IOP levels and optic nerve head flow values in normal and in hypertensive subjects, while there is no linear or significative correlation in the advanced glaucoma group (Table 1).

It was concluded that in some way, hypertensive patients were able to regulate their blood flow in the presence of ocular hypertension, while glaucomatous patients were not able to preserve adequate flow values in presence of ocular hypertension. This fact may be due to the role played by the autoregulation system [1–7].

## Results

The results presented in this preliminary report belong to the Brimonidine, Brinzolamide and Betaxolol groups, after one month of treatment.

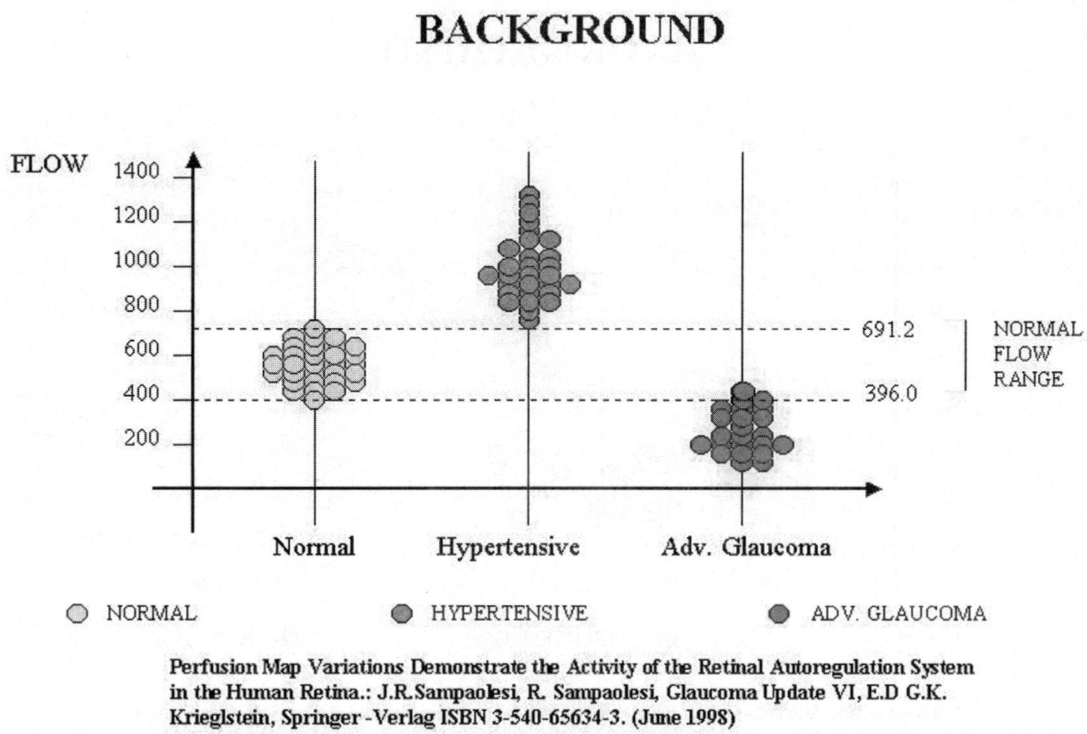


*Figure 1.*

The Brimonidine Group showed decreased flow values after 1 month under therapy, both in the hypertensive and preperimetric subgroups, while the perimetric subgroup flow values remain almost unchanged (Figure 2). Note that the decrease of flow in T1 is to the normal flow range, because flow values were over the normal range at T0.

The mean IOP reduction in Brimonidine Group was 6.083 mm Hg. The reduction of flow values in the first two groups is statistically significant (Table 2) and an example of this reduction can be appreciated in Figure 3.

The Brinzolamide Group showed decreased flow values after 1 month under therapy, both in the hypertensive and preperimetric subgroups, while the perimetric subgroup flow values remain almost unchanged (Figure 4). Note that the decrease of flow in T1 is to the normal flow range, because flow values were over the normal range at T0.

The mean IOP reduction in Brinzolamide Group was 5.111 mm Hg. The reduction of flow values in the first two groups is statistically significant (Table 3) and an example of this reduction can be appreciated in Figure 5.

The Betaxolol Group showed no significative change in flow values after 1 month under therapy, neither in the hypertensive or the preperimetric subgroups, while the perimetric subgroup flow values also remain unchanged (Figure 6). Note that flow values in T0 are over the normal range, and that they remain the same in T1.

The mean IOP reduction in Betaxolol Group was 2.375 mm Hg. There was no reduction of flow values in the first two groups. The flow values found for each subgroup in T0 and T1 are not statistically different (Table 4). There is even a slight increase in optic nerve head flow in the hypertensive and preperimetric subgroups, but this difference is also not statistically significant. An example of a patient treated with Betaxolol is shown in Figure 7.

**Discussion**

Antiglaucomatous agents that significantly reduce IOP seem to reduce optic nerve head flow to normal values.

They may be able to turn off the regulation system in hypertensive and preperimetric eyes by decreasing the IOP levels.

# BACKGROUND

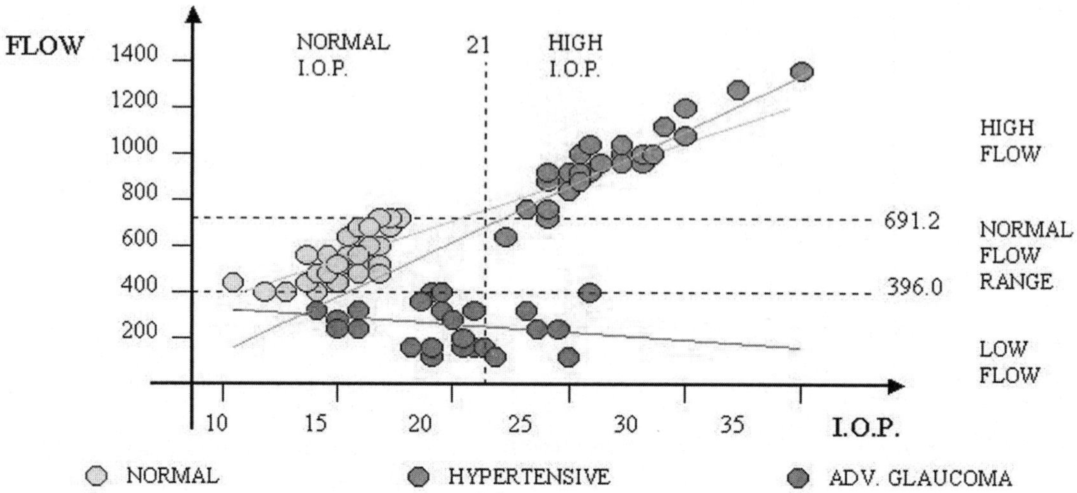


**Perfusion Map Variations Demonstrate the Activity of the Retinal Autoregulation System in the Human Retina.: J.R.Sampaolesi, R. Sampaolesi, Glaucoma Update VI, E.D G.K. Krieglstein, Springer-Verlag ISBN 3-540-65634-3. (June 1998)**

*Figure 2.*

# BACKGROUND

| | NORMAL | HYPERTENSIVE | ADV.GLAUCOMA |
|---|---|---|---|
| LINEAR REGRESSION | P < 0.001 | P < 0.001 | N.S. |
| RUNS TEST | Linear Correlation | Linear Correlation | Non Linear Correlation |
| r | 0.63 | 0.92 | 0.21 |
| $r^2$ | 0.40 | 0.86 | 0.04 |
| Linear regression analysis according to the least square method ( r & $r^2$ ). | | | |

**Perfusion Map Variations Demonstrate the Activity of the Retinal Autoregulation System in the Human Retina.: J.R.Sampaolesi, R. Sampaolesi, Glaucoma Update VI, E.D G.K. Krieglstein, Springer-Verlag ISBN 3-540-65634-3. (June 1998)**

*Figure 3.*

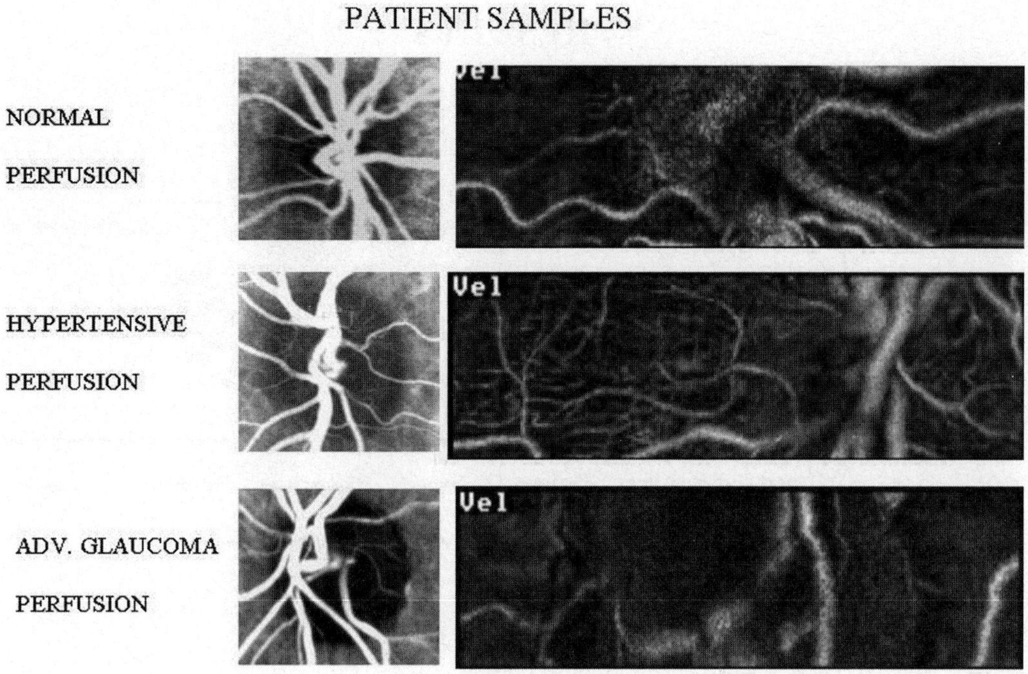


*Figure 4.*

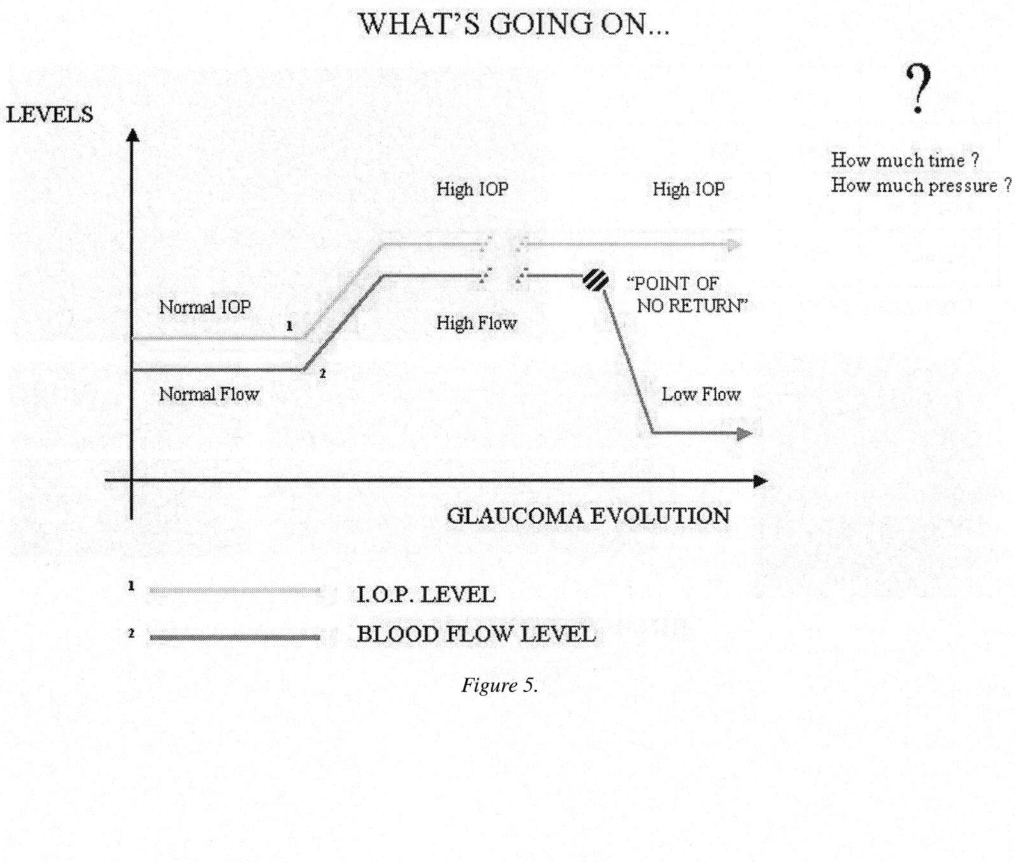


*Figure 5.*

## RESULTS

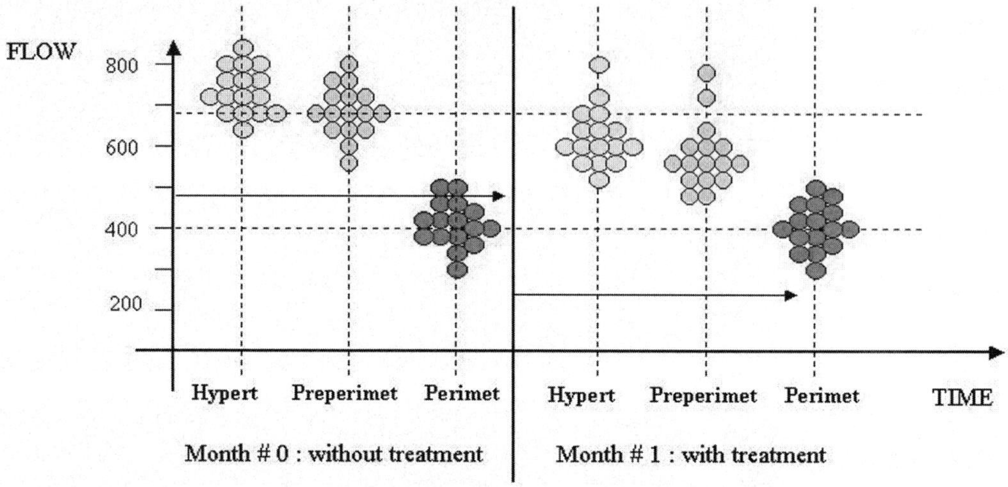


**BRIMONIDINE GROUP**

*Figure 6a.*

## RESULTS

| GROUP | T 0 | T 1 | ST. |
|---|---|---|---|
| Hypert. | 682 | 621 | ** |
| Preperim. | 671 | 550 | ** |
| Perimet. | 394 | 401 | ns |

IOP REDUCTION : 6.083 mm HG.

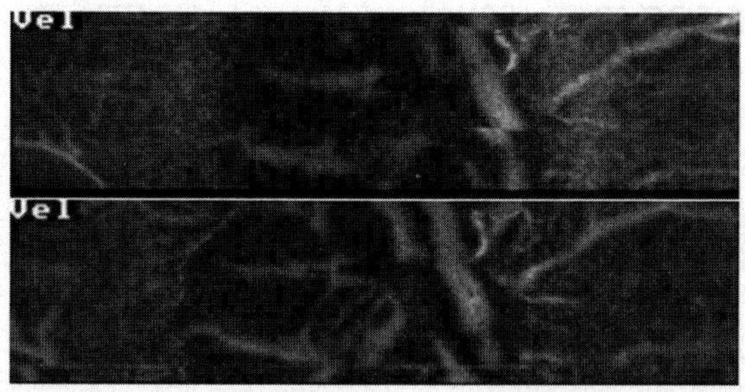


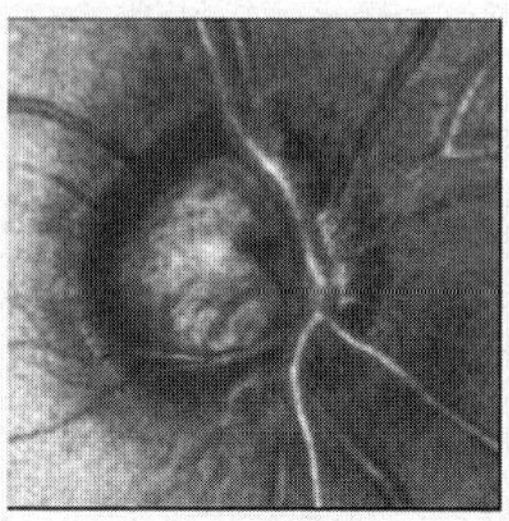

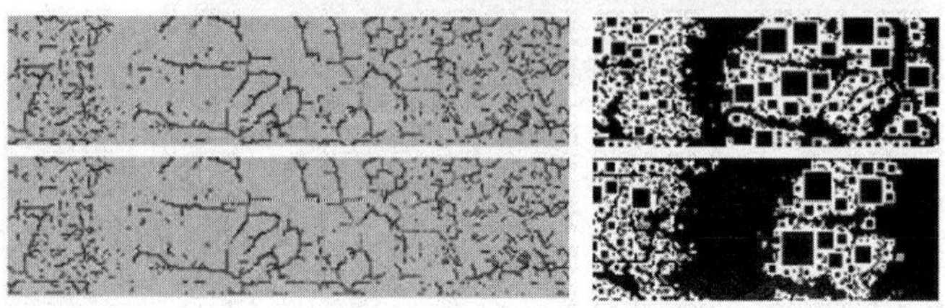

**BRIMONIDINE GROUP**

*Figure 6b.*

## RESULTS

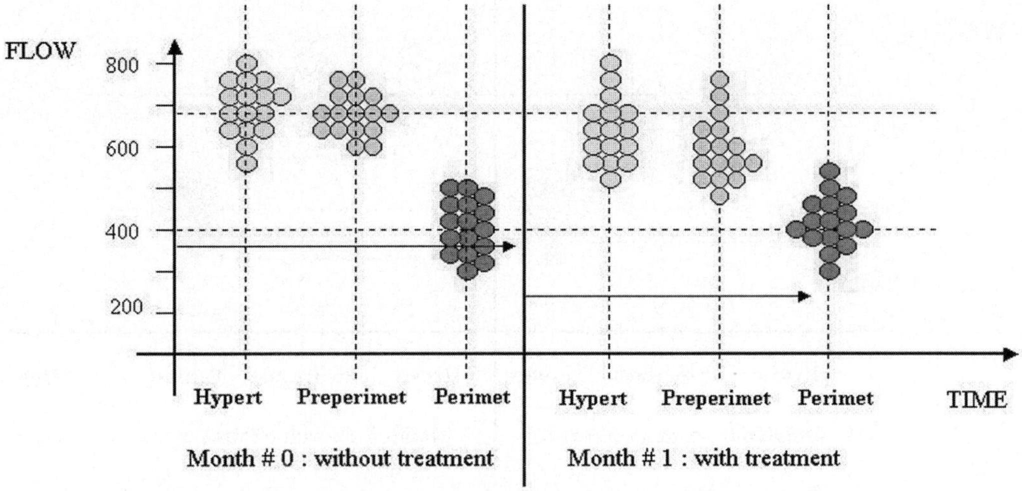


**BRINZOLAMIDE GROUP**

*Figure 7a.*

## RESULTS

| GROUP | T 0 | T 1 | ST. |
|---|---|---|---|
| Hypert. | 640 | 617 | * |
| Preperim. | 628 | 590 | * |
| Perimet. | 400 | 408 | ns |

IOP REDUCTION : 5.111 mm HG.

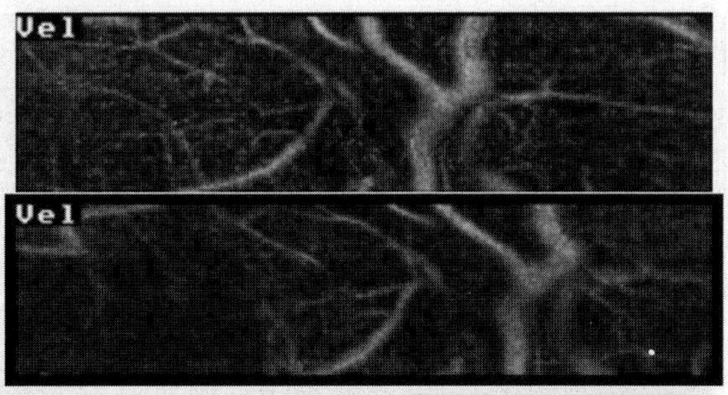


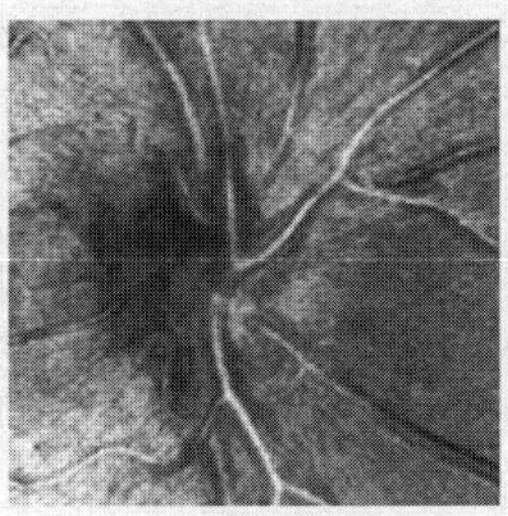

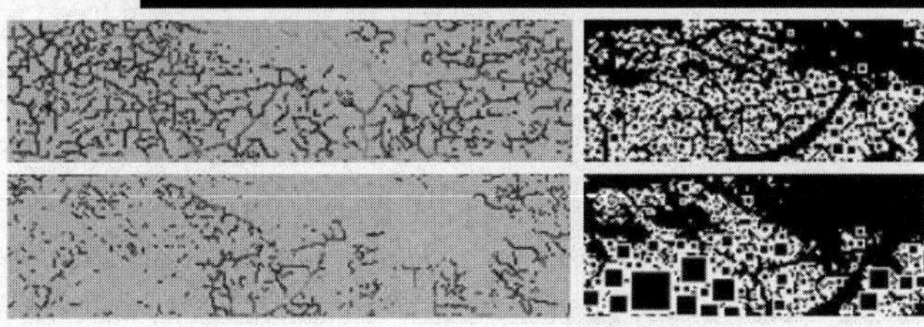

**BRINZOLAMIDE GROUP**

*Figure 7b.*

## RESULTS

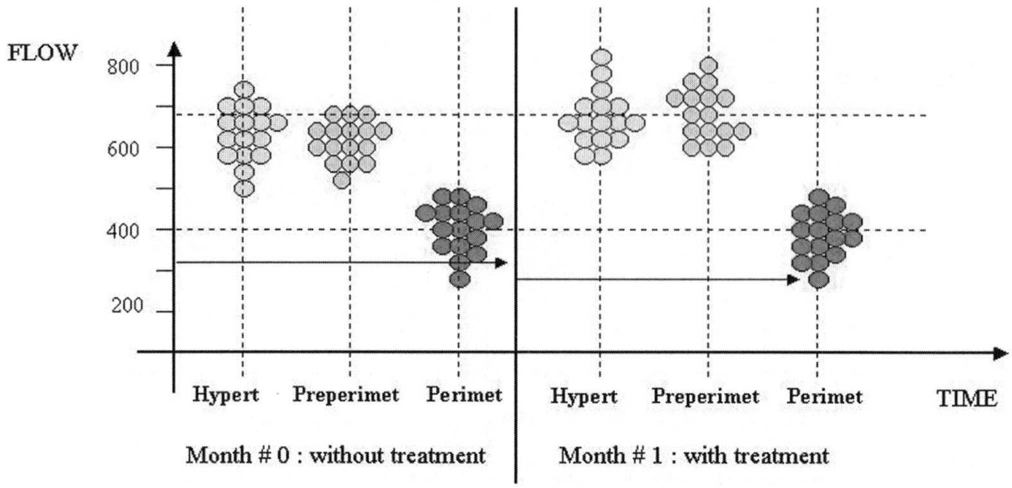


**BETAXOLOL GROUP**

*Figure 8a.*

## RESULTS

| GROUP | T 0 | T 1 | ST. |
|---|---|---|---|
| Hypert. | 612 | 687 | ** |
| Preperim. | 621 | 678 | ** |
| Perimet. | 390 | 389 | ns |

IOP REDUCTION : 2.375 mm HG.

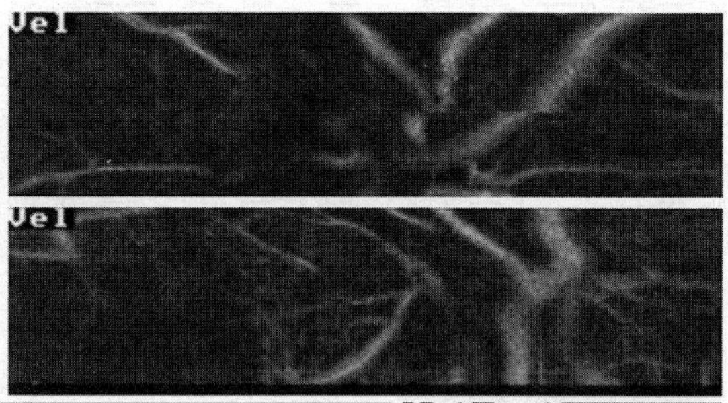


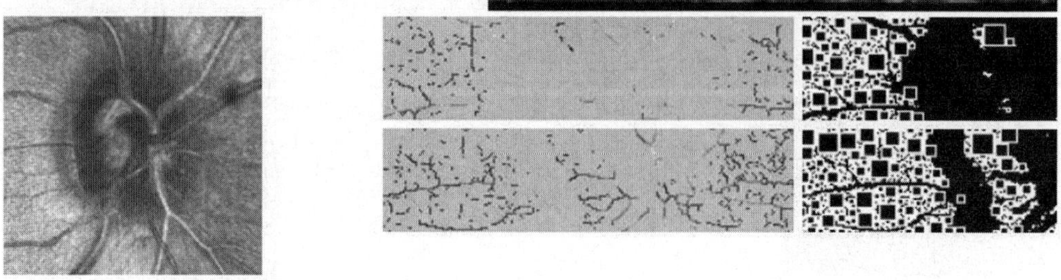

**BETAXOLOL GROUP**

*Figure 8b.*

In all the cases the flow showed no significant reduction, the intraocular pressure was not significantly reduced, or even changed.

It is unclear if treated patients with high flow values are really patients with improved optic nerve head flow or are just patients with IOP not well or enough regulated.

IOP reduction may goes together with flow reduction to normal values. When flow remains high, it may be because the IOP is not well enough reduced, and the autoregulation has to increase flow in order to keep optic nerve head blood supply at normal values.

In perimetric glaucomas, no significant changes were observed in optic nerve head blood flow values.

From previous data it was suggested that these patients are not able to increase blood flow in response of ocular hypertension.

This phenomenum may be related to the fact that advanced glaucomas seem no to be able to autoregulate their blood flow. Anatomopathologically there are findings which support these results[8]. There are also proofs of this with ICG.

Further investigations are needed to confirm the effect of these and other drugs on the optic nerve head flow, and the response of the autoregulation system.

## References


1. Michelson G, Schmauss B, Langhans MJ, Harazny J, Groh MJM. Principle, validity and reliability of scanning laser Doppler flowmetry. J Glaucoma 1996; 5: 99–105.
2. Pournaras CJ. Autoregulation of ocular blood flow, in HJ Kaiser, J Flammer, P Hendrikson (eds.), Ocular Blood Flow. Glaucoma Meeting 1995, Karger, Basel, 1996: 40–50.
3. Haefliger IO, Anderson DR. Pericytes and capillary blood flow modulation, in HJ Kiser, J Flammer, P Hendrickson (eds.), Ocular Blood Flow. Glaucoma Meeting 1995, Karger, Basel, 1996: 74–78.
4. Petrig BL, Riva CE. Optic nerve head laser Doppler flowmetry: Principles and computer analysis, in HJ Kiser, J Flammer, P Hendrickson (eds.), Ocular Blood Flow. Glaucoma Meeting 1995, Karger, Basel, 1996: 120–127.
5. Pillunat LE, Anderson DR, Knighton RW, Joos KM, Feuer WJ. Effect of increased intraocular pressure on optic nerve head blood flow, in HJ Kiser, J Flammer, P Hendrickson (eds.), Ocular Blood Flow. Glaucoma Meeting 1995, Karger, Basel, 1996: 138–144.
6. Sampaolesi J, Sampaolesi R. Perfusion-map variations demonstrate the activity of the retinal autoregulation system in the human retina, in Krieglstein (ed.), Glaucoma Update VI. Verlag, Springer, 1999: 95–102.
7. Michelson G, Harazny J, Langhans M, Groh M, Dichtl A, Horn F, Retinal circulation in glaucoma, in SM Drance (ed.), Vascular Risk Factors and Neuroprotection in Glaucoma. Update 1996. Kugler Publications, Amsterdam/New York, 1996: 217–220.
8. Sampaolesi R, Sampaolesi J. Confocal tomography of the retina and the optic nerve head. City Druck, Bergheimer Str. 119, D-69115 Heidelberg, 1999.

International Ophthalmology **23:** 369–371, 2001.
J.R. Sampaolesi (ed.), Laser Scanning: Update 1, 187–189.




# Vascular blood flow in different optic nerve head neuropathies

Juan Sampaolesi & Roberto Sampaolesi
*Fundación Argentina Oftalmológica Juan Sampaolesi and Department of Ophthalmology, School of Medicine, University of Buenos Aires, Argentina*

## Abstract

*Purpose:* To analyze the vascular features of different optic neuropathies and their blood flow values, by scanning laser Doppler flowmetry. To show, through these examples, the analysis possibilities of optic nerve head blood flow with a new software version. *Material and methods:* We analyzed normal optic nerve heads, and others, affected with glaucoma (localized nerve fiber defect), glaucoma (diffuse atrophy), megalopapilla, anterior ischemic optic neuropathy (A.I.O.N), and ischemics disorders (Carotid Obstruccion). These diseases were analyzed with the Heidelberg retina flowmeter, by scanning laser Doppler flowmetry, with a new software AFPPIA (automatic full field percusión image analyzer). *Results:* Through the different options of analysis the quantitative and qualitative results are shown in each pathology. He values of each case were compared with the reference values of a normal control group and the significance of the same were studied. *Conclusion:* The new automatic full field perfusion image analyzer seems to be a very useful tool to the evaluation of different optic nerve head disorders concerning blood flow. We still have to study larger groups in a prospective way and in relation with other conventional examinations in order to determine the results statistically.

## References

D.C.: **Reflective Image:** is the image obtained without analysis, similar to the one obtained with confocal tomography.

P.M.: **Perfusion Map:** is the image obtained with the flowmeter. Every bright point is related to circulation.

I.S.: **Intercapillar Space:** the less the intercapillar space, the smallest square. Big squares identify great intercapillar spaces. (qualitative analysis)

Z.I.: **Ischemic Zones:** bright zones identify ischemic areas, were the distance for oxygen diffusion is increased.

E.L.: **Capillaries Distribution:** show the distribution of capillaries in the analyzed area.

R.F.: **Cumulative Curve:** graphic with frequence distribution. It compares temporal flow (FT: yellow), nasal flow (FN: red) and rim flow (NRF: green).

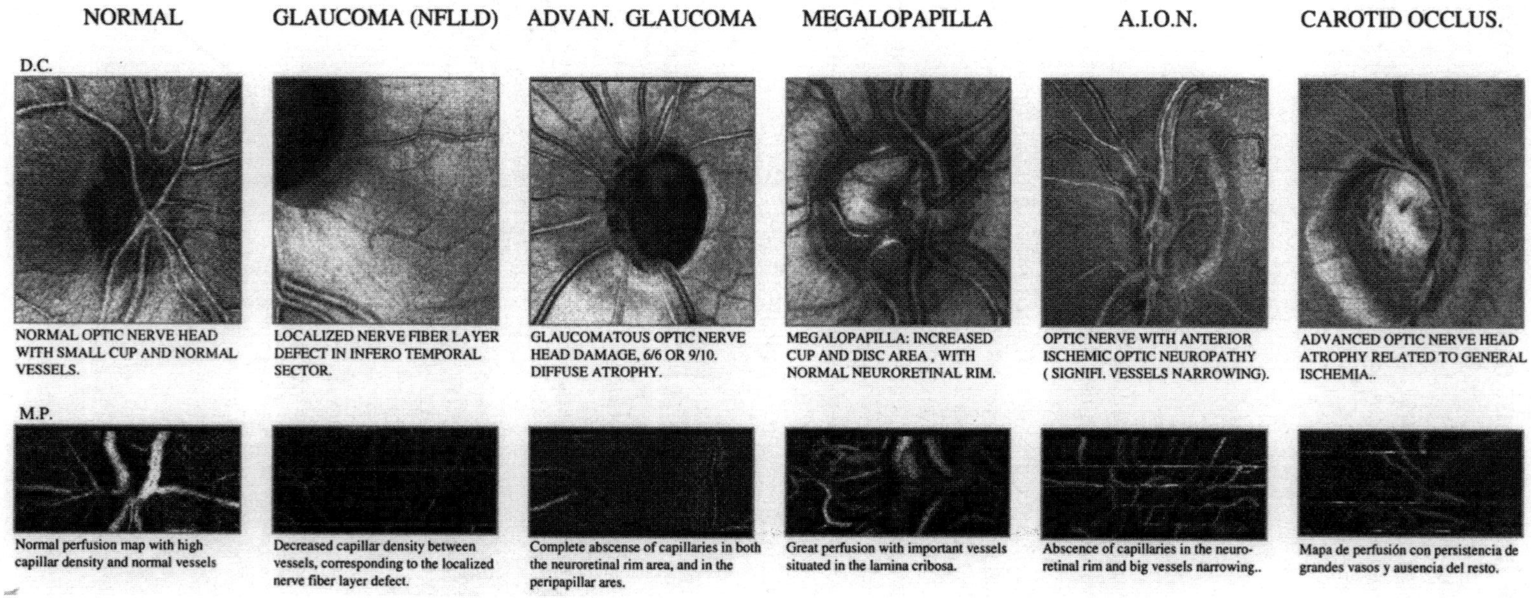
NORMAL
GLAUCOMA (NFLLD)
ADVAN. GLAUCOMA
MEGALOPAPILLA
A.I.O.N.
CAROTID OCCLUS.
D.C.
NORMAL OPTIC NERVE HEAD WITH SMALL CUP AND NORMAL VESSELS.
LOCALIZED NERVE FIBER LAYER DEFECT IN INFERO TEMPORAL SECTOR.
GLAUCOMATOUS OPTIC NERVE HEAD DAMAGE, 6/6 OR 9/10. DIFFUSE ATROPHY.
MEGALOPAPILLA: INCREASED CUP AND DISC AREA , WITH NORMAL NEURORETINAL RIM.
OPTIC NERVE WITH ANTERIOR ISCHEMIC OPTIC NEUROPATHY ( SIGNIFI. VESSELS NARROWING).
ADVANCED OPTIC NERVE HEAD ATROPHY RELATED TO GENERAL ISCHEMIA..
M.P.
Normal perfusion map with high capillar density and normal vessels
Decreased capillar density between vessels, corresponding to the localized nerve fiber layer defect.
Complete abscense of capillaries in both the neuroretinal rim area, and in the peripapillar ares.
Great perfusion with important vessels situated in the lamina cribosa.
Absence of capillaries in the neuro-retinal rim and big vessels narrowing..
Mapa de perfusión con persistencia de grandes vasos y ausencia del resto.

I.S.

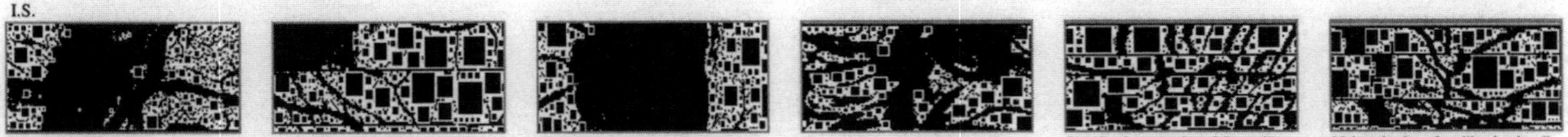

Very short interacapillar space due to high capillar density in a normal optic nerve head.

Increased interacapillar space in a localized nerve fiber defect (big squares)

Decreased perfusion in neuroretinal rim, with big squares in the peripapillary area.

Great perfusion in the lamina cribos and normal density in neuroretinal rim area.

Significative reduction of the capillar density in prepapillary capillary plexus.

Main reduction of capillar density all arround the central vessels. General ischemic disease.

Z.I.

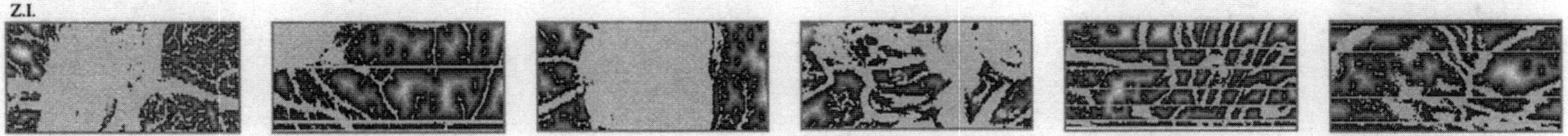

Absence of ischemic areas

Temporal inferior ischemic area in correlation with localized nerve fiber defect

General ischemia in advanced glaucoma

Poor image due to the large disc area in megalopapilla.

Central ischemic areas, with normal peripheric areas..

Generalized ischemia.

E.L.

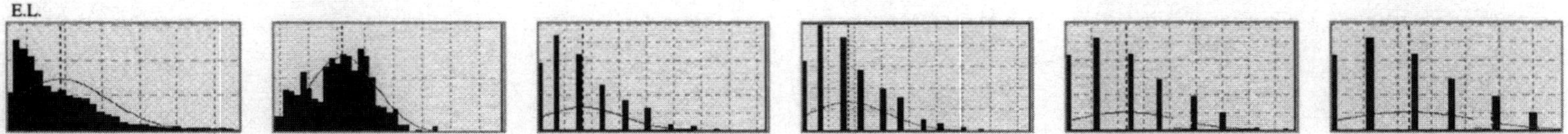

Normal Histogram.

Histogram in localzed nerve fiber defect.

Histogram of advanced glaucoma

Histogram in megalopapilla.

Abnormal Histogram in AION.

Completly abnormal Histogram

R.F.

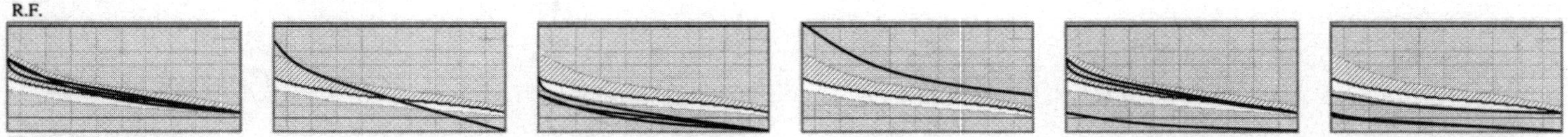

FT, FN, y NRF are included in the normal range.

Part of the curve is normal while the sencond part is below the normal range.

Advanced Glaucoma : TF, NF, and NRF are decreased.

Megalopapilla: big vessels measured in the lamina cribosa show very high flow values.

AION:In this case, NRF is significantly decreased, while TF and NF are still normal.

Carotid occlusion: TF, NF and NRF are decreased and all the values are below the normal range.

*International Ophthalmology* **23:** 373–379, 2001.
*J.R. Sampaolesi (ed.), Laser Scanning: Update 1*, 191–197.




# Optic nerve head behavior in Posner-Schlossman syndrome

V. Darchuk, J. Sampaolesi, O. Lopez Mato, C. Nicoli & R. Sampaolesi
*Department of Ophthalmology, School of Medicine, University of Buenos Aires, Fundación Argentina, Oftalmológica Juan Sampaolesi and Instituto de la Visión, Argentina*

## Abstract

*Purpose:* To analyze the effect of glaucomatocyclitic crisis on the optic nerve head, as well as its consequent structural and hemodynamic changes. To evaluate the surgical indication for Posner-Schlossman syndrome. *Materials and methods:* Four Posner-Schlossman syndrome cases were evaluated for an analysis of the phenomena occurring in the optic disc by means of retinal confocal tomography and Scanning Laser Doppler Flowmetry, and their correlation with visual function as measured with automated computerized perimetry. In all cases the images were obtained in the mediate period after the attack, while in case 1, measurements were performed in the immediate and mediate periods before, during and after the attack. *Results:* Significant differences were found between optic nerve head parameters and retinal flow measurements in the different phases evaluated. No permanent optic nerve head damage was demonstrated in any of the cases studied, except for case 4 (Posner-Schlossman syndrome associated with traumatic glaucoma). All the variables analyzed returned to normal values after remission of the acute rise in intraocular pressure. *Conclusions:* During the attacks of ocular hypertension the optic nerve head experiences significant morphologic changes and hemodynamic variations; nevertheless since they are transient, they fail to cause permanent damage. There was only one case, in which there was optic disc damage and visual field loss. It can be concluded that surgery should be restricted to those cases with severe and disabling symptoms (relative surgical indication) or to cases with progressive optic neuropathy with visual field loss, when the syndrome is associated with glaucoma (absolute surgical indication).

## Introduction

Four cases were evaluated in order to analyze the behavior of the optic nerve head when exposed to ocular hypertensive attacks in glaucomatocyclitic crisis. The optic nerve head was examined before, during and after each attack, both in the immediate and mediate periods. Each evaluation consisted of optic disc photographs, then confocal tomography, and finally, optic nerve head flowmetry and computerized perimetry, with a purpose to analyze whether the hypertensive attacks characterizing this syndrome have an impact on optic nerve head and visual function, as well as their possible implications. The roll of surgical management of this syndrome is thus reviewed.

## Materials and methods

The patients studied experienced a mean of 20 attacks during the mean follow-up period of 8 years, presenting with the characteristic symptoms of the syndrome, and always in the same eye. The contralateral unaffected eye was also evaluated using the same methods and simultaneously with the other eye. Of the total 4 cases studied three had Posner-Schlossman syndrome alone and in the fourth case it was associated with secondary glaucoma.

Optic nerve head topography was prospectively evaluated using optic disc confocal tomography. Images were acquired before, during and after each attack (after one week and 4 months) in case 1. In cases 2, 3 and 4 the topographs were obtained in the mediate period after the attacks. Imaging was obtained in all cases using the Heidelberg Retina Tomograph (wavelength: 638 nm), version 2.0, with two meas-

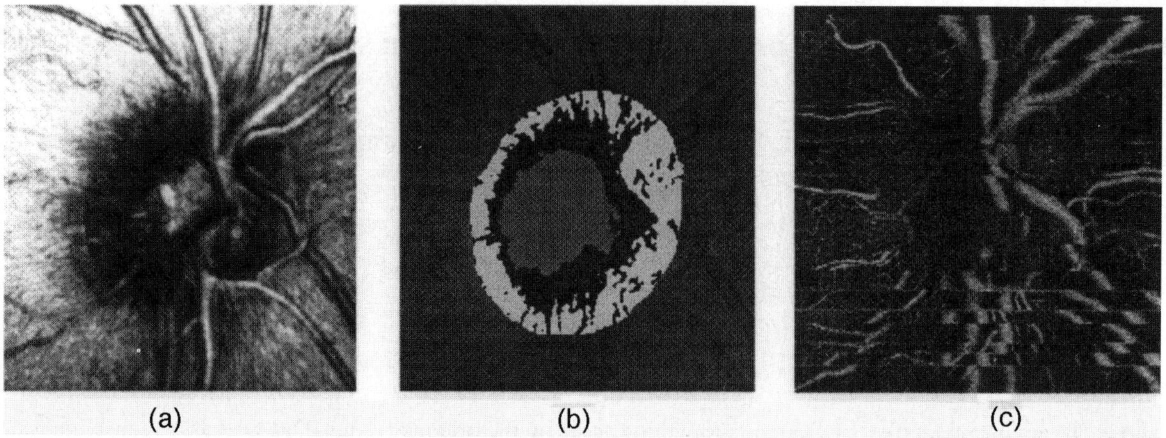


*Figure 1.* (Case 1) Before attack: (a) reflective image; (b) stereometric analysis; (c) perfusion map.

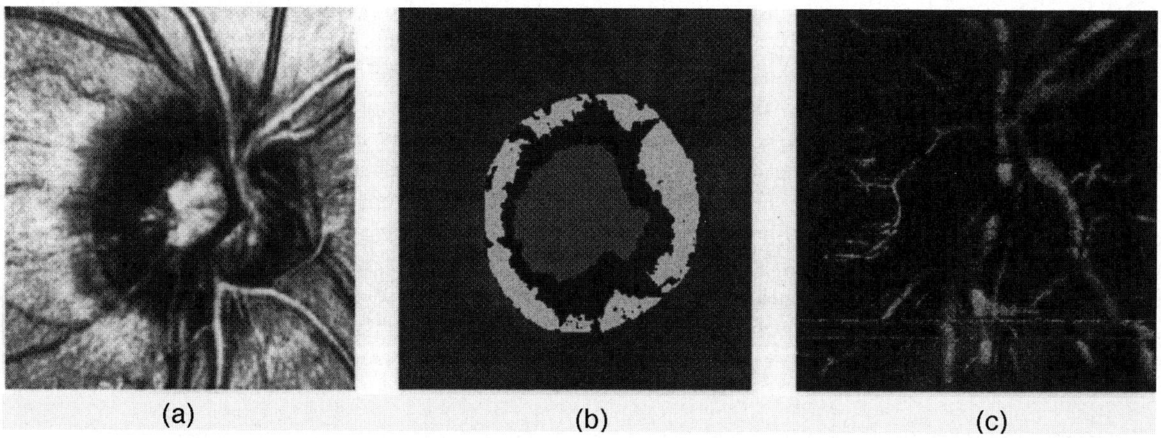


*Figure 2.* (Case 1) During attack: (a) reflective image; (b) stereometric analysis: (c) perfusion map.

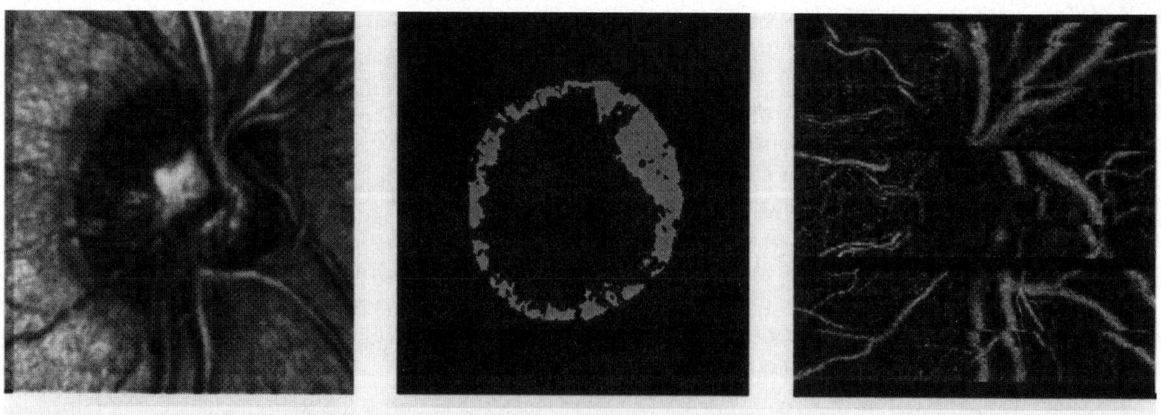


*Figure 3.* (Case 1) After attack (immediate): (a) reflective image; (b) stereometric analysis; (c) perfusion map.

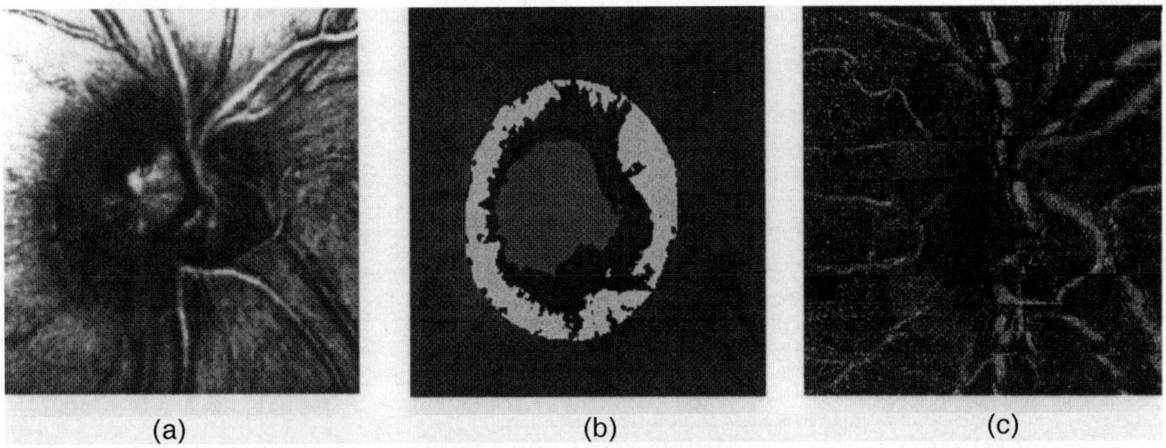

*Figure 4.* (Case 1) After attack (mediate): (a) reflective image; (b) stereometric analysis; (c) perfusion map.

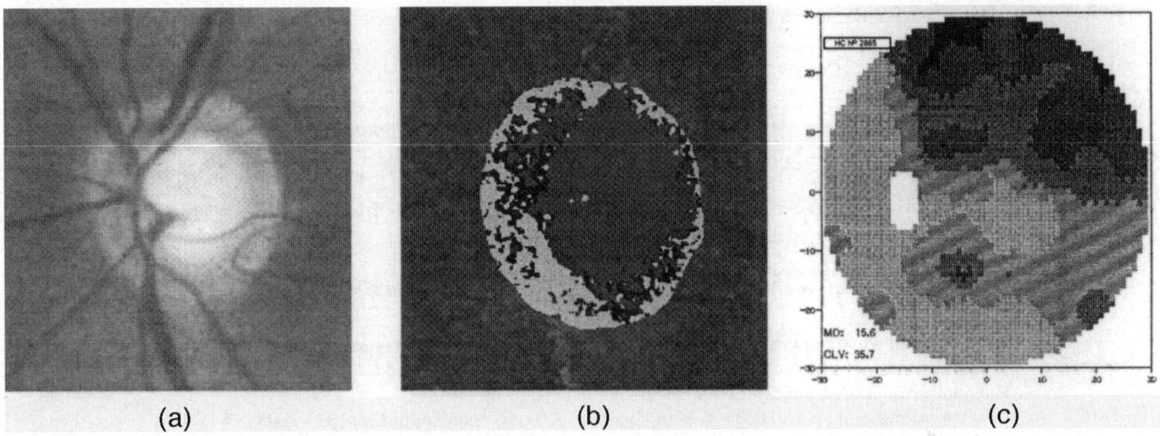


*Figure 5.* (Case 4) (a) fundus photography; (b) stereometric analysis; (c) computed visual field.

urements using 10 and 20 degrees of field [1]. Triple measurements were performed in each phase of the attack, of which a mean with a standard deviation of n < 10 was obtained. Each patient was followed through three different attacks in the previous 6 months.

Flowmetry was performed using the Heidelberg Retina Flowmeter (wavelength: 638 nm) [2]. Perfusion maps were analyzed with the Automatic Full Field Perfusion Image Analyzer (AFFPIA), version 3.2.3 [3]. Five images of each phase in glaucomatocyclitic crisis were acquired in order to cover the whole disc area, improve the image obtained with the tomograph and thus correlate both studies. Fundus photographs were taken with a fundus camera in order to observe the ophthalmoscopic appearance of the optic disc and look for any changes that may occur during the hypertensive attack.

Visual field examinations were performed with automated computerized perimetry (Octopus 500 EZ) in order to correlate eventual optic nerve damage with visual field condition and its possible functional changes during the attacks.

Hypotensive therapy was administered after diagnosis and acquisition of the examinations mentioned above. Upon examination, intraocular pressure (IOP) readings ranged from 50 to 60 mmHg during the hypertensive attack.

The mean IOP readings at each phase of glaucomatocyclitic crisis were as follows:

- Before the attack: 14 mmHg.
- During the attack: 60 mmHg.
- After the attack (immediate period) 15 mmHg.
- After the attack (mediate period) 17 mmHg.

*Table 1.* Values of each parameter (ordinates) at the different phases (abscissas) in case 1

| Parameters | Control group | Before attack | During attack | After attack (immediate) | After attack (mediate) | Units |
|---|---|---|---|---|---|---|
| Disc area | 1.60 to 2.51 | 2.181 | 2.186* | 2.186* | 2.182* | $mm^2$ |
| Rim volume | 0.32 to 0.65 | 0.321 | 0.227*** | 0.177*** | 0.312* | cmm |
| Rim area | 1,37 to 2.10 | 1.631 | 1.593** | 1.687** | 1.730** | $mm^2$ |
| Cup volume | 0.00 to 0.12 | 0.107 | 0.192*** | 0.144** | 0.119* | cmm |
| Cup area | 0.06 to 0.60 | 0.550 | 0.693*** | 0.600** | 0.608** | $mm^2$ |
| Third moment | –0.29 to –0.15 | –0.295 | –0.260** | –0.374*** | –0.275* | – |
| RNFL CSA | 0.65 to 1.10 | 0.841 | 0.813* | 0.758** | 1.024** | $mm^2$ |

Parameters listed: Disc area, rim volume, rim area, cup volume, cup area, third moment and RNFL cross-section area (CSA).
The number of asterisks indicates statistical significance level:
***Very significant ($p < 0.001$); **significant ($p < 0.05$); * non-significant ($p > 0.05$).
Statistical significance has been established in relation to the control group [4].

*Table 2.* Values of each parameter (ordinates) at the different phases (abscissas) in case 1

| Parameters | Control group | Before attack | During attack | After attack (immediate) | After attack (mediate) | Units |
|---|---|---|---|---|---|---|
| Temporal flow | 380 to 600 | 494.3 | 361.9*** | 490.4* | 498.8* | relative |
| Rim flow | 300 to 650 | 417.1 | 257.3*** | 424.9* | 421.3* | relative |
| Nasal flow | 370 to 620 | 399.6 | 203.0*** | 398.4* | 412.1* | relative |
| DNC (10–20) | 40 to 70 | 60 | 34*** | 60* | 62* | um |
| Mean of squ. | 15 to 30 | 28.4 | 34.8** | 28.2* | 26.5* | um |
| Ischemia | No | No | Yes | No | No | |

Parameters shown: Temporal flow, rim flow, nasal flow, distance to nearest capillaries (DNC), mean of squares (mean of squ.) and presence of ischemia in the perfusion map.
The number of asterisks indicates statistical significance level:
***Very significant ($p < 0.001$); **significant ($p < 0.05$); *non-significant ($p > 0.05$).
Statistical significance has been established in relation to the measurements performed before the attack.

## Results

When comparing the optic nerve head topography during the attacks and at the intervals between them, the tomographs revealed significant changes. There were also significant changes between the findings before, after and during hypertensive attacks. Interestingly, the images obtained in the mediate period before and after the attacks were identical, and showed no significant changes. They, in turn, were significantly different from those acquired during the attack and at the immediate period after it.

During the attack, Rim Volume and Rim Area decreased, while Cup Volume and Cup Area increased and the Third Moment revealed a positive trend. The RNFL Cross Section Area showed and initial decrease, which was followed by a return to its original values. The Disc Area (total disc area) remained unchanged and the remaining parameters showed no significant variations Table 1). In all the parameters of case 4 there were highly significant and permanent optic nerve changes.

Flowmetry revealed normal values before and after the attack, during which the flow was reduced at the peripapillary temporal and nasal sectors, with a further decrease at the level of the neuroretinal rim (NRR). Perfusion maps evidenced a decrease in the blood supply to these areas, with a significant increase in intercapillar distance (longer oxygen diffusion distance). There was also an increase in the intercapillar space, which is indicative of a decreased number of capillaries showing active circulation upon measurement, particularly at the level of the prepapillary capillary plexus (Table 2). Case 4 had a general decreased flow, with marked reduction at the level of the NRR accompanied with permanent hemodynamic changes, within the mediate period after the crisis.

*Table 3.* Values of each index (ordinates) at the different phases (abscissas) in case 1

| Parameters | Control group | Before attack | During attack | After attack (immediate) | After attack (mediate) | Units |
|---|---|---|---|---|---|---|
| MD | –2.0 to 2.0 | 0.6 | 1.0* | 1.5* | 1.4* | dB |
| CLV | 0.0 to 4.0 | 3.8 | 2.6* | 1.2* | 1.1* | dB2 |
| RF | 0.0 to 10.0 | 1.8 | 2.5* | 2.3* | 3.6* | % |

Parameters listed: Mean defect (MD), corrected loss variance (CLV) and reliability factor (RF).
*Statistically non-significant ($p > 0.05$) (relative to the values obtained at the examination performed before the attack).
Reference values [5].

These variations are clearly reflected by the images obtained during each phase (case 1). In the imaging acquired *before the attack* the appearance of the optic nerve head falls within the lower normal limits (Figure 1a), with a considerable flat (in green) and tilted NRR (in blue) (Figure 1b). Flowmetry evidences normal vascular and capillary meshworks, with a preserved prepapillary capillary plexus (Figure 1c). The images obtained *during the attack* reveal that the disc margins, as well as those of the cup become lighter, with visualization of the lamina cribrosa (Figure 2a). The Cup Area increases at the expense of the Rim Area, which decreases (Figure 2b). A substantially decreased number of active capillaries at the area of the NRR and peripapillary retina is evidenced by flowmetry images (Figure 2c). The images obtained *in the immediate period after the attack*, show little optic disc changes (Figure 3a), however, the stereometric analysis evidences an increased area of the tilted NRR at the expense of a reduced area of the flat NRR (Figure 3b). Flowmetry also reveals the slow flow restoration occurring at the level of the capillaries, mainly at the peripapillary level (Figure 3c). Images obtained *in the mediate period after the attack* show how the lamina cribrosa becomes less visible and the margins of both disc and cup start to become dimmer again (Figure 4a). The decreased areas of the cup and tilted NRR, with the consequent increased area of the flat NRR are evident (Figure 4b). Flowmetry shows the normalization of the flow, which may even be increased, with an evident recirculation at the level of the prepapillary capillary plexus (Figure 4c).

No changes were observed between the visual field indices obtained before, during and after the attack. In addition, no diffuse defects or local scotomas developed, the slight variations found were short-term fluctuations (Table 3). The only case presenting with a significant visual field defect was the one in which the syndrome was associated with secondary glaucoma (case 4).

Table 4 shows the close similarity in the values of the parameters of the first three cases, and their significant difference with case 4 (syndrome associated with secondary glaucoma) in the examinations performed at the mediate and immediate periods after the attack. No significant differences were found between the first three cases and the control group, while there were very significant differences with the fourth case ($p < 0.001$).

Op tic nerve head parameters, flow measurements and visual field indices of the contralateral eyes yielded similar values in all the measurements, with no statistically significant differences ($p > 0.05$) between all cases.

## Conclusion

The results presented have demonstrated that though the structural topography of the optic nerve head changes significantly during hypertensive attacks, these changes are transient and wholly reversible. Though similar to those observed in glaucomatous optic neuropathy, they differ in their transient nature and reversibility.

The variations found between the examinations at the different phases, might be due to the *baropressive* effect of ocular hypertension on the optic nerve head, and it is presumed that no degenerative or trophic change is involved. This *baropressive* phenomenon also influences retinal microcirculation, thus causing decreased blood flow at this level.

The acute rise in IOP which characterizes glaucomatocyclitic crisis sets into action the retinal autoregulation system (RAS), which elevates capillary flow in response to an IOP rise, in order to maintain the

*Table 4.* Optic nerve parameters, hemodynamic values and visual field indices found in the mediate period after the attack in all four cases studied

| Parameters | Control | Case 1 | Case 2 | Case 3 | Case 4 | Units |
|---|---|---|---|---|---|---|
| Disc area | 1.60 to 2.51 | 2.182* | 2.147* | 1.604* | 2.463* | $mm^2$ |
| Rim volume | 0.32 to 0.65 | 0.312* | 0.419* | 0.320* | 0.193* * * | cmm |
| Rim area | 1.37 to 2.10 | 1.730* | 1.587* | 1.479* | 1.011** | $mm^2$ |
| Cup volume | 0.00 to 0.12 | 0.119* | 0.100* | 0.013* | 0.477* * * | cmm |
| Cup area | 0.06 to 0.60 | 0.608* | 0.561* | 0.125* | 1.451* * * | $mm^2$ |
| Third moment | –0.29 to –0.15 | –0.275* | –0.218* | –0.320* | –0.066* * * | – |
| RNFL CSA | 0.65 to 1.10 | 1.024* | 1.282* | 1.084* | 0.690** | $mm^2$ |
| Flujo Temp. | 380 to 600 | 498.8* | 471.5* | 467.3* | 318.2** | relative |
| Flujo ANR | 300 to 650 | 421.3* | 432.8* | 448.6* | 175.4* * * | relative |
| Flujo nasa1 | 370 to 620 | 412.1* | 474.5* | 437.9* | 340.2** | relative |
| D.N.C.(10–20) | 40 to 70 | 62* | 68* | 75* | 25* * * | um |
| Mean of squ. | 15 to 30 | 26.5* | 24.2* | 22.1* | 45.1** | um |
| Ischemia | No | No | No | No | No | – |
| MD | – 2.0 to 4.0 | 1.4* | 1.1* | 2.2* | 15.6* * * | dB |
| CLV | 0.0 to 4.0 | 1.1* | 2.6* | 4.1* | 35.7* * * | dB2 |
| RF | 0.0 to 10.0 | 3.6* | 3.1* | 1.4* | 2.8* * * | % |

Notice that cases 1, 2 and 3 are pure glaucomatocyclitic crises while in case 4 the glaucomatocyclitic crisis is associated with glaucoma.
The number of asterisks indicates the different statistical levels:
***Very significant ($p < 0.001$); **significant ($p < 0.05$); *non-significant ($p > 0.05$).
Statistical significance has been established in relation to the control group, while the significance levels of flow values have been established in relation to the different cases.

adequate blood supply. But the IOP rises up to such an extent that the RAS comes to a point where it losses its ability to maintain the adequate retinal flow. At this moment there is high ocular hypertension simultaneously with extremely decreased flow values. When the IOP becomes normal, the flow rises again until it returns to normal values.

Glaucomatocyclitic crisis differs from simple chronic glaucoma, where there is sustained medium-intensity ocular hypertension (depending on each case), in that the retinal circulation is exposed to a sudden, and mainly, extensive IOP rise. The magnitude of this rise is thought to exceed the regulation capacity of the SAR, which thus collapses, and there is a consequent low flow at the retinal level.

This halt in the perfusion system is transient and, therefore, though there is a low supply of oxygenated blood to the retinal fibers during it, this does not seem to stop circulation totally, besides, it is not sufficiently long so as to produce ischemia leading to necrosis. The flow is wholly restored after the attack (both in the mediate and immediate periods).

Strikingly, though the flow is immediately restored after remission of the IOP rise, the disc topography parameters take almost 4 months to return to the values found before the crisis.

All these findings, correlated to the coexistence of a normal visual function, definitely show that Posner-Schlossman syndrome produces optic nerve head changes during hypertensive attacks, but they are transient, harmless and reversed after it subsides.

After this syndrome was described for the first time [6–9], there were very few papers reporting the effects of hypertensive attacks on the posterior segment, and particularly on the optic nerve head. Some authors have reported that there are no significant changes in the optic nerve and that, consequently, no visual field manifestations develop either [6, 10–12]. Nevertheless, other prospective studies have demonstrated otherwise, and in these cases, damage was found mainly in the affected eye, and less frequently in the contralateral eye [6, 13-16]. This syndrome has even been reported to serve as a simple glaucoma model for the description of progressive optic nerve damage [17].

Pederson and Herschler were the first to report the cup reversibility in adults after the remission of hypertensive attacks [11]. Later on, Park and Hong evaluated these changes objectively with optic nerve confocal tomography [12].

We have analyzed four patients with Posner-Schlossman syndrome. In three of these cases the presence of normal optic nerve head and visual field after a follow-up of 8–10 years was established. In the fourth case, the syndrome was associated with trauma on the same left eye, after which he had elevated IOP during the intervals between attacks, with the consequent optic nerve head damage (Figures 5a and 5b) and subsequent glaucomatous visual field defects (Figure 5b).

Our experience has shown that hypotensive and anti-inflammatory therapy is sufficient, with no need for surgery except for one precise indication, when the syndrome is associated with glaucoma, and diagnosis of progressive optic neuropathy and visual field loss is well established.

The results demonstrate the marked difference between two different entities (with their corresponding prognoses): Posner-Schlossman syndrome alone and the same syndrome in association with glaucoma.

To our knowledge, this is the first paper to report optic nerve changes as revealed by tomography, and including a comprehensive analysis of all the optic nerve parameters as well as of the circulation condition during an acute IOP rise in glaucomatocyclitic crisis.

## Acknowledgement

Research work was supported by Allergan and Fundación Argentina Oftalmológica Juan Sampaolesi.

## References

1. Burk ROW et al. Analysis of three dimensional optic disk topography by laser scanning tomography. In: Nasemann JE, Burk ROW, eds. Scanning laser ophthalmoscopy and tomography, Quintessenz, Munchen, 1990; pp. 161–176.
2. Michelson G, et al. Principle, validity and reliability of scanning laser doppler flowmetry. J Glaucoma 1996; 5: 99–105.
3. Michelson G, et al. Automatic full field analysis of perfusion images gained by scanning laser Doppler flowmetry. Br J Ophthalmol 1998; 82(11): 1294–1300.
4. Sampaolesi R, Sampaolesi J. Tomografía confocal de la retina y del nervio óptico. Arch Oftal Bs As 1997; LXX(5): 77–86.
5. Sampaolesi R. Visual field indices. In: Sampaolesi R, ed. Glaucoma Edit Med Panamericana, 1991, pp. 265–271.
6. Terrien F, Veil P. Certain so-called primary glaucomas. Bull Et mem Soc Franc d'ophth 1929; 42: 349.
7. Kraupa, E. Ocular hypertension in acute angioneurosis of ciliary body ('glaucoma allergicum'): its relationship to cyclitic and heterochromic glaucoma. Arch f Augenh 1935; 109: 416.
8. Kronfeld PC. Gonioscopic correlates of responsiveness to miotics. Arch Ophthalmol 1944; 32: 447.
9. Posner A, Schlossman A. Syndrome of unilateral recurrent attacks of glaucoma with cyclitis symptoms. Arch Ophthalmol 1948; 39: 517–535.
10. Gerhard JF. A propos de l'evolution du syndrome de Posner-Schlossman. Bull Soc Ophthalmol Fr 1981; 81(8–9): 657–659.
11. Pederson JE, Herschler J. Reversal of glaucomatous cupping in adults. Arch Ophthalmol 1982; 100: 426–431.
12. Park KH, Hong Ch. Reversal of optic disc topography in patients with glaucomatocyclitic crisis after remission of attack. J Glaucoma 1998; 7: 225–229.
13. Varma R et al. Surgical treatment of acute glaucomatocyclitic crisis in a patient with primary open-angle glaucoma. Am J Ophthalmol 1998; 105(1): 99–101.
14. Mattson R. Ueber glaukomatozylklitische Krisen. Acta Ophthalmol 1954; 32: 523–533.
15. Kass MA et al. Glaucomatocyclitic crisis and primary open angle glaucoma. Am J Ophthalmol 1973; 75: 668.
16. Raitta C, Vannas A. Glaucomatocyclitic crisis. Arch Ophthalmol 1997; 95: 608.
17. Iwata K et al. Die beginnende Fundus infolge rezidivierender kleiner Krisen beim Posner-Schlossman-Syndrom – ein Modell fur das Glaucoma Simplex. Klin Mbl Augenheilk 1982; 180: 20–26.

*International Ophthalmology* **23:** 381–383, 2001.
*J.R. Sampaolesi (ed.), Laser Scanning: Update 1*, 199–201.




# A.I.O.N.: Vascular findings with Scanning Laser Doppler Flowmetry

Martinez Haydée & Juan R. Sampaolesi
*Department of Ophthalmology, School of Medicine, University of Buenos Aires and Fundación Argentina Oftalmológica Juan Sampaolesi*

**Abstract**

*Purpose:* To find the vascular effects of the anterior ischemic optic neuropathy on the optic nerve head. *Material and methods:* Scanning Laser Doppler Flowmetry was performed on 60 normal volunteers, also on 33 patients with Anterior Ischemic Optic Neuropathy (A.I.O.N). Patients were examined with the Heidelberg Retina Flowmeter, by Scanning Laser Doppler Flowmetry, (680 nm), and Flow, Volume and Velocity were assessed in each report. Perfusion maps have been reconstructed with the new SLDF software, version 3.2, called AFFPIA. (Automatic Full Field Perfussion Analizer). Visual acquity, visual field, fundus examination and Doppler studies were also recorded. *Results:* In this preliminary report, it was found that eyes with AION present a decreased flow in the neurorretinal rim area, corresponding to the prepapillary capilar plexus. The flow values were almost normal, (but also decreased) in the peripapillary area, both temporal and nasal. Rim flow shows a significant difference with control group values ($p < 0.001$) while temporal and nasal peripapillary flow values demonstrate no significant differences. (N.S.) *Conclusion:* Prepapillary capilar plexus seems to be the most affected zone by anterior ischemic optic neuropathy, while the peripapillary zone seems to remain almost normal. *Discussion:* H.R.F is a good tool in order to study blood flow at the optic nerve head in ischemic disorders.

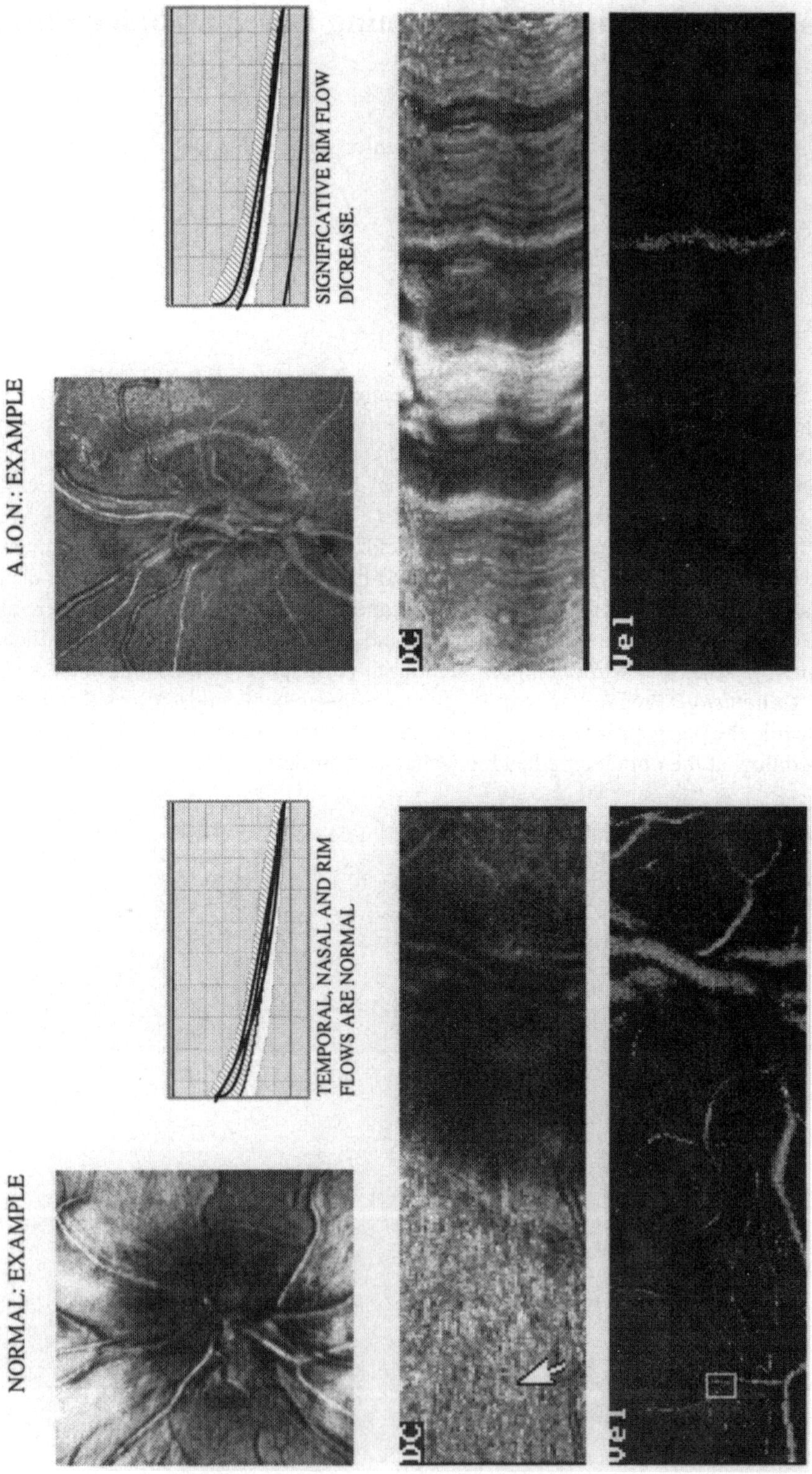
NORMAL: EXAMPLE
TEMPORAL, NASAL AND RIM FLOWS ARE NORMAL
DC
Ve1
A.I.O.N.: EXAMPLE
SIGNIFICATIVE RIM FLOW DICREASE.
DC
Ve1

| RESULTS: | CONTROL GROUP | A.I.O.N. | STATISTICS |
|---|---|---|---|
| RIM FLOW | **600.01** | **203.97** | P < 0.001 |
| TEMPORAL FLOW | **584.20** | **394.48** | P > 0.05 (NS) |
| NASAL FLOW | **567.30** | **389.52** | P > 0.05 (NS) |
| Kruskal Wallis Population Test. | | | |

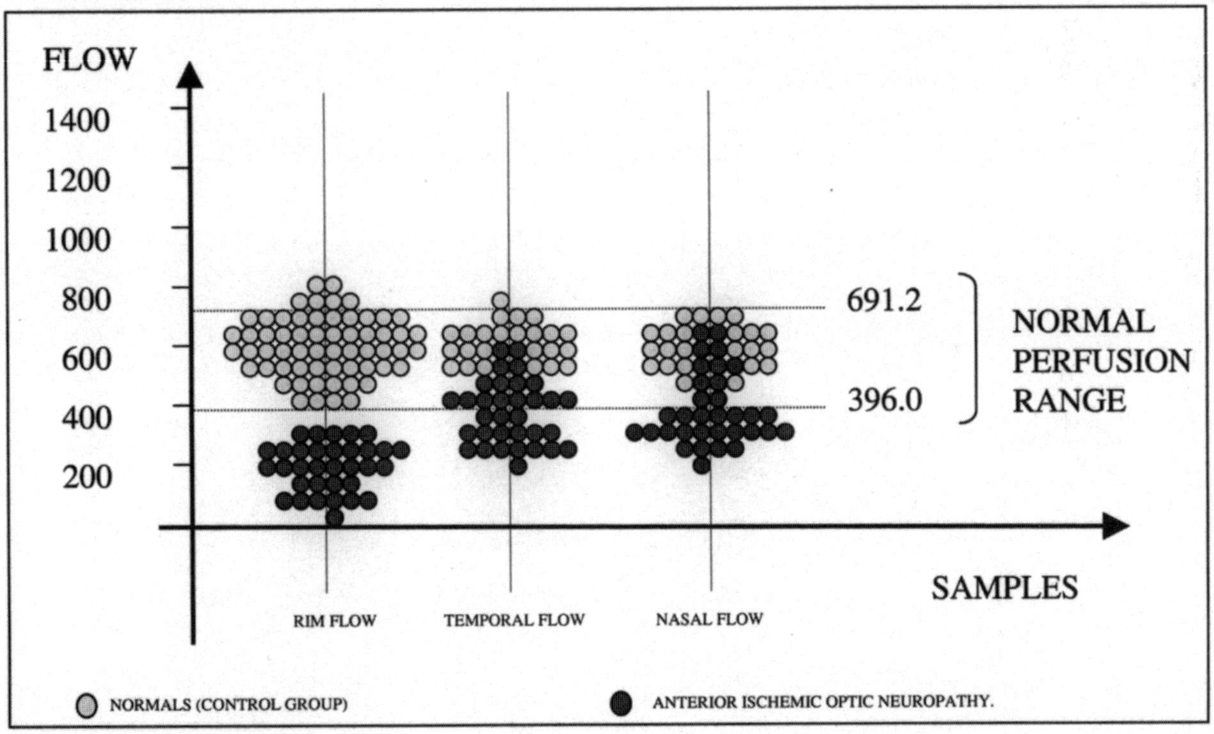

International Ophthalmology **23:** 385–393, 2001.
J.R. Sampaolesi (ed.), Laser Scanning: Update 1, 203–211.




# Imaging the microcirculation of untreated and treated human choroidal melanomas *

A.J. Mueller[1], D.-U. Bartsch[2], U. Schaller[1], W.R. Freeman[2] & A. Kampik[1]
[1]*Augenklinik der Universität, München, Germany;* [2]*University of California San Diego, Shiley Eye Center, 9415 Campus Point Dr. 0946, La Jolla, CA 92093-0946, USA*

*Key words:* choroidal melanoma, confocal scanning laser ophthalmoscope, fluorescein angiography, indocyaninegreen angiography, microcirculation pattern

## Abstract

*Introduction*: Histologically demonstrable microcirculation patterns (microcirculation pattern) of human choroidal melanomas have prognostic significance for the patient. We report on our experience in imaging these microcirculation pattern *in vivo* using simultaneous confocal Fluorescein (FA)- and Indocyaninegreen (ICG) angiography before and after brachytherapy. *Patients and methods*: The simultaneously procured confocal FA- and ICG angiograms of 50 patients with untreated choroidal melanomas were studied for the visibility of microcirculation pattern. Patients were also followed with simultaneous FA/ICG after brachytherapy. *Results*: Confocal FA disclosed signs of tumor vascularization in 12 (24%) of the 50 examined patients but microcirculation pattern only in 3 patients (6%). In contrast, simultaneously obtained confocal ICG disclosed microcirculation pattern in 47 patients (94%). In 10 (77%) of the 13 patients the tumor microcirculation changed considerably after brachytherapy: Distortion, thickening, thinning, as well as complete obliteration of vessels could be observed. *Conclusion*: Histologically demonstrated microcirculation pattern can be imaged *in vivo*. This offers the possibility to assess the likely biologic behavior of the individual tumor without the need for obtaining a cytologic or histologic specimen via enucleation or fine-needle biopsy. Confocal ICG angiogiography images microcirculation pattern better than FA which can be explained by the different absorption-, fluorescence- and exudation-characteristics ICG. Follow-up with confocal ICG of choroidal melanomas after brachytherapy shows different features and allows for visualization of the microcirculation reaction to the treatment which might be a useful tool for studying the effects of future anti-angiogenesis based tumor therapies.

## Introduction

Recently, a large retrospective histopathological study identified 9 different microcirculation patterns (silent, normal, straight, parallel with and without crosslinking, arcs with and without branches, loops, networks) and correlated the presence of these microcirculation patterns (microcirculation pattern) in histological sections with data about death from metastatic disease [1–6]. This study was confirmed by other independent groups [7–9]. Two of these microcirculation pattern (networks, parallel with crosslinking) showed a very strong correlation with metastatic disease. In a Cox proportional hazard model of various prognostically significant factors the presence of networks was the single most important factor for developing metastatic disease. In contrast, two other microcirculation pattern (silent, parallel without crosslinking) were correlated with a more favorable outcome for the patient.

We demonstrated previously that microcirculation pattern that were so far only be described histologically can be imaged by indocyanine green angiography [10]. We also confirmed the presence of various ICG angiographically imaged microcirculation pattern in direct angiographic/histopathologic correlations in hu-

* Supported by Deutsche Forschungsgemeinschaft (DFG) Mu1204/1-1,1-2, Whitaker Foundation (D.- U.B.), and Stern Foundation (W.R.F.)

mans and in newly developed human melanoma animal models that were specifically designed for this purpose [10, 11]. We also demonstrated that the identification of microcirculation pattern using confocal ICG is superior to fluorescein angiograms using a conventional acquisition technique with a fundus camera [12]. It is well known that conventional fluorescein angiography using a fundus camera does not show a pathognomonic fluorescence pattern in choroidal melanomas [13–17] nor are microcirculation pattern consistently seen. Size and pigmentation of the tumor has considerable influence on the appearance of the fluorescein angiogram. The effect of the tumor on adjacent ocular structures, particularly the retinal pigment epithelium, also contributes significantly to its fluorescein angiographic appearance. If the overlying retinal pigment epithelium is completely intact, a fairly normal fluorescein angiogram may result. Overlying retinal vessels are usually visible easily during all phases. However, the so called double or tumor circulation (simultaneous visualization of retinal and choroidal circulation) is often difficult to image. Usually it is only visible when the tumor has broken through Bruch's membrane and/or in the very early phase of the FA. In later phases leakage of dye may mask these phenomenon relatively early. Clearly, the absence of this finding does not indicate lack of extensive tumor vasculature.

Indocyanine green angiography is a method which was primarily developed for imaging choroidal vessels *in vivo* [18–22]. Because the near-infrared light used for indocyaninegreen angiography penetrates the pigmented layers of the fundus more easily than the short wavelength light used in fluorescein angiography, indocyaninegreen angiography has been primarily used to study leakage of dye in choroidal neovascularization during the late phase and thereby to detect changes in the vascular permeability of these vessels. Some of the early studies also have indicated a possible role for indocyanine green angiography in investigating choroidal masses [23, 24], but this was not further investigated. Recently our group described the confocal scanning laser technology to perform indocyaninegreen angiography [25] and later simultaneous FA/ICG angiography [26]. This technique uses sensitive digital image acquisition and processing. The horizontal image resolution has considerably improved to below 20 $\mu$m. The size of the reported histologic microcirculation pattern fall in this range and thus, we hypothesized that they are therefore imageable with the above described technique.

We now report our experience in using a simultaneuos fluorescein-/indocyanine green angiograph to study microcirculation pattern in choroidal melanomas. We also followed the tumors with the same technique after brachytherapy.

## Patients and methods

We report our findings from 50 patients with newly diagnosed prominent choroidal masses diagnosed as choroidal melanomas. Diagnosis of choroidal melanoma was established by indirect ophthalmoscopy based on the characteristic appearance of this tumor and confirmed using standardized ultrasound A- and B-scan. Only tumors with a maximum apical height of more than 1.5 mm and low or low to medium internal reflectivity (2 to 60% retinal spike height) according to standardized ultrasound A-scan were enrolled. Thereafter simultaneous fluorescein/indocyanine green angiography was obtained using a confocal scanning laser ophthalmoscope (Heidelberg Retina Angiograph (HRA), Heidelberg Engineering, Heidelberg (Germany)). The instrument has been described in detail previously [26].

The optics of this instrument allowed for spherical aberration compensation between $-12$ and $+12$ diopters (D). By automatically adjusting the focal plane, confocal serial optical sectioning were obtained (Figure 1). This was performed to facilitate visualization of deep tumor vessels. The simultaneous fluorescein/indocyanine green angiograms were compared for the visibility of the so called double or tumor circulation. Tumor circulation was defined by the presence of tubular structures that were filling with dye during the early (arterial) phase and that appeared empty in the late phase of the angiogram. In addition it was required that the filling pattern of these vessels was separate from the that of the overlying retinal circulation.

If tumor vessels were visible in the fluorescein or indocyanine green angiogram the microcirculation pattern of the tumor circulation was classified in a second step. The angiographic appearance of microcirculation pattern was classified according to the histologic description: normal, silent, straight, parallel with and without crosslinking, arcs with and without branching, loops, networks [1, 2]. If a tumor did not show any sign of localized or diffused hyperfluorescence within the tumor borders the microcirculation pattern was classified as silent. However, some tumor

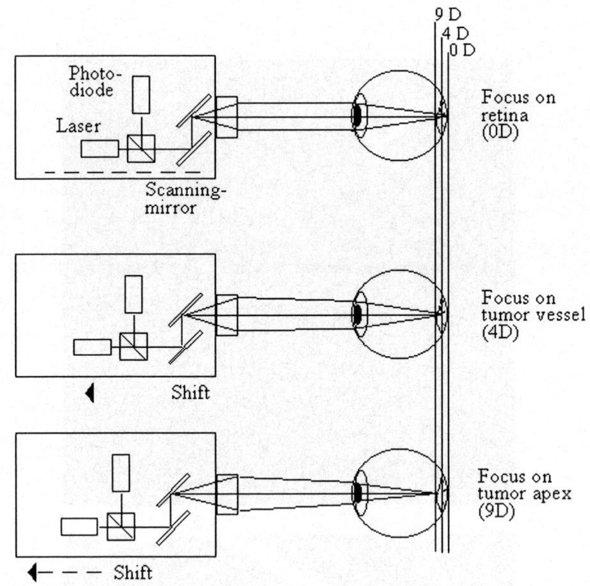


*Figure 1.* Schematic drawing illustrating an optical sectioning series in a tumor of 3 mm height in an emmetropic eye. In this series, the scanning laser beam would be rapidly and consecutively focused in 1 D steps from 0 D (tumor basis) to 9 D (tumor apex). The following steps are shown: focusing on the plane of adjacent retina (top, 0 D), on a tumor vessel (middle, 4 D) and on the tumor apex (bottom, 9 D).

angiograms disclosed multiple microcirculation pattern in different as well as in the same confocal optical sections which is described histologically. Thus, it was possible to classify multiple microcirculation pattern within one tumor.

If the tumor was treated using brachytherapy, the patient was followed regularly using the same technique.

## Results

The fluorescein and indocyanine green angiograms of 50 patients with untreated choroidal melanomas all of whom had a confocal series were investigated in this study. Three patients (5%) had an amelanotic melanoma. The other tumors were pigmented. The median maximal apical tumor height according to standardized A-scan was 3.5 mm with a 95% CI of 2.0 to 7.7 mm. The median maximal diameter of the tumor basis was 10 mm with a 95% CI of 4.5 to 14 mm. The internal reflectivity according to standardized A-scan was low in 34 tumors (68%) and low to medium in the other 16 (32%) tumors.

Tumor vessels in the confocal fluorescein angiograms could be detected in 12 (24%) of the 50 patients. In contrast, confocal indocyanine green angiography revealed tubular structures within the melanoma which were identified as tumor vessels based on their angiographic appearance in 47 (94%) of the studied 50 patients. The difference between the visibility of tumor vessels in the confocal fluorescein (24%) and indocyanine green angiograms (94%) was statistically highly significant ($p = 0.001$, Fisher's exact test).

Traces of microcirculation pattern could be detected in 3 (6%) of the 50 examined patients in the confocal fluorescein angiogram (Figure 2). In all 3 cases it was only possible because the images were taken in the very early arterial phase and in all three cases the microcirculation pattern was masked in later frames due to leakage of the dye. In contrast, microcirculation pattern were identified in the confocal indocyanine green angiograms of all 47 patients in which tumor vessels were identified. Moreover, this was easily possible throughout the whole arterial venous phase and not limited to the early arterial phases (Figure 3). Additionally, all 3 angiograms which did not show any tumor vessels could be classified as having a silent microcirculation pattern.

In summary, in 13 areas the microcirculation was classified as "normal", in 10 as "silent", in 21 as "straight", in 3 as "parallel without crosslinking", in 12 as "parallel with crosslinking", in 7 as "arcs without branching", in 7 as "arcs with branching", in 6 as "loops" and in 6 more areas as "networks" (Figure 4). None of the cases with multiple microcirculation pattern contained the "silent" or "parallel without crosslinking" pattern together with the "parallel with crosslinking" pattern or "networks".

Thirteen of these patients were followed after brachytherapy. The age of the patients ranged from 42 to 82 years (mean 65.4 years). Of these 9 patients were female (69%) and 4 male (31%). Seven tumors were located in the right and 6 in the left eye.

Maximum apical tumor height of this subgroup was median 5.5 mm with a 95% confidence interval of 3.4 to 7.7 mm according to standardized A-scan ultrasound. All tumors showed a low or low to medium internal reflectivity according to standardized A-scan. Before radiation brachytherapy signs of tumor microcirculation could be imaged with indocyanine green angiography in all 13 patients (100%). After radiation brachytherapy we observed changes of the microcirculation in 10 patients 77%). An angiographic response

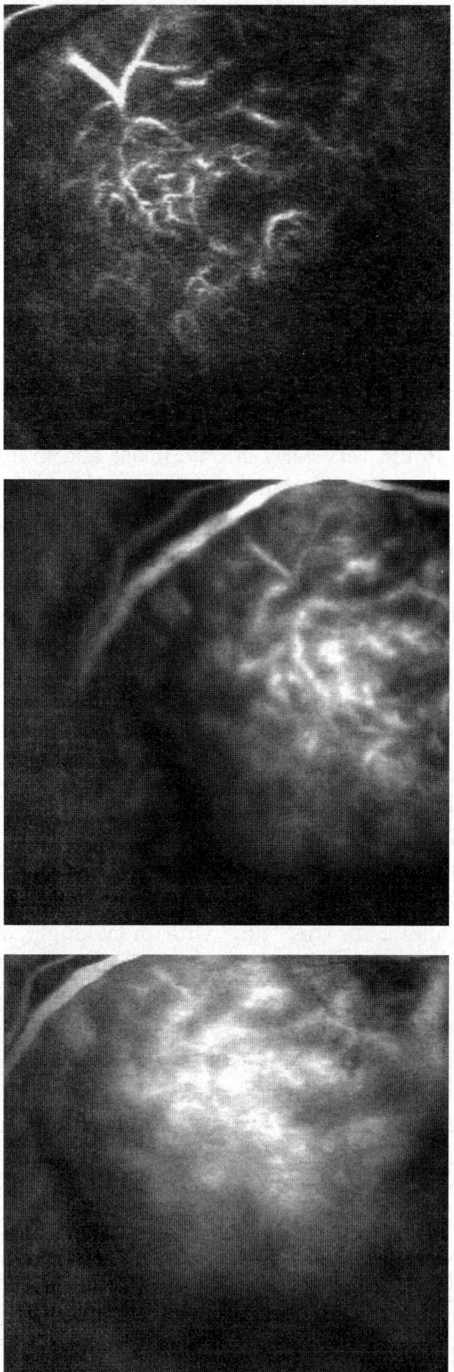

*Figure 2.* Serial confocal fluorescein angiography of a choroidal melanoma. While the early phase (top) clearly reveals signs of a double circulation, the tumor microcirculation is masked in the middle (center) and late phase (bottom) due to diffuse leakage of dye.

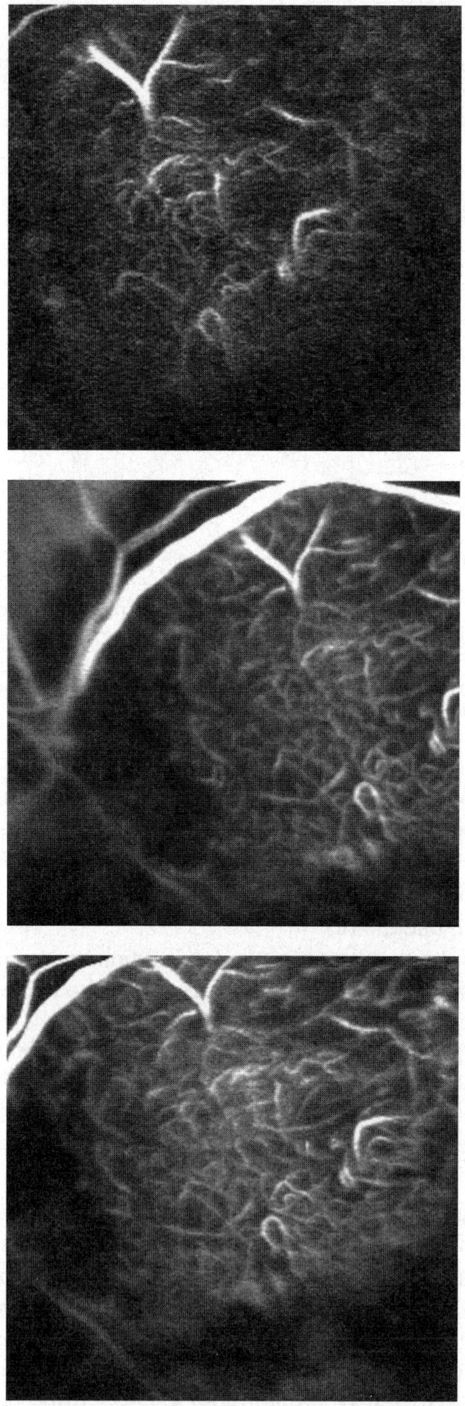

*Figure 3.* Confocal indocyanine green angiogram of the same tumor as in Figure 2. The microcirculation pattern can be consistently identified throughout the early (top), middle (center) and late (bottom phase.)

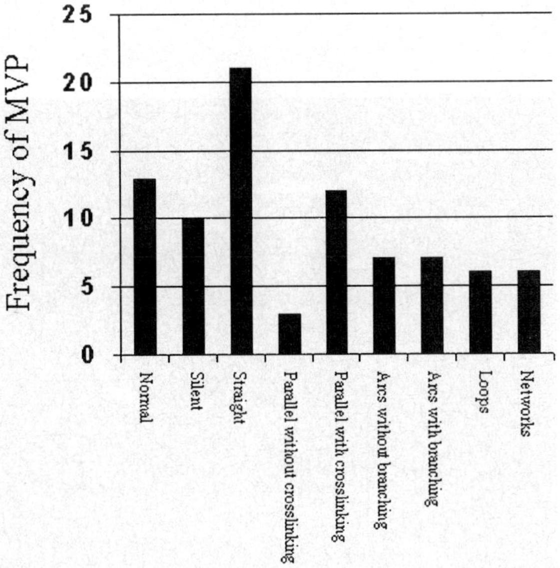


*Figure 4.* Distribution of identified microcirculation pattern. All histopathologically described microcirculation pattern could be identified.

was defined as distortion, thickening, thinning, or obliteration of previously imaged tumor vessels (Figure 5). In 4 patient (30%) we detected thickening of tumor vessels. Thinning was observed in 4 patients (30%). Obliteration of vessels was observed in 6 patients (46%). In 4 patients (31%) we observed distortion of vessels compared to the location before treatment. In 3 of these patients (30%) we saw only one of the above defined changes of the microcirculation after radiation brachytherapy, in 5 patients (38%) we found 2, and in 2 more patients (20%) we found 3 different changes of the microcirculation as described before.

In 12 patients (92%) we saw a clinical tumor regression after one year. Maximum apical tumor height was median 2.9 mm with a 95% confidence interval of 1.5 to 4.5 mm according to standardized A-scan ultrasound. 8 tumors (61%) showed a medium or medium to high internal reflectivity according to standardized A-scan. 4 tumors (30%) showed a low to medium internal reflectivity, in one tumor (7%) the internal reflectivity remained low within the observation period. In 1 patient we observed tumor revascularisation in a previously avascular area (Figure 6). Subsequently the tumor increased in height, which was documented with standardized A-scan followed by enucleation of the eye. In this case the ICG-angiographically documented changes preceeded echographically detected regrowth by 2 months.

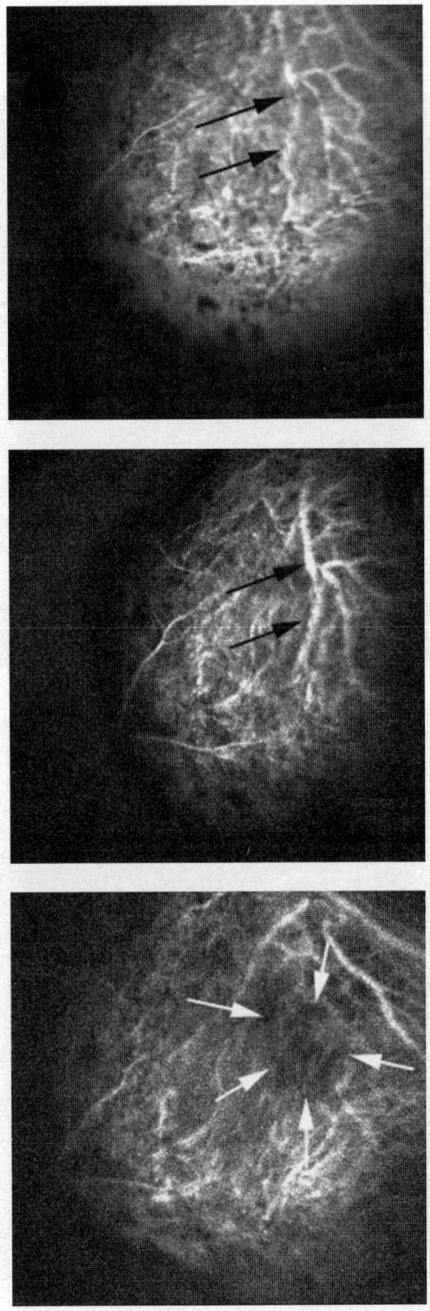

*Figure 5.* Female patient with a choroidal melanoma located in the upper temporal quadrant of the left eye. The maximum apical tumor height is measured with 4.9 mm according to standardized ultrasound. Top: Indocyanine green angiography before treatment. Multiple dilated tumor vessels with anastomoses are imaged. Note the localization of the vessel marked with black arrows. Center: Indocyanine green angiography 5 weeks after brachytherapy using I-125. The tumor vessel marked with black arrows is slightly distorted but still perfused. Bottom: Indocyanine green angiography at 9 weeks after brachytherapy. The vessel marked with black arrows in the previous images is not perfused any more (white arrows).

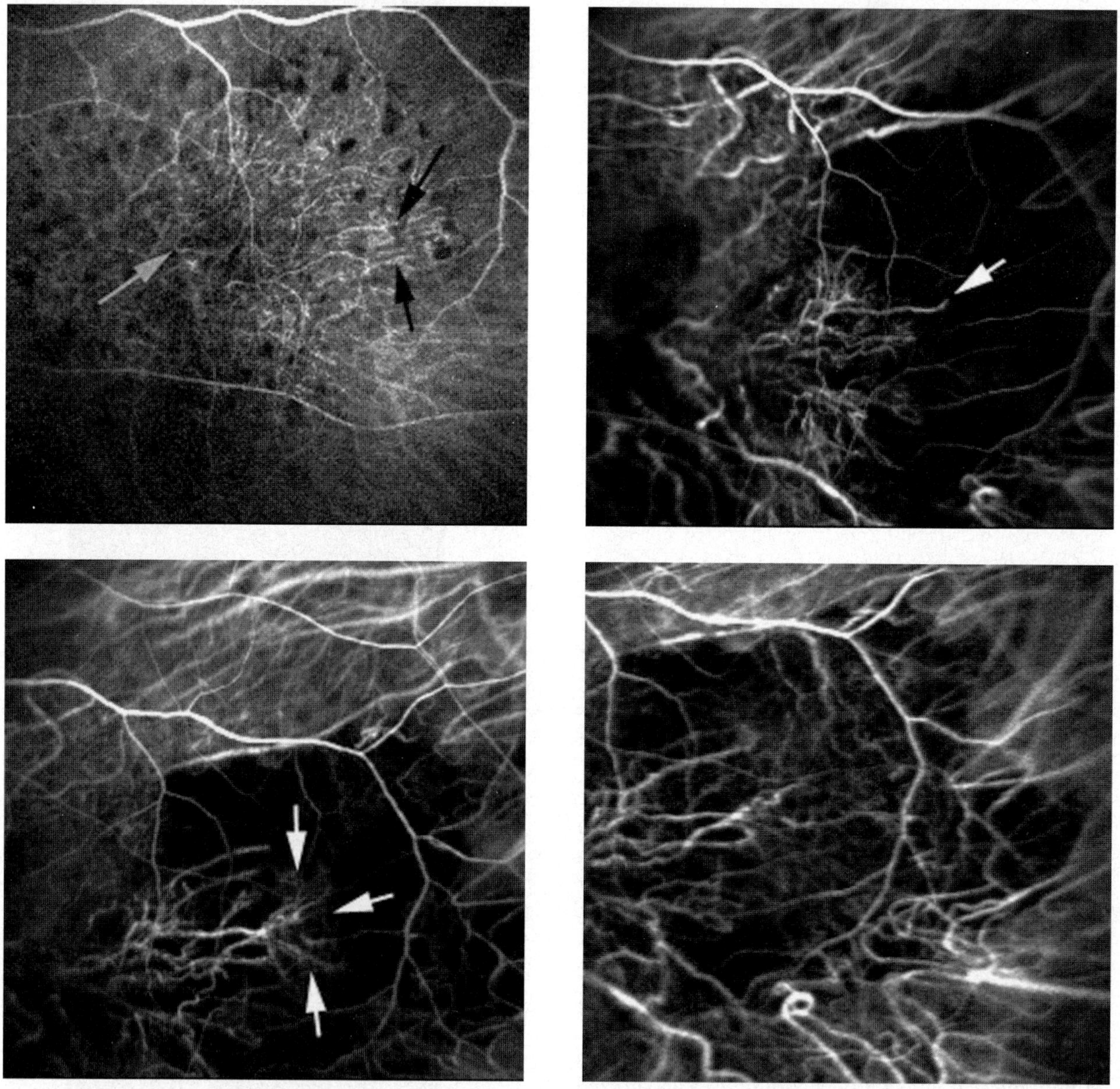

*Figure 6.* Male patient with a choroidal melanoma located temporal to the macula of the left eye. Maximum apical tumor height is measured with 3.4 mm according to standardized ultrasound. Top left: Indocyanine green angiography before treatment. Two microcirculation patterns are imaged in this confocal optical section using indocyanine green angiography: parallel without crosslinking (black arrows) and arcs (grey arrow). Top right: Indocyanine green angiography 7 months after local brachytherapy. An area of complete avascularity is seen. Within this area a partially perfused vessel trunk (white arrow) originating from a rest plexus of tumor vascularization is imaged. Bottom left: Indocyanine green angiography 15 months after local brachytherapy. A fan-shaped revascularization (white arrows) originates from the vessel trunk noted in the previous image. Bottom right: Indocyanine green angiography 23 months after local brachytherapy. The revascularization has considerably enlarged. The patient was enucleated 2 months later due to echographic documented tumor regrowth.

In 3 patients (23%) we did not observe any change of tumor microcirculation after radiation brachytherapy. However, we found a tumor regression in the standardized A-scan in.

## Discussion

The accuracy of diagnosing a choroidal melanoma has considerably increased over the last 25 years [27]. Several well established treatment options are currently available, most of them capable of sufficiently controlling the primary tumor. However, it appears that none of these approaches is able to prevent late metastasis. Because none of the currently available treatment options seems to influence the rate of metastasis, it is desirable to identify clinically patients with high and low risk of developing metastases. For patients with low risk of developing metastases frequent clinical observation might be the best management of the tumor, especially if the tumor is small and does not show any growth. In contrast, several publications during the last years have suggested a possible role of active immunotherapy for the prevention or adjuvant chemotherapy for the treatment of overt metastases [28, 29]. Some of the used substances are toxic and certainly not suitable for all patients with choroidal melanomas. Thus, it is also desirable to identify clinically patients with high risk of developing metastatic disease who might qualify for these new therapies. Although highly desirable the role of tumor markers for identifying patients with a high risk for developing metastatic disease is currently under investigation but still remains to be established [30].

Recently it has been shown that certain histologically identified microcirculation pattern are an independent risk factor for the metastatic behavior of choroidal melanomas. Especially the microcirculation pattern "parallel with cross-linking" and "networks" indicate a high probability for metastatic disease whereas the presence of "silent" and "parallel without crosslinking" microcirculation pattern indicated a better prognosis for the patients [1–6]. Because this study was clearly confirmed by other independent histopathology labs [7–9] these important findings motivated us to determine whether it is possible to image and identify prognostically significant microcirculation pattern of choroidal melanomas *in vivo* using conventional fluorescein and indocyanine green angiography with a new confocal scanning ophthalmoscope [25]. We have found that confocal indocyanine green angiography is capable of detecting microcirculation pattern and superior to fluorescein angiography using a conventional fundus camera. Since the commercial availability of a simultaneous fluorescein/indocyanine green angiograph [26] we now routinely use this instrument to study microcirculation pattern in human choroidal melanomas.

When comparing the visibility of tumor vascularization in fluorescein with confocal indocyaninegreen angiograms, both taken simultaneously with the same confocal scanning laser ophthalmoscope, we found that tumor vessels were imaged in 47 (94%) of the examined 50 patients using confocal indocyanine green angiography. In the remaining 3 patients the tumor appeared indocyanine angiographically as empty. In these patients the microcirculation pattern was classified as silent. Hence, indocyanine green angiographically determined counterparts of histologically described microcirculation pattern could be determined in all patients in various confocal optical sections. Specifically we were able to image "parallel with crosslinking" in 12 areas and "networks" in 6 areas. The microcirculation pattern "silent" was imaged in 10 areas and "parallel without crosslinking" in 3 areas. All 4 microcirculation pattern have been demonstrated to be highly correlated with a specific risk for the patient of developing metastatic disease. In none of our cases high risk microcirculation pattern ("networks", "parallel with crosslinking") were seen together with low risk microcirculation pattern ("silent", "parallel without crosslinking").

In contrast, simultaneously taken confocal fluorescein angiography was able to reveal signs of tumor vessels in only 12 (24%) of the studied 50 patients. Moreover, in only 3 of the latter cases, traces of microcirculation pattern could be identified in the confocal fluorescein angiograms. However, this was only possible in the very early arterial phase of the angiogram. In the later phases the microcirculation pattern was masked due to leakage of the dye.

It has been previously attempted to image the microcirculation of choroidal melanomas using fluorescein angiography after brachytherapy [15, 17]. These studies have found that the most reliable angiographic signs of successful radiotherapy are occlusion of tumor and choroidal vessels, avascular areas within the radiated zone, and decreased or no leakage into the tumor remnants. Histologically, cellular necrosis, fibrosis of the tumor stroma, accumulation of pigmented macrophages, dose related vacuolization of tumor cells and inflammation are described in eyes enucleated after radiation therapy with helium ion, proton beam or radioactive plaques [31–34]. Vascular changes in the tumor were also noted and consisted of degeneration, hyalinization, thickening and necrosis of vessel walls. Vascular occlusions are described as well [31–33, 35].

In our sample of 13 patients that were followed after brachytherapy we found changes of the tumor microcirculation in 10 patients (77%). We could observe thickening, thinning, obliteration, and distortion of the vessels. These changes did not preceed the regression of the mass. We also observed regrowth of new vessels in one case into a previously avascular area. Later on in the follow-up of this patient standardized A-scan revealed regrowth of the lesion and the eye was enucleated. In this case ICG-documented revascularization indicated a very early and sensitive sign of tumor regrowth even before this was detectable clinically or by standardized ultrasound. It might be argued that this revascularization is due to the formation of a choroidal neovascularisation. However, the fact that tumor regrowth was documented makes it more likely that these new vessels are linked to the tumor regrowth.

Interestingly, 3 patients (23%) showed no changes of the microcirculation despite a documented tumor regression. At the present time we don't know the significance of this finding. We are continuing to follow these patients, because persistent functioning microcirculation of a tumor might reflect a bad prognostic sign for regrowth. In the latter cases, we should observe regrowth during further follow-up. Our analysis of time to regression demonstrates that radiation therapy does not appear to have an effect on the microcirculation before the tumor begins to regress as measured by ultrasound. This is an important finding because some authors speculate that the principle effect of radiation therapy is to destroy the tumor vascularization which subsequently causes the tumor to shrink. At this point we do not know if the observed changes are a result of the radiation or merely an accompanaying effect of tumor regression. However the observation that the changes in angiography did not preceed the sonographically measured regression would argue against a direct effect of radiation on the tumor microcirculation.

In summary, our study demonstrates that indocyanine green angiography using a confocal scanning laser ophthalmoscope is superior to fluorescein angiography even when using the same instrument in imaging microcirculation pattern. The angiographically seen microcirculation patterns appear to be identical to patterns that have been identified histologically and that have prognostic significance in choroidal melanomas. Thus, this minimally-invasive technique has the potential to be used as a "non-invasive biopsy". This offers the possibility to determine the likely biologic behavior of a choroidal melanoma without the removal of tissue. Follow-up of these tumors after treatment with the same technique provides the examiner with additional information about the biologic behavior of primary choroidal melanomas after radiation brachytherapy. This method may also play a future role in monitoring patients with choroidal melanomas who will undergo anti-angiogenesis therapy. Although this is not a standard treatment yet there are currently 20 compounds of anti-angiogenesis factors under clinical investigation [36] and it is likely that some of these will enter phase III trials. In this scenario confocal ICG could be helpful in monitoring the anti-angiogenetic effect *in vivo*.

## References


1. Folberg R, Pe'er J, Gruman L, et al. The morphologic characteristics of tumor blood vessels as a marker of tumor progression in primary human uveal melanoma: a matched case-control study. Hum Pathol 1992; 23: 1298–1305.
2. Folberg R, Rummelt V, Parys, et al. The prognostic value of tumor blood vessel morphology in primary uveal melanoma. Ophthalmology 1993; 100: 1389–1398.
3. Folberg R, Mehaffey M, Gardner L, et al. The microcirculation of choroidal and ciliary body melanomas. Eye 1997; 11: 227–238.
4. Rummelt V, Folberg R, Rummelt C, et al. Microcirculation architecture of melanocytic nevi and malignant melanomas of the ciliary body and choroid. Ophthalmology 1994; 101: 718–727.
5. Rummelt V, Gardner L, Folberg R, et al. Three-dimensional relationships between tumor cells and microcirculation with double cyanine immunolabeling, laser scanning confocal microscopy, and computer-assisted reconstruction: an alternative to cast corrosion preparations. J Histochem Cytochem 1994; 42: 681–686.
6. Rummelt V, Folberg R, Woolson R, et al. Relation between the microcirculation architecture and the aggressive behavior of ciliary body melanomas. Ophthalmology 1995; 102: 844–851.
7. Char D, Kroll S, Crawford J, et al. Uveal melanoma cycling, vascular patterns and prognosis. Inv Ophth Vis Sci 1994; 35: 17–21.
8. Sakamoto T, Sakamoto M, Yoshikawa H, et al. Histologic findings and prognosis of uveal malignant melanoma in japanese patients. American Journal of Ophthalmology 1996; 121: 276–283.
9. Seregard S, Spanberg B, Juul C, Osskarsson M. Prognostic accuracy of the mean of the largest nucleoli, vascular pat-

terns, and PC-10 in posterior uveal melanoma. Ophthalmology 1998; 105: 485-491.

10. Mueller AJ, Bartsch DU, Folberg R, et al. Imaging the microvasculature of choroidal melanomas with confocal indocyanine green scanning laser ophthalmoscopy. Arch Ophthalmol 1998; 116: 31–39.
11. Mueller AJ, Folberg R, W. F, et al. Evaluation of the human choroidal melanoma rabbit model for microvascularization patterns. Exp Eye Res 1999; 68: 671–678.
12. Mueller AJ, W. F, Folberg R, et al. Evaluation of microvascularization pattern visibility in human choroidal melanomas comparing simultaneously taken confocal Fluorescein with Indocyanine green angiograms. Graefe's Arch Clin Exp Ophthalmol 1999; 237: 448–456.
13. Edwards MG, Schachat AP. Tumors in the posterior pole. International Ophthalmology Clinics 1995; 35: 123–135.
14. Flindall R, Gass J. A histopathologic fluorescein angiographic correlative study of malignant melanomas of the choroid. Can J Ophthalmol 1971; 6: 258–267.
15. Lommatzsch P, Ballin R, Helm W. Fluorescein angiography in the follow-up study of choroidal melanoma after 106Ru/106Rh plaque therapy. Retina 1987; 7: 148–155.
16. Pettit T, Barton A, Foos R, Christensen R. Fluorescein angiography of choroidal melanomas. Arch Ophthalmol 1970; 83: 27–38.
17. Tarkkanen A, Laatikainen L. Fluorescein angiography in the long-term follow-up of choroidal melanoma after conservative treatment. Acta Ophthalmol 1985; 63: 73–79.
18. Flower R. Infrared absorption angiography of the choroid and some observations on the effects of high intraocular pressures. Am J Ophthalmol 1972; 74: 600–614.
19. Flower R, Hochheimer B. Clinical infrared absorption angiography of the choroid. Am J Ophthalmol 1972; 73: 458–459.
20. Flower R. Injection technique for indocyanine green and sodium fluorescein dye angiography of the eye. Invest Ophthalmol Vis Sci 1973; 12: 881–895.
21. Flower R, Hochheimer B. Indocyanine green dye fluorescence and infrared absorption choroidal angiography performed simultaneously with fluorescein angiography. Johns Hopkins Med J 1976; 138: 3–42.
22. Hochheimer B. Angiography of the retina with Indocyanine green. Arch Ophthalmol 1971; 86: 564–565.
23. Bischoff P, Flower R. Ten years experience with choroidal angiography using indocyanine green dye: a new routine examination or an epilogue? Doc Ophthalmol 1985; 60: 235–239.
24. Chopdar A, Turk A, Hill D. Fluorescent infra-red angiography of the fundus oculi using indocyanine green dye. Trans Ophthalmol Soc Uk 1978; 98: 142–146.
25. Bartsch DU, Weinreb R, Zinser G, Freeman W. Confocal scanning infrared laser ophthalmoscopy for indocyanine green angiography. Am J Ophthalmol 1995; 120: 642–651.
26. Freeman WR, Bartsch DU, Mueller AJ, et al. Simultaneous indocyanine green and fluorescein angiography using a confocal scanning laser ophthalmoscope. Archives of Ophthalmology 1998; 116: 455–463.
27. Albert D, Marcus D. Accuracy of diagnosis of choroidal melanomas in the collaborative ocular melanoma study. Arch Ophthalmol 1990; 108: 1268–1273.
28. Mitchell M, Harel W, JK-M, et al. Active specific immunotherapy of melanoma with allogeneic cell lysates. Rationale, results, and possible mechanisms of action. Ann NY Acad Sci 1993; 690: 153.
29. Quan W, Mitchell M. Phase II trial of carbetimer in metastatic melanoma. Invest New Drugs 1993; 11: 231–233.
30. Schaller U, Mueller AJ, Bosserhoff A, et al. Melanoma inhibiting activity (MIA): Evaluierung eines neuen tumorassozierten Antigens als Serummarker für uveale Melanome. Ophthalmologe, in press.
31. Crawford J, Char D. Histopathology of uveal melanomas treated with charged particle radiation. 1987; 94: 639–643.
32. MacFault P, Morgan G. Histopathological changes in malignant melanomas of the choroid after cobalt plaque therapy. Brit J Ophthalmol 1977;61:221-8.
33. Messmer E, Bornfeld N, Foerster M, et al. Histopathologic findings in eyes treated with a ruthenium plaque for uveal melanoma. Graefes Arch Clin Exp Ophthalmol 1992; 230: 391–396.
34. Shields C, Shields J, Karlsson U, et al. Enucleation after plaque radiotherapy for posterior. Ophthalmol 1990; 97: 1665–1670.
35. Seddon J, Gragoudas E, Albert D. Ciliary body and choroidal melanomas treated by proton beam irradiation. Arch Ophthalmol 1983; 101: 1402–1408.
36. Casey R, Li W. Factors controlling ocular angiogenesis. Am J Ophthalmol 1997; 124: 521–529.

*International Ophthalmology* **23:** 395–398, 2001.
*J.R. Sampaolesi (ed.), Laser Scanning: Update 1*, 213–216.



# Oral fluorescein angiography with scanning laser ophthalmoscope

C.J. Argento, O.R. Lopez Mato & M.D. Martinez-Cartier
*Instituto de la Visión, Buenos Aires, Argentina*

**Abstract**

*Objective:* To study the efficacy of oral fluorescein angiography compared to intravenous (IV) fluorescein angiography in several retinal diseases. *Patients and methods:* We performed oral fluorescein angiography with a confocal SLO (Heidelberg Retina Angiograph-HRA) in fourteen psychologically challenged individuals and children, all of whom explicitly refused venipuncture. Patient's diagnosis included diabetes, central serous choroidopathy, toxoplasmosis and choroidal neovascularization. A dose of 30 mg/kg body weight of fluorescein was given. *Results:* We obtained images of adequate quality to allow interpretation in all cases. *Conclusions:* Even though the images were not nearly as good as when obtained after IV injection, we conclude that oral fluorescein angiography with SLO is an appropriate alternative in those cases where a venipuncture is contraindicated by medical or psychological reasons.

## Introduction

Intravenous (IV) fluorescein angiography (FA) is a well established procedure which has been performed with excellent results for almost 40 years. However, oral FA has not gained wide acceptance because of its technical limitations, mainly low image resolution when obtained with conventional fundus cameras. With this technique early fluorescein stages are not detected and in most cases do not allow visualization of fine vessel detail necessary to perform photocoagulation, being suited only for conditions in which coarse late leakage determination is useful for confirmation of diagnosis (i.e., cystoid macular edema, central serous choroidopathy, diabetic retinopathy with neovascularization).

Several complications are related to IV FA; they can range from minor ones like nausea and vomiting, to severe ones like bronchospasm and anaphylaxis. It has been reported that oral FA is associated with fewer adverse reactions compared to IV FA, as is the case with several other drugs. Besides, in a number of cases oral FA is the only alternative because of inaccesible veins or lack of patient cooperation.

## Patients and methods

Fourteen subjects underwent oral fluorescein angiography with a confocal SLO (Heidelberg Retina Angiograph-HRA, Heidelberg Engineering, Germany) most of them psychologically challenged individuals (9 patients) whom refused a venipuncture, 3 children and 2 normal volunteers. Image frames of 30° and 20° were taken with a resolution of 512 × 512. There were 9 females and 5 males, with ages ranging from 6 to 74 years. A dose of 30 mg/kg body weight of fluorescein was given in all cases.

All studies were performed at Instituto de la Visión between August 1998 and June 1999 by a single operator. All patients fasted at least 3 hours before the procedure. Pupillary dilation was performed with tropicamide 1% (Alcon Mydril®), 1 drop every 15 minutes, four drops total. Patients drunk orange juice mixed with 30 mg/kg body weight of 10% fluorescein (Fluoresceina Loa®). Images were obtained at 5 minutes intervals beginning at 10 minutes after drinking the fluorescein mixture. Pictures so gathered were stored at a hard disk for subsequent analysis.

Angiograms were graded according to the method described by Watson into four categories: 3 = strong fluorescence; 2 = moderate fluorescence; 1 = faint fluorescence; 0 = no useful fluorescence. Grade 3 were

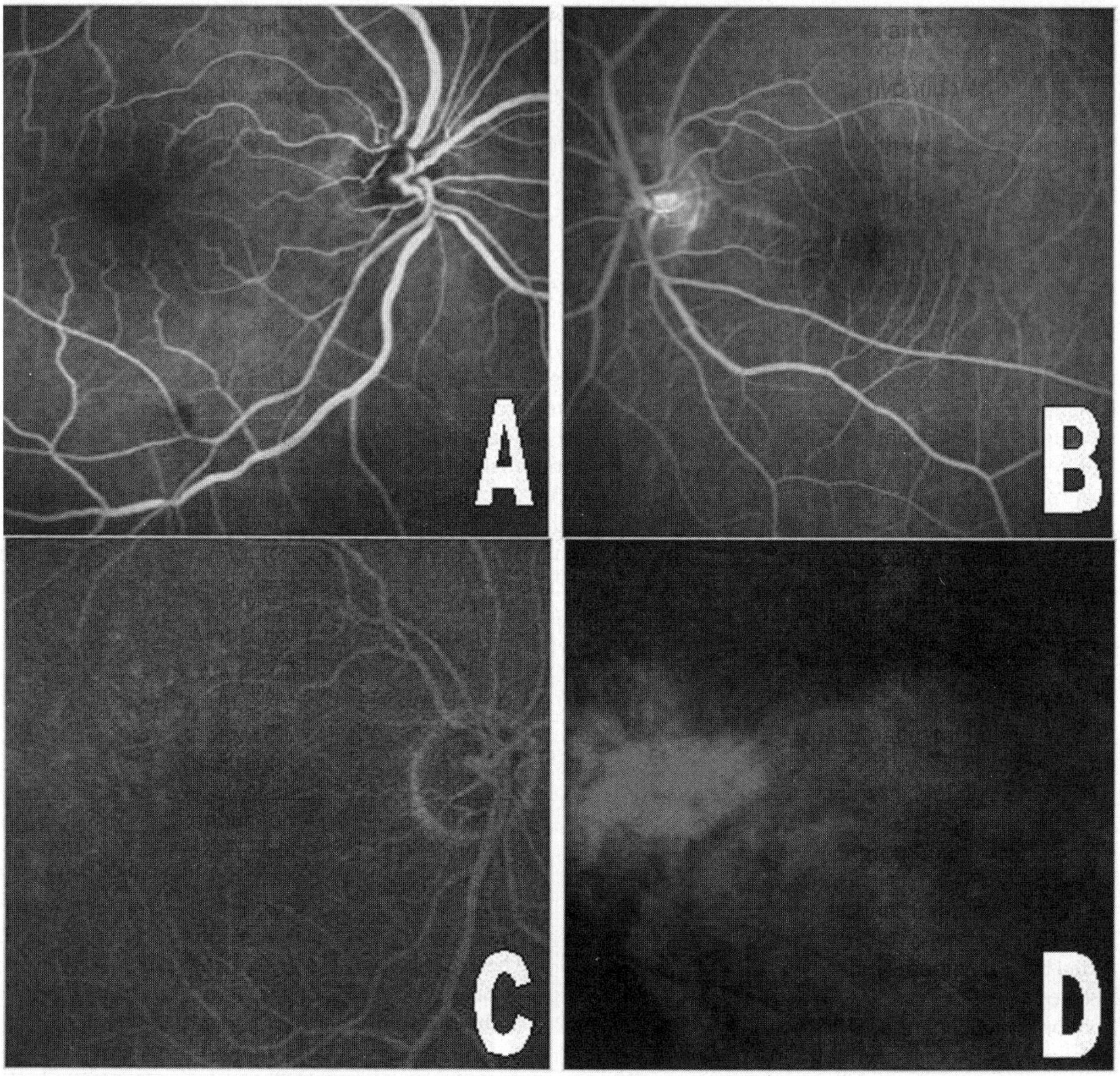


*Figure 1.* Image grading: (a) grade 3; (b) grade 2; (c) grade 1; (d) grade 0.

considered excellent and grade 2 were considered adequate for interpretation; on the other hand, in grade 1 only the optic nerve, macula and prominent leakage would be noted and in grade 0 no useful information would be obtained.

## Results

We obtained grade 3 images in four cases, grade 2 images in eight cases and grade 1 cases in the two remaining cases. No patient experienced adverse reactions to oral FA. Early fluorescence was noted at the first imaging session at 10 minutes, but the best images were obtained at 20 minutes in most cases. We did not encounter a single case of inaccessible veins, although this may represent an indication for oral FA.

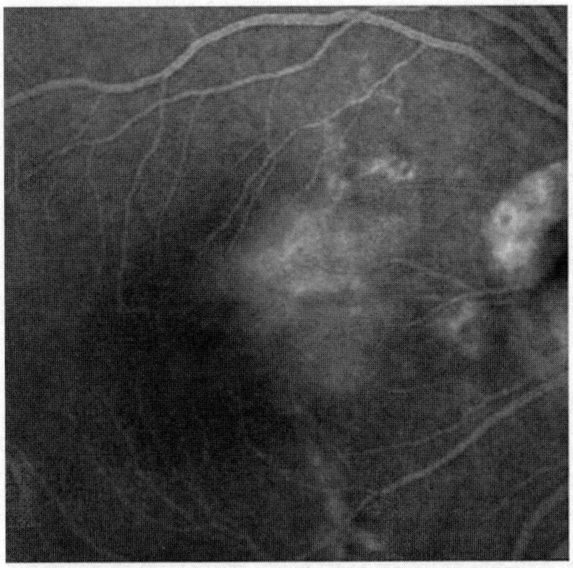

*Figure 2.* Image taken at 25 minutes. It shows right macula of a patient with active toxoplasmosis retinochoroiditis.

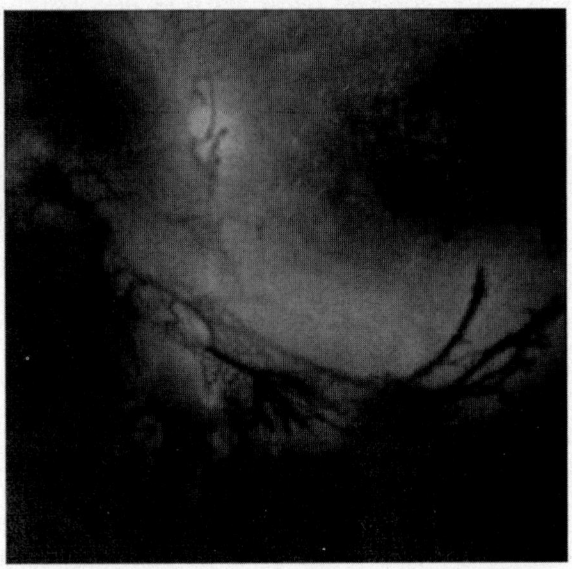

*Figure 3.* Image taken at 35 minutes. Neovascular net silhouette can be seen along the inferior temporal arcade. Prominent leakage from retinal neovascularization is shown.

## Discussion

Conventional FA by IV injection results in good images in most cases because fluorescein is a very efficient molecule in terms of light emission. Confocal SLO are more sensitive than fundus cameras to gather faint fluorescence traditionally associated with oral

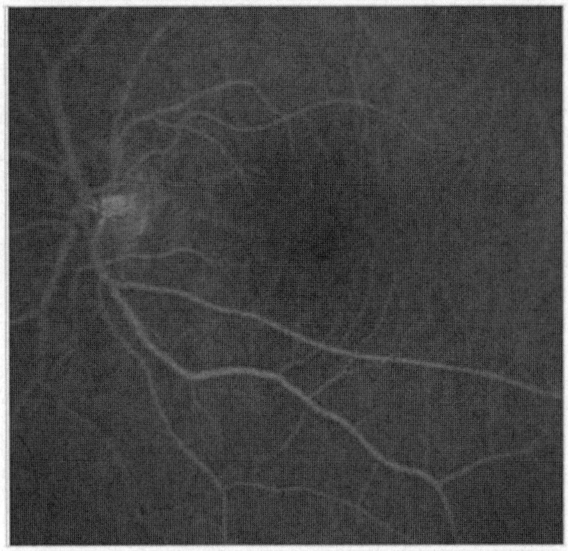

*Figure 4.* Thirty degree image taken at 20 minutes from a normal volunteer.

FA. We agree than IV FA is superior and less time consuming than oral FA; we only performed this study when there was no other alternative (12 cases out of 850). Patients who were very anxious about IV injection were particularly grateful of given the possibility of avoiding venipuncture, as well as some children. Most of adult patients involved in this study were taking antidepressants or tranquilizers.

Even though the number of patients in this series is limited, we did not observe any adverse effect, compared to 8% incidence of mild adverse effect in our series (unpublished data, prospective study on IV FA and ICG complications).

Oral FA studies in the past were done using fundus cameras and concluded that this modality was useful only when late low quality images were acceptable, as in patients with cystoid macular edema or diabetic retinopathy with neovascularization. We found that with a SLO this is not necessarily the case and were able to obtain good to excellent images in 85% of patients.

## References


1. Gass JD. Stereoscopic Atlas of Macular Diseases: Diagnosis and Treatment, Mosby-Yearbook, 1997.
2. Gómez-Ulla F, Malvar A, Parafita M, et al. Oral fluorescein angiography and fluoroscopy: Determination of plasma fluorescein levels and clinical application. Optom Vis Sci 1992; 69: 986–990.

3. Martínez-Cartier et al. Estudio prospectivo sobre complicaciones de la angiografia con fluoresceína e indocianina verde (resultados preliminares).
4. Watson AP, Rosen ES. Oral fluorescein angiography: Reassessment of its relative safety and evaluation of optimum conditions with use of capsules. Br J Ophthalmol 1990; 74: 458–461.
5. Yannuzzi LA, Rohrer KT, Tindel LJ, et al. Fluorescein angiography complication survey. Ophthalmology 1986; 93: 611–617.

International Ophthalmology **23:** 399–404, 2001.
J.R. Sampaolesi (ed.), Laser Scanning: Update 1, 217–222.




# Basic investigations for 2-dimensional time-resolved fluorescence measurements at the fundus

D. Schweitzer, A. Kolb, M. Hammer & E. Thamm
*Department of Experimental Ophthalmology, University of Jena, Germany*

## Introduction

Fluorescence measurements are considered as a tool in the discovering of the pathomechanism of ocular diseases especially in age-related macular degeneration. The presentation of fluorescence images is a kind of 2-dimensional functional diagnosis. For an interpretation of fluorescence images, it is necessary to know, which substances are detected at different fundus sites. For that reason, it was the goal of the investigation is to find out a method, which allows both the discrimination of different fluorophores and the 2-dimensional distribution of autofluorescent substances at the fundus. The main problem in fluorescence measurements at the fundus is the very weak detectable signal, which cannot be increased because of the limitation of the excitation by the maximal permissible exposure. As the eye is a moving object, an averaging of a number of fluorescence images to improve the signal to noise ratio is possible up to a certain degree, but it requires additional expenditures. A further limitation for fluorescence measurements is given by the spectral transmission of the ocular media, which allows spectral investigations only in a wavelength-range between about 400 nm and 900 nm.

## Discrimination of fluorophores

In the literature fluorescent substances in biological tissue are characterised by excitation and emission spectra, as well as by the decay time [1–4]. Some of these substances describe distinctive stages of metabolism in the tissue. But only few of them can be excited at the fundus. As the cornea has an absorption edge at 350 nm and the crystalline lens at 400 nm, only fluorophores of the fundus are detectable, having an excitation range higher than 400 nm. In Table 1, the wavelengths for excitation and emission as well as the decay time of some groups of intrinsic fluorophores are given. Only the heavy typed fluorophores in the table are detectable at the fundus in principle. The most expected fluorophores at the fundus are the products of the lipid-peroxidation and the flavines.

A discrimination of fluorophores is possible according to the spectral range of the excitation, to the emission spectra or to the decay time of fluorescence after short time excitation.

Taking into account the transmission of the ocular media, the excitation spectra of the fluorophores at the fundus exhibit in the visible range no large differences. That means, a discrimination of fundus fluorophores by their excitation spectra is very difficult.

One possibility for the discrimination of fluorophores is the measurement of the emission spectra by imaging spectrometry [5, 6]. A line of 1.5 mm at the fundus is excited in different spectral ranges. Using matched cut off filters, the fluorescence light will be dispersed by a spectrograph. At the exit of the spectrograph, the spectrally dispersed fluorescence light of different excited locations is simultaneously detectable by an intensified CCD matrix. As an example, in Figure 1, the excitation ranges and the corresponding fluorescence spectra for fundus investigations are given. The spectral characteristic of the detection arrangement was compensated. The autofluorescence spectra after short-wave and long-wave excitation were investigated in age-related macular degeneration. It can be demonstrated a different behaviour of the integral normalised long-wave autofluorescence in case of early or in late AMD. In contrast, the short-wave excited autofluorescence is not different between both stages of AMD [7].

Despite a discrimination of fluorophores is possible by imaging spectrometry, the detection of the 2-dimensional distribution of different fluorophores is very time consuming.

*Table 1.* Groups of natural fluorophores

| Autofluorescent substances | Excitation $\lambda_{max}$ (nm) | Emission $\lambda_{max}$ (nm) | Decay time (ns) |
|---|---|---|---|
| Aminoacids | 248–289 | 282–359 | 2.6–3.6; 6.4–6.8 |
| Nucleotides | 260–350 | 313–510 | < 0.02; <1; 6.3 |
| Photoproducts of Tryptophans | 300–370 | 385–490 | 1.2; 3.7; 6.1; 8.6 |
| Connecting tissue | 280–**450** | 310–530 | |
| Lipid-peroxidation (Lipofuscines) | 261–**500** | 323–672 | 0.06; 0.2–0.32; 1.2–1.7; 3.1; 4.8; 6.5 |
| Melanin | 350–**500** | 440–490 | 0.2; 0.7; 2.3; 7.8 |
| Eosinophile | 370–**500** | 440–550 | |
| Hemsynthesis | **400–575** | 613–705 | 0.7; 2; 3; 4.5; 17; 20 |
| Hem-decomposition | **475** | 525–575 | 0.02–0.09; 1–2 |
| NAD | 260 | 445 | |
| NADH + $H^+$ | 290, 340 | 440, 450, 470 | 0.2–0.3; 1.4–2.4; 6.0–8.0 |
| FAD | **450; 488** | 500–650 | 0.2–0.3; 1; 1.96 |
| FMN | **436** | 500 | 5 |

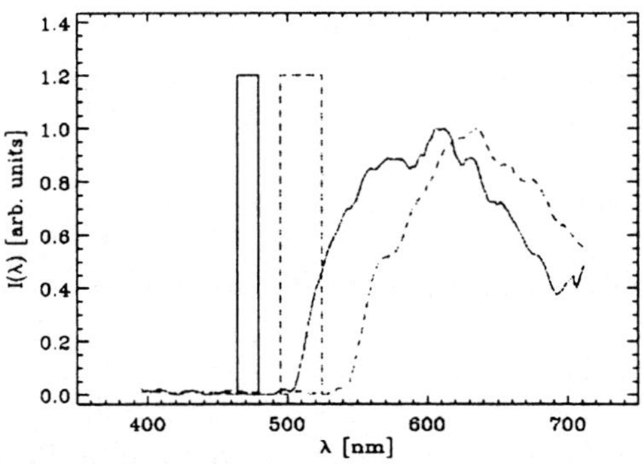


*Figure 1.* Excitation and emission range: —, short-wave excitation; - - - - - -, long-wave excitation.

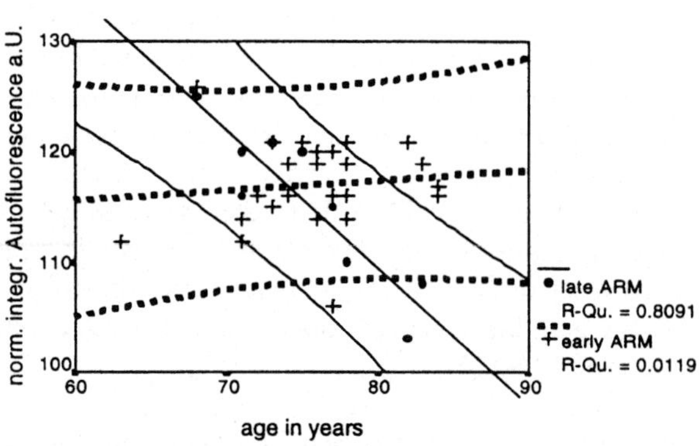


*Figure 2.* Age-dependence of integral normalised long-wave autofluorescence in dry and wet AMD.

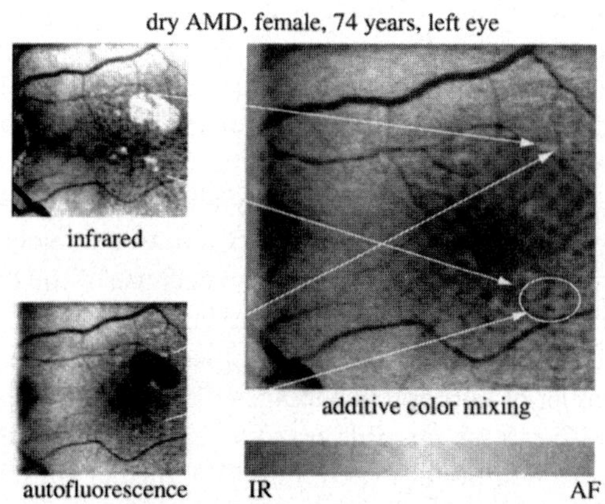


*Figure 3.* Infrared and autofluorescence images in AMD.

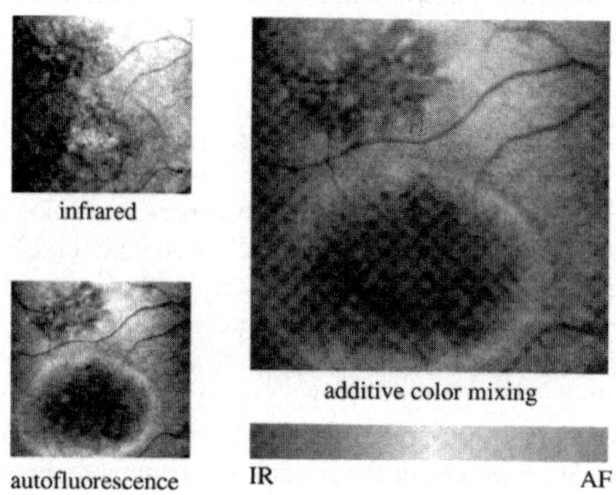


*Figure 4.* Infrared and autofluorescence images in AMD.

The most known technique for the measurement of autofluorescence images is performed by laser scanning ophthalmoscopes [8]. To get a sufficient signal to noise ratio, the integral autofluorescence is detected without any spectral dispersion. That means, a discrimination of different fluorophores is not possible using this technique. An improvement for the registration of fluorescence image is reached, when infrared images are detected simultaneously with fluorescence images (Figures 3, 4). In this case, landmarks in the infrared images can be used for the averaging procedure of fluorescence images.

Furthermore, infrared images exhibit additional information. The interpretation of autofluorescence images is more easy, if an additive colour mixing between infrared and autofluorescence images will be performed [9]. More detailed information are visible in the fluorescence images. Despite this new combination, a discrimination of different fluorophores is not possible using the conventional laser scanner ophthalmoscopy.

Corresponding to the state of the art, different fluorescence spectra can be detected by imaging spectrometry, but a mapping presentation is to time consuming. On the other hand, the integral fluorescence measurement by the SLO technique allows a fluorescence mapping, but a discrimination of fluorophores is not possible.

## Time-resolved fluorescence measurement at the fundus

If the decay time of the fluorescence after short time excitation is measurable, the discrimination between fluorophores and also the presentation of the 2-dimensional distribution of different fluorophores as a "tau-mapping" should be possible. Methods for the measurement of the decay time of fluorescence can be classified according to two principles. In the frequency domain, the whole object will be excited by Intensity-modulated light. The decay time is calculated from the phase shift or the demodulation between excitation and emission light. This principle can be applied, if the fluorescence light is sufficient bright. In the time domain, the fluorescent sample will be excited by a short pulse and the responded fluorescent light will be detected at different times after excitation.

The decision about the optimal measuring principle can be done according to the expected number of fluorescence photons. Starting from the maximal permissible exposure, let us consider a pulse of 5 pJ energy with a repetition rate of 77 MHz in the plane of the cornea. This pulse is attenuated by the transmission of the ocular media. Considering the quantum efficiency of the fluorophores at the fundus, the wavelength range of the emission, the solid angle of the detection, the transmission of the measuring device and of the cut off filter as well as the quantum efficiency of the detector, and the energy of a fluorescent photon, the number of the detected photons can be calculated. Under these assumptions, only 0.06 Photons can be detected as response of an excitation pulse of 5 pJ. The considered values, taken in this calculation, are given in Equation (1).

$$\begin{aligned} N_{\mathrm{PMT}} & \\ &= \frac{E_{\mathrm{Ex}} \cdot \tau_{\mathrm{Ex}} \cdot \phi_F \cdot \Delta\lambda \cdot \Omega \tau_{\mathrm{Em}} \cdot \tau_{\mathrm{SLO}} \cdot \tau_F \cdot \phi_{\mathrm{PMT}}}{h \cdot \nu_{\mathrm{Em}}} = 0.06 \\ &< \text{required max. } N_{\mathrm{TCSPC}} = 0.1 \end{aligned} \tag{1}$$

where $E_{\mathrm{Ex}}$ = energy of an excitation laser pulse in the cornea plant = 5 pJ, $\tau_{\mathrm{Ex}}$ = ocular transmission for excitation light = 0.9, $\phi_F$ = quantum yield of fluorophor = 100 nJ $\mathrm{nm}^{-1}$ $\mathrm{sr}^{-1}$/J [10], $\Delta\lambda$ = wavelength-range of emission = 100 nm, $\Omega$ = solid angle of detection = 0.02 sr, $\tau_{\mathrm{Em}}$ = ocular transmission for emitted light = 0.9, $\tau_{\mathrm{SLO}}$ = transmission of the laser scanner ophthalmoscope = 0.9, $\tau_F$ = transmission of the cut off filter = 0.8, $\phi_{\mathrm{PMT}}$ = quantum yield of the photomultiplier = 0.1, $\nu_{rmEm}$ = frequency of the emitted light ≡600 nm, $h$ = Planck's quantum.

This very low expected number of fluorescent photons excludes the application of tau mapping in the frequency domain. In contrast, these conditions are optimal for the application of the time correlated single photon counting (TCSPC), which is the most sensitive detecting principle.

The basic unit of the time correlated single photon counting technique is a time to amplitude converter. The excitation pulse, hitting the object, starts simultaneously the ramp of the time to amplitude converter. The ramp is stopped by the response fluorescence photon. According to the reached amplitude, one count is added in a register, which corresponds to the amplitude. After a sequence of measurements, the content of the registers represents the time dependent fluorescence intensity after short pulse excitation.

We have developed an experimental arrangement of a laser scanner system for the 2-dimensional measurement of the time resolved fluorescence (Figure 4). An active mode locked A laser is used as the light source in a conventional laser scanner ophthalmo-

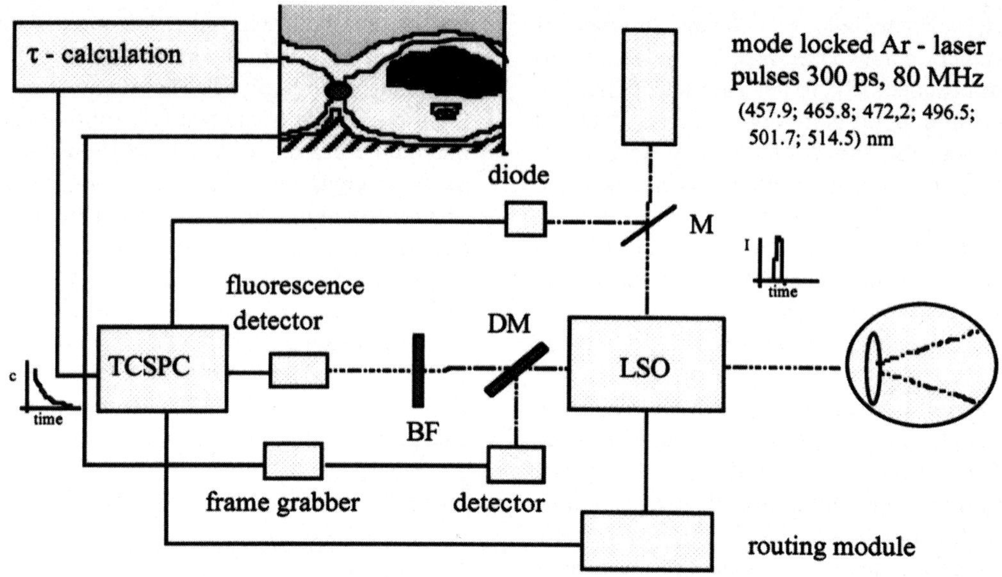


*Figure 5.* Scheme of the laser scanner ophthalmoscope for the 2-dimensional detection of the decay time.

scope (cSLO, Carl Zeiss, Germany). The pulsed laser beam scans the sample (e.g., fundus) at selectable wavelengths (457.9; 465.8; 472.7; 496.5; 501.7; 514.5 nm) with a repetition rate of 77.3 MHz. The pulse duration is <350 ps. One part of the excitation pulse starts the ramp in the time to amplitude converter. The main part of the pulse excites the fluorescence at the fundus. Only one detected fluorescence photon is necessary for stopping the ramp. If the ramp is stopped, one count is added in the register, which corresponds to the reached amplitude. The scanned field of 20 degree at the fundus is detected in a matrix of 128 × 128 pixel, corresponding to a lateral resolution of 50 $\mu$m. The synchronisation between the scanning process and the time resolved fluorescence measurement is realised by a routing module. As only about 300 photons are required for the calculation of the mono-exponential decay time, a complete 20 degree image will be measured within less than 3 seconds. Eye movements will be compensated using the simultaneously detected reflection image. Some technical parameters are given in Table 2.

As an example, the content of the time registers is demonstrated for the fluorescence measurement of a structured sample of Rhodamin 6 g. The measuring conditions corresponded to *in vivo* investigations at the fundus. The detected 300 photons allow the determination of the decay time within an error of 5%. For the comparison measurement of the decay time, a streak camera was used. The measuring time of the streak camera was in the minute range. Because of the very low number of photons, required for the determination of a monoexponential decay time, an image of 20 degree can be detected in less than 3 s. During this time a sufficient fixation of the patient's eye should be possible. An image registration for the improvement of the signal to noise ratio can be performed using the simultaneously acquired reflection images.

*Table 2.* Parameters of the experimental set-up for tau-mapping

| Scanned field | 20 degree |
|---|---|
| Lateral resolution | 50 $\mu$m |
| Radiation energy/pulse | <5 pJ |
| Mean radiation power | <390 $\mu$W |
| Required photons/pixel | ? 300 |
| Time/pixel | 160 $\mu$s |
| Acquisition time/image | <2.6 s |
| Decay time error (mono.-exp.) | 5% |

A very important advantage of decay time measurements for the discrimination of fluorophores is, that the determination of the decay time is independent of the intensity of the fluorescence light. It could be demonstrated, that the contrast in a tau-image is constant, despite the fluorescence intensity is not homogeneous in the field. That means, the tau mapping is insensitive to the shading.

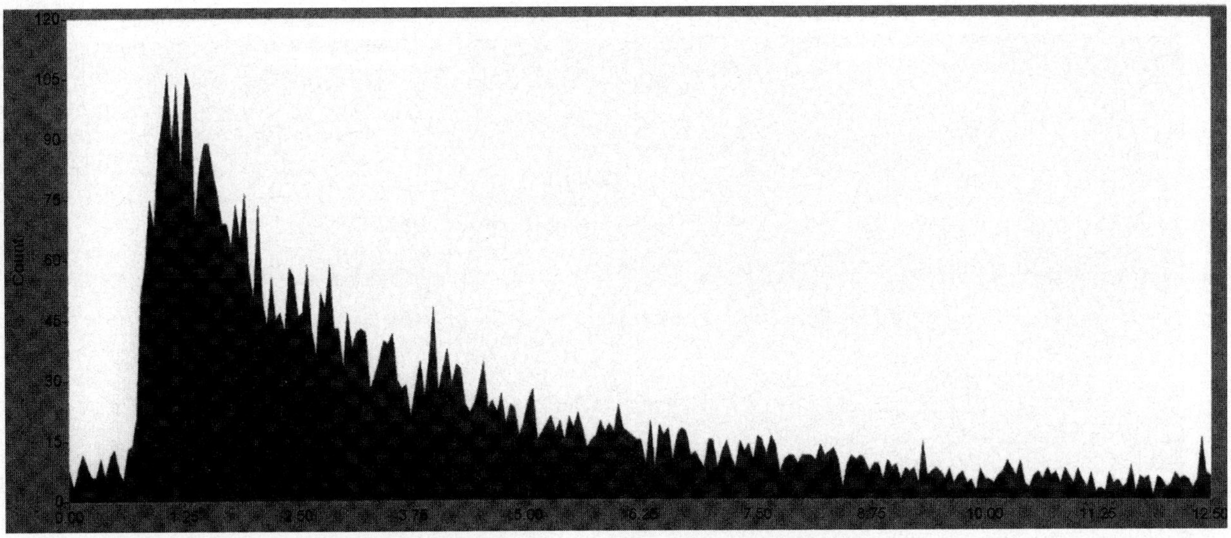


*Figure 6.* Content of the registers (about 300 photons) in the time correlated single photon counting technique (TCSPC) for the calculation of the decay time.

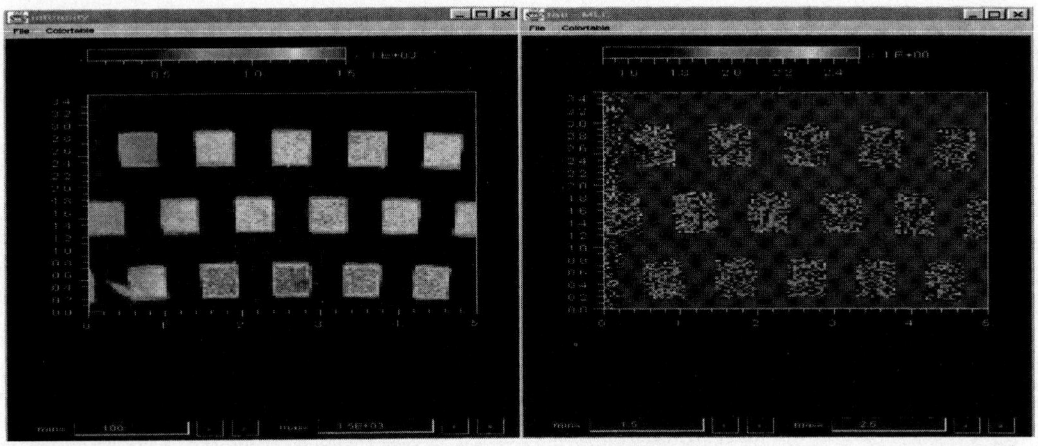

*Figure 7.* Comparison between locally resolved integral fluorescence intensity (left) and fluorescence decay mapping (right) of a structured sample, containing Rh 6 G.

The determination of the decay time images of the fundus requires algorithms, which calculate the decay time exact also from a very low number of photons. The time for the calculation of the decay time image should be as short as possible. In principle both the least square fit and the maximum likelihood algorithm are suited for the calculation of the decay time. The results of a test measurement under the conditions of the living human eye show, that the maximum likelihood algorithm [10] delivers a more homogeneous distribution of the decay time than the least square fit.

The life-time measurement of the autofluorescence is practically limited to monoexponential processes. In case of multiexponential decay behaviour, the duration of the measurement must be prolonged by about a factor of 100. Such a long time can not expect of patients. If in case of multiexponential decay the time constants differ in a large range from each other, an approximation by a monoexponential decay is also possible, when the time ranges are equivalent adapted.

## Summary

The decay time is characteristic for several natural fluorophores. The determination of the decay time is independent of the fluorescence intensity. As a consequence, a short living weak fluorescence should be

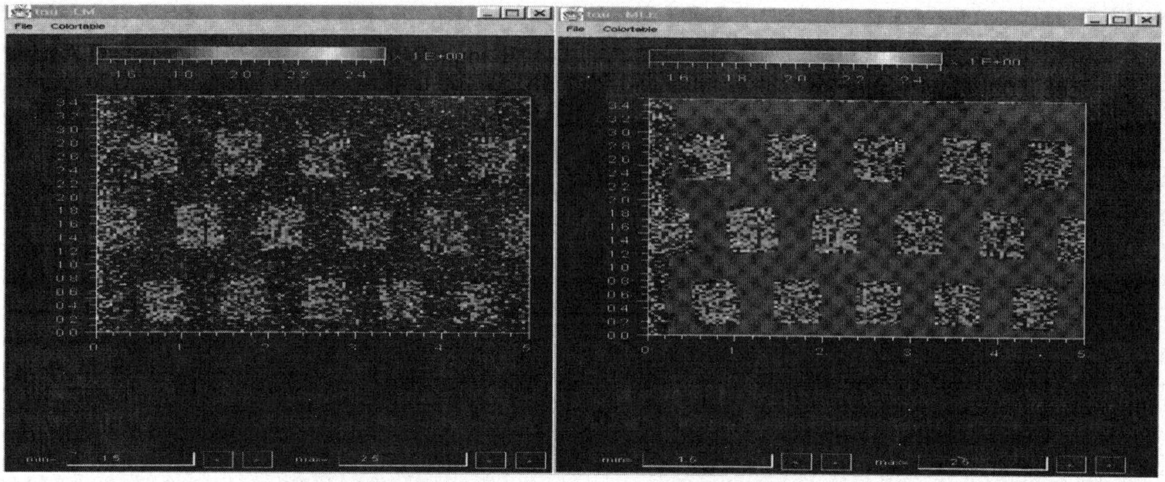

*Figure 8.* Comparison of the least square algorithm (left) and the maximum likelihood method (right) for the calculation of the decay time from noisy signals.

detectable also if it is covered by a intensive long-living fluorescence. As the decay time is influenced by the embedding matrix, information about the cellular stage might be possible.

The laser scanning technique in combination with the time correlated single photon counting technique seems to be the optimal method for the discrimination of different fluorophores at the fundus according to the decay time. Fields of an equal decay time are presented as a tau mapping. An experimental set up was developed. Until now, only basic experiments were done on structured fluorescent tests, but under the conditions of the living eye. The results are promising. For the separate detection of the most important short lifetime of 120 ps of A2E (excitation at 413 nm, emission 450–600 nm) as a putative precursor of lipofuscin in age-related macular degeneration [11], a light source for pulses in the range of about 50 ps is required.

## References


1. Richards-Kortum R, Sevick-Muraca E. Quantitative optical spectroscopy for tissue diagnostics. Annu Rev Phys Chem 1996; 47: 555–606.
2. Wolfbeis OS. Fluorescence of organic natural products, in SG Schulman (ed.), Molecular Luminescence Spectroscopy. Volume 1. J. Wiley and Sons, New York, 1993: 167–370.
3. Bottiroli GA, Croce AC, Locatelli D, Marchesini R, Pignoli E, Tomatis S, Cuzzoni C, Dipalma S, Dalfante M, Spinelli P. Natural fluorescence of normal and neoplastic human colon: A comprehensive *ex vivo* study. Lasers Surg Med 1995; 16: 48–60.
4. Wagnieres GA, Star WM, Wilson BC. *In vivo* fluorescence spectroscopy and imaging for oncological applications. Photochemistry and Photobiology 1998; 68(5): 603–632.
5. Schweitzer D, Hammer M, Scibor M. Imaging spectrometry in ophthalmology – principle and applications in microcirculation and investigation of pigments. Ophthalmic Res 1996; 28(Suppl. 2): 37–44
6. Hammer M, Schweitzer D, Leistritz L, Scibor M, Donnerhacke KH, Strobel J. Imaging spectrometry of the human ocular fundus *in vivo*. Journal of Biomedical Optics 1997; 2(4): 418–425.
7. von Rückmann A, Fitzke FW, Bird AC. Distribution of fundus autofluorescence with a scanning laser ophthalmoscope. Br J Ophthalmol 1995; 79: 407–412.
8. Schweitzer D, Lang GE, Remsch H, Beuermann B, Hammer M, Thamm E, Spraul ChW, Lang GK: Vergleich zwischen Patienten mit altersabhängiger Makulopathie, deren Kindern und Augengesunden durch Messung von Sauerstoffsättigung, Xanthophyll und natürlicher Fluoreszenz am Fundus. Der Ophthalmologe (In press).
9. Schweitzer D, Kolb A, Hammer M, Thamm E. Imaging spectrometry, 2-D static fluorometry, and 2-D time resolved fluorometry – a comparison of methods for the determination of the autofluorescence of the fundus. Investigative Ophthalmol Vis Sci 1999; 40(4): S 918.
10. Delori FC. Spectrometer for non-invasive measurement of intrinsic fluorescence and reflectance of the ocular fundus. Applied Optics 1994; 33(31): 7439–7452.
11. M. Köllner, J. Wolfrum. How many photons are necessary for fluorescence-lifetime measurements? Chemical Physics Letters 1992; 200(1, 2).
12. Cubeddu R, Taroni P, Hu DN, Sakai N, Nakanishi K, Roberts JE. Photophysical studies of A2E, putative precursor of lipofuscin, in human retinal pigment epithelial cells. Photochem Photobiol 1999; 70: 172–175.

International Ophthalmology 23: 405–408, 2001.
J.R. Sampaolesi (ed.), Laser Scanning: Update 1, 223–226.




# Differential imaging in scanning laser ophthalmoscopy

Ross A. Ashman, Fred Reinholz & Robert H. Eikelboom
*Lions Eye Institute, and Centre for Ophthalmology and Visual Science, University of Western Australia, Perth, Western Australia*

*Key words:* differential imaging, multiple wavelength

**Abstract**

Differential imaging may be useful for understanding pathologies of the choroid. We use a prototype scanning laser ophthalmoscope capable of simultaneous multiple wavelength imaging to record fundus images. The advantages of simultaneous acquisition compared to serial acquisition are, reduced image capture times, and image registration is not required.

The system is run under non-confocal conditions in using slit apertures with a width of up to 600 $\mu$m. The laser lines launched into the SLO were the 633 nm line of a HeNe- laser, and the 815 nm line from a tunable (740 nm to 840 nm) cw Ti:Sapphire-laser. The difference in absorption at these two wavelengths is used to produce a differential image. We compare conventional imaging with differential spectral imaging for viewing the choroidal vessels. Qualitative results show the contrast of choroidal vessels in differential imaging is improved compared with normal imaging in the region of the optic disk and for the level of pigmentation in the subjects examined.

## Introduction

The choroid is an interesting region because of the association of choroidal new vessels (CNV) with various retinal diseases. It has been shown that infrared (IR) imaging improves the ability to detect areas suspicious of CNV over indocyanine green (ICG) and fluorescein [1] through increased contrast of choroidal vessels.

It is known that choroidal patterns can also be seen with red illumination [2]. Imaging with 633 nm the choroidal vessels appear lighter than the surrounding fundus. Imaging with longer wavelengths, 780–820 nm, the vessels appear dark against a bright background. The change in visibility of the choroidal vessels against their background with wavelength may be used to increase their contrast by imaging differentially.

Differential techniques such as differential optical absorption spectroscopy and Tunable Infrared Laser Differential Absorption Spectroscopy (TILDAS) [3] have been used in such areas as pollution monitoring and astronomy. Differential spectroscopy is used to improve the sensitivity of the detecting instrument. Until now such differential techniques have not been used in ophthalmology. We compare conventional imaging with differential spectral imaging for viewing the choroidal vessels.

## Method

An existing scanning laser ophthalmoscope set-up for recording multiple wavelengths simultaneously was used [4]. The laser lines used include the 633 nm line of a HeNe- laser, and the 815 nm line from a tunable (740 nm to 840 nm) cw Ti:Sapphire-laser. Intensities of the different components are adjusted using laser intensity stabilisers (noise eaters).

All optical elements in the common part of the illuminating and detection beam path are silver mirrors. This ensures minimum loss over the imaging wavelength range, no chromatic aberrations can occur, and focusing effects every wavelength in the same way. In the detection beam path the different wavelengths are separated using dichroic filters.

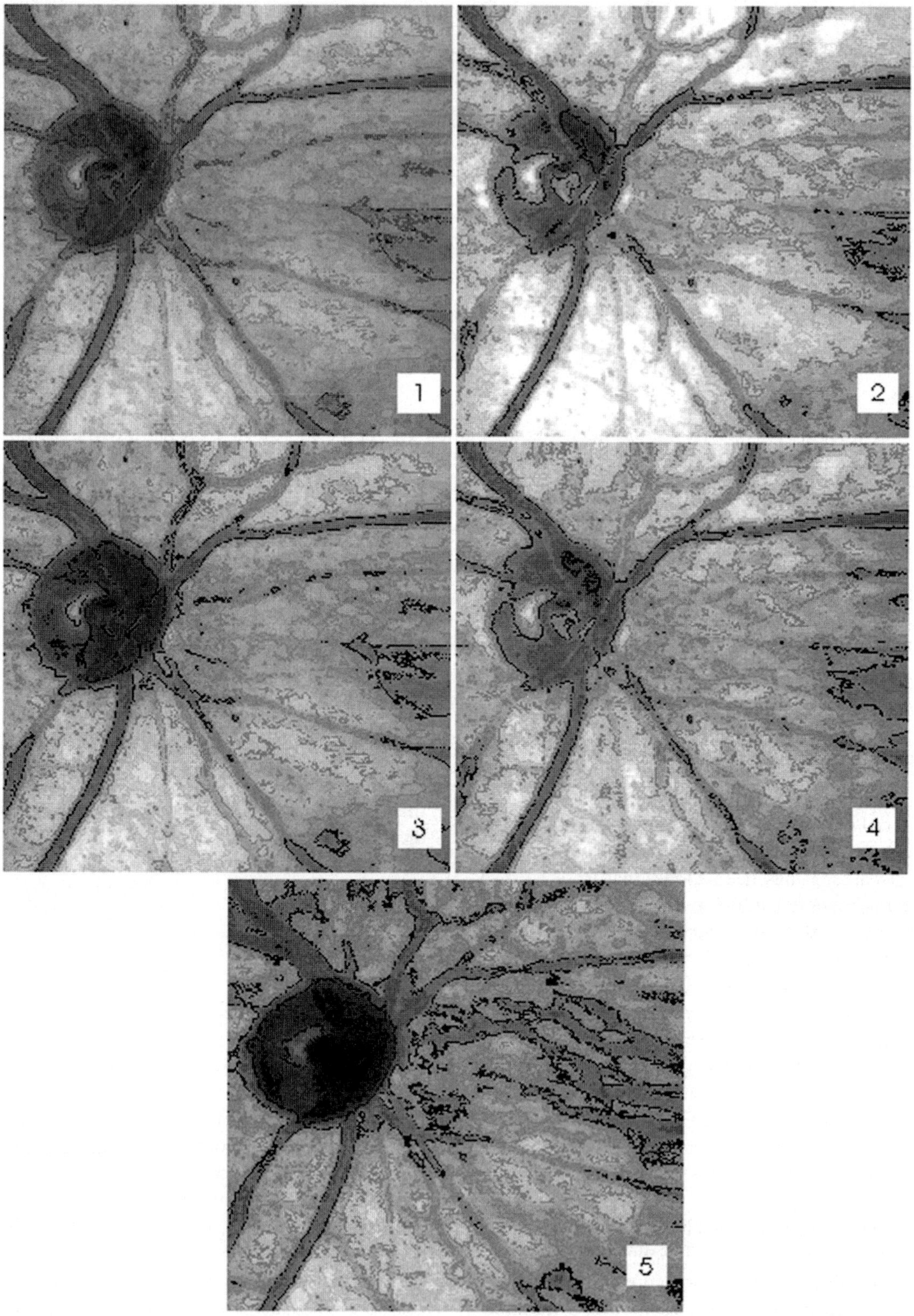


*Figure A.* Series of images produced for subject (A). Image (1) is a two colour image; (2) grey scale; (3) 815 nm frame; (4) 633 nm frame; (5) differential image.

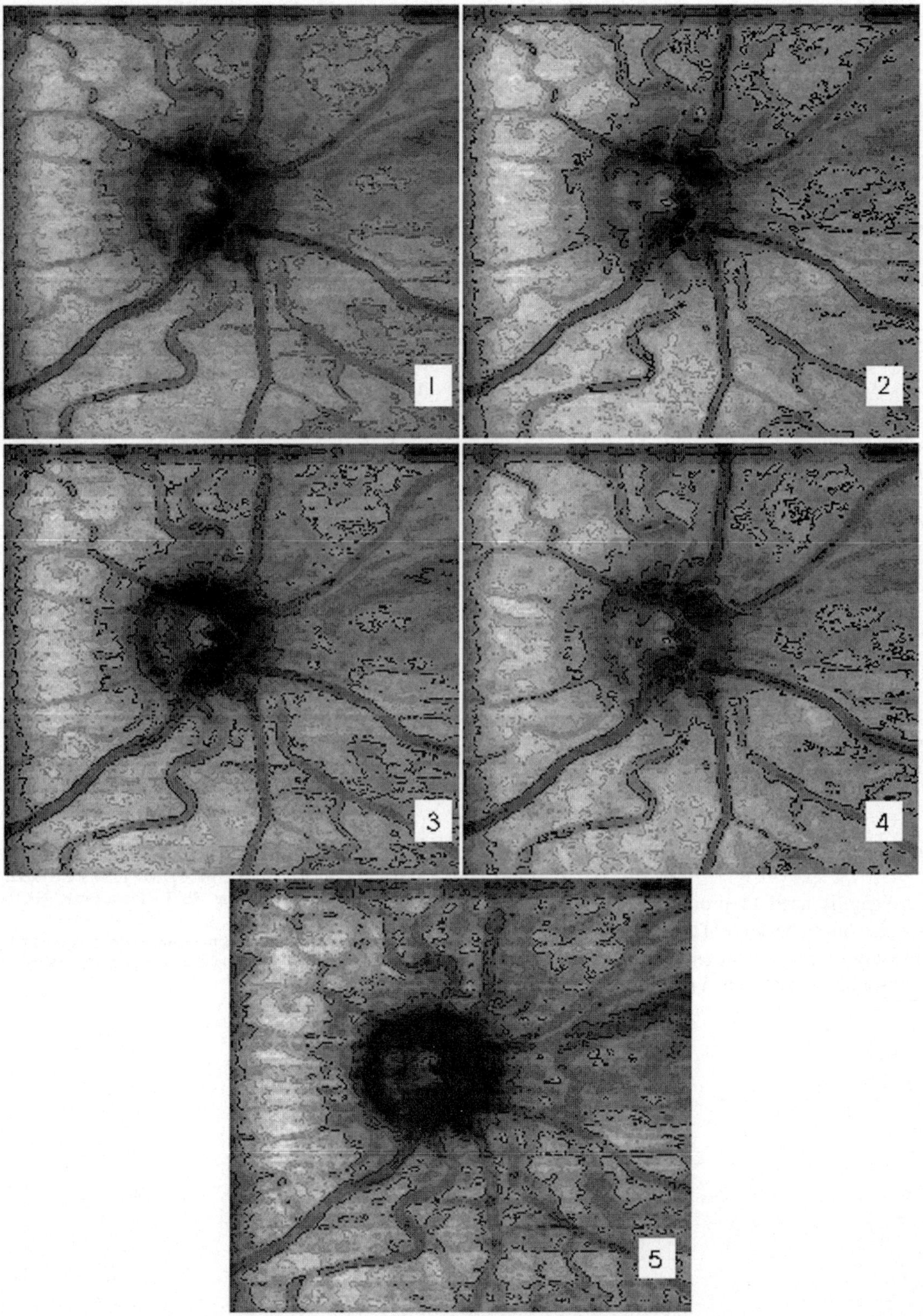


*Figure B.* Series of images produced for subject (B). Image (1) is a two colour image; (2) grey scale; (3) 815 nm frame; (4) 633 nm frame; (5) differential image.

After the separation each beam is focused individually onto a slit aperture in front of a avalanche photo diode. The system is run under non-confocal conditions. Images are recorded with a size of $512 \times 512$ pixels, an 8 bit dynamic range in each channel, and the frame rate for simultaneous imaging was 7 Hz.

Subjects' pupils were dilated with Mydriacyl eye drops. Subject (A) had blue/grey irises and subject (B) had hazel/green irises. The differential image is produced after acquisition. The 633 nm frame is subtracted with saturation from the 815 nm frame with an intensity ratio of two.

## Results

In the two colour images (Figure A1 and B1) it is possible to distinguish choroidal vessels by their red appearance against a yellow background. This can also be seen in grey scale (Figure A2 and B2) image. The 815 nm images clearly show large vessels (Figure A3 and B3). With 633 nm the vessels are harder to distinguish individually but give an overall mottled appearance to the image (Figure A4 and B4). A comparison of monochromatic imaging (Figure A2 and B2) and differential imaging (Figure A5 and B5) shows that with the latter, contrast of the choroidal vasculature is greatly improved, particularly the smaller choroidal vessels.

## Conclusion

The wavelengths used to produce differential SLO images were chosen because of their ability to improve the contrast of retinal structures of interest, in particular the choroidal vessels. When illuminating the ocular fundus with 633 nm laser light the choroidal vessels appear as regions of high intensity in a dark surrounding background. The reflectance of blood is greater than the reflectance of the surrounding melanin.

Conversely when illuminating with 815 nm or infrared light the choroidal vessels appear dark against a bright background. At longer wavelengths the reflection of surrounding melanin is greater than that of blood. It is the change in absorption of choroidal blood and melanin with wavelength that is used to enhance the contrast of the choroidal vessels in differential imaging.

The contrast of choroidal vessels in differential imaging is improved compared with normal imaging in the region of the optic disk and for the level of pigmentation in the subjects examined. Differential imaging may provide a means of imaging and monitoring choroidal vessels without the need for contrast enhancing dyes. However further study is required to determine its usefulness in other regions of the retina and on a greater subject pigmentation range.

## References


1. Elsner AE, Hartnett, ME, Weiter J, Buzney SM, Burns S. Advantages of infrared imaging in detecting choroidal new vessels. OSA Technical Digest (Vision science and its applications). 1996; 1: 102–105, 1996, (Abstract).
2. Elsner AE, Burns S, Weiter J, Delori FC. Infrared imaging of sub-retinal structures in the human ocular fundus. Vision Research 1996; 36: 191–205.
3. Kolb CE, McManus BJ, Nelson DD, Zahniser ZS. Tunable infrared laser differential absorption spectroscopy (TILDAS) monitors for atmospheric trace gases. Lasers and Electro-Optics Society Annual Meeting. IEEE. Piscataway, NJ, USA, pp. 921996.
4. Reinholz F, Ashman F. Simultaneous three wavelength imaging with a scanning laser ophthalmoscope. Cytometry 1999; 37: 165–170.

*International Ophthalmology* **23:** 409–412, 2001.
*J.R. Sampaolesi (ed.), Laser Scanning: Update 1*, 227–230.




# Occult CNV imaging with scanning laser ophthalmoscope

Carlos Argento, Omar Lopez Lopez & Mauricio Martinez Cartier
*Instituto de la Visión, Marcelo T. de Alvear 2261, 1122 Buenos Aires, Argentina*

## Introduction

Fluorescein angiography provides useful images of the retinal vessels in most instances, but choroidal vessel assessment may be hindered because of inherent properties of fluorescein (i.e., rapid difussion through the choriocapillaris and blockage by pigments like xanthophylle, melanin or hemoglobin).

Indocyanine green angiography have some theoretical as well as practical advantages compared to fluorescein: infrared emission allowing greater transmission through the RPE; greater protein binding (98% vs 80%), resulting in less difussion through the choriocapillaris.

The Heidelberg Retina Angiograph (HRA) is a confocal scanning laser ophthalmoscope that allows simultaneous FA and ICG angiography with excelent image contrast, since confocal optics eliminates rays that originate at points other than the focal plane. HRA angiography is a safe and confortable technique, allowing better delineation of vessel arquitecture of occult CNV compared to conventional fundus cameras [1], greater sensibility and specificity in early diagnosis of vitreoretinal interphase syndrome [2], dual FA and ICG angiography [3], imaging through small pupils or media opacities.

For simultaneous FA and ICG angiography we inject 500 mg of fluorescein and 12.5 mg of ICG in bolus. For FA imaging a 488 nm argon laser is used and for ICG angiography a 795 nm diode laser is used. Digital images are seen on real time on a high resolution monitor. We present 4 case reports in which the ability of the HRA to present simultaneous images was helpful for diagnosis and management of patients with exudative AMD.

Case 1: 75-year-old male that complained of decrease on his visual acuity and metamorphopsia. Best corrected visual acuity was 20/80. On funduscopy an elevated lesion with hemorrhage is evident on the macula (Figure 1a). FA angiography reveals diffuse hyperfluorescene but ICG shows well defined hyperfluorescence corresponding to the feeder vessel (Figure 1b, c).

Case 2: 82-year-old female presenting with 20/200 vision OD. On fundus examination a large RPE detachment and retinal folds are evident (Figure 2a). On FA diffuse hyperfluorescence corresponding to type I occult CNV is evident; ICG reveals well defined neovascular net at site of RPE notch (Figure 2b, c).

Case 3: 76-year-old male what underwent cataract extraction a few years ago presented with metamorphopsia OD (Figure 3a). Best corrected visual acuity was 20/100. FA reveals diffuse hyperfluorescence (Figure 3b); ICG allows visualization of well defined neovascularization (Figure 3c, d).

Case 4:[1] 72-year-old female that complained of decrease in visual acuity OD. Red free image shows lesion consistent with CNV. FA shows ill-defined hyperfluorescence, but ICG clearly reveals neovascular net and its feeder vessel. Late phase FA shows prominent leakage.

## Conclusion

SLO allows better delineation of choroidal vessel architecture mainly because of its fast-frame characteristics and confocal optics. We performed approximately 200 simultaneous FA and ICG angiograms on patients with exudative AMD and observed well defined vessels or hyperfluorescence on 45% of patients. On the other hand Yanuzzi et al. reported a 12% incidence of classic CNV with conventional fundus cameras which increased to 39% using ICG videoangiography [4].

[1] Unfortunately, the figures for Case 4 did not reach us in time.

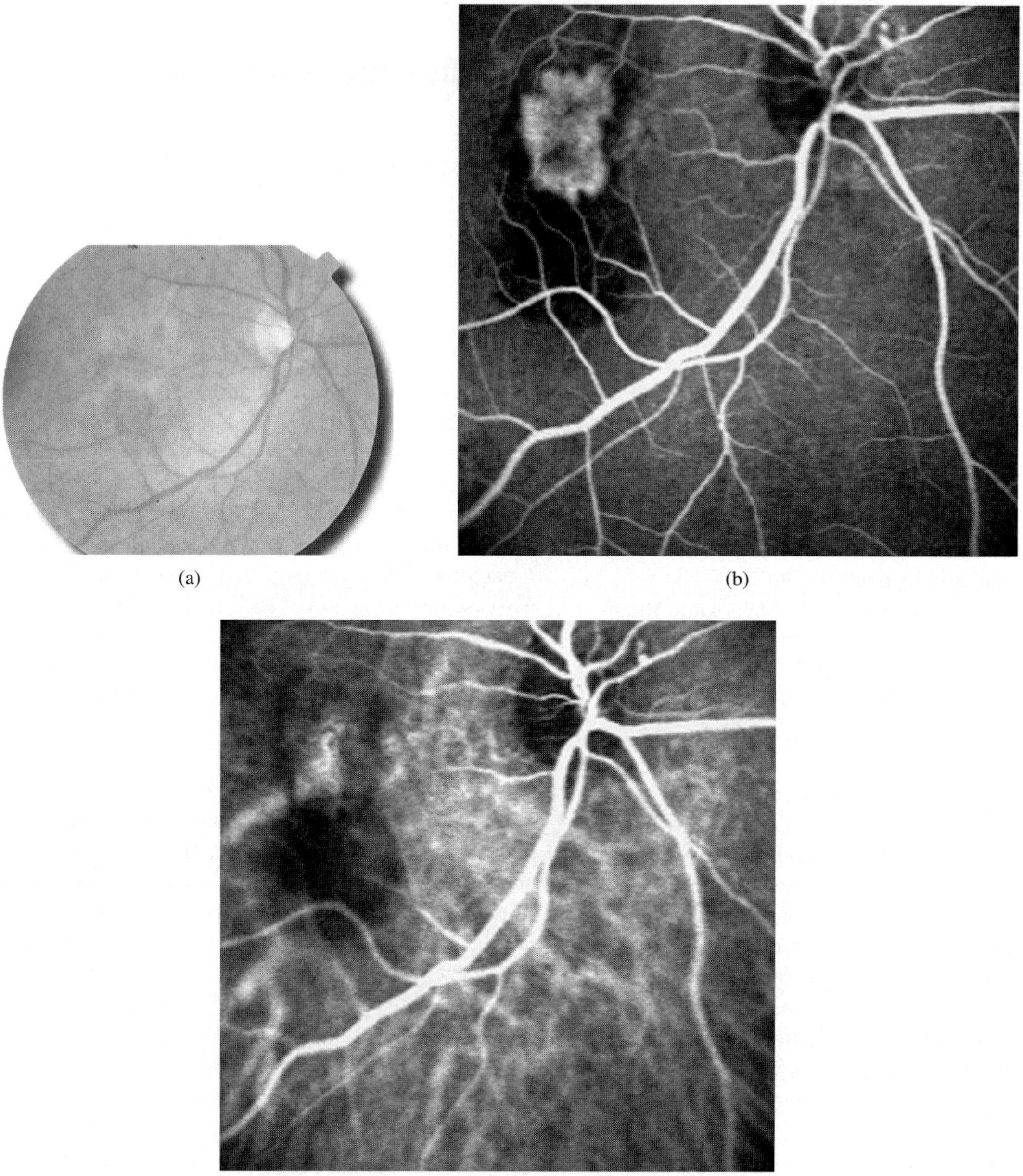

(a) (b) (c)

*Figure 1.*

## References


1. Gelisken et al. Indocyanine green videoangiography of occult choroidal neovascularization: A comparison of scannning laser ophthalmoscope with high-resolution digital fundus camera. Retina 1998; 18: 37–43.
2. Varano et al. Scanning laser ophthalmoscopy in the early diagnosis of vitreoretinal interface syndrome. Retina 1997; 17: 300–305.
3. Holz et al. Simultaneous confocal scanning laser fluorescein and indocyanine green angiography. American Journal of Ophthalmology 1998; 125: 227–236.
4. Yannuzzi et al. Digital indocyanine green videoangiography and choroidal neovascularization. Retina 1992; 12: 191–223.

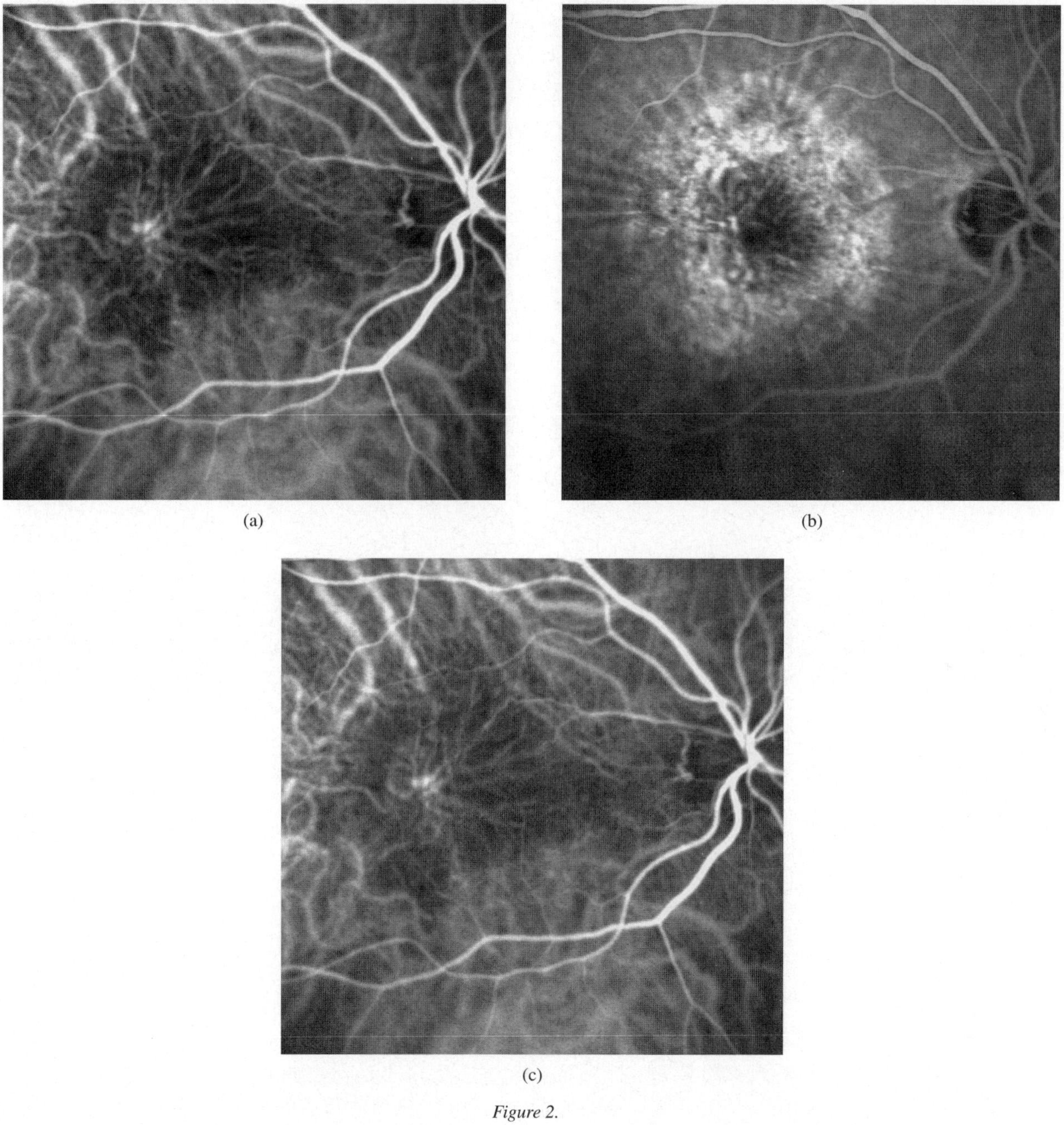

(a) (b) (c)

*Figure 2.*

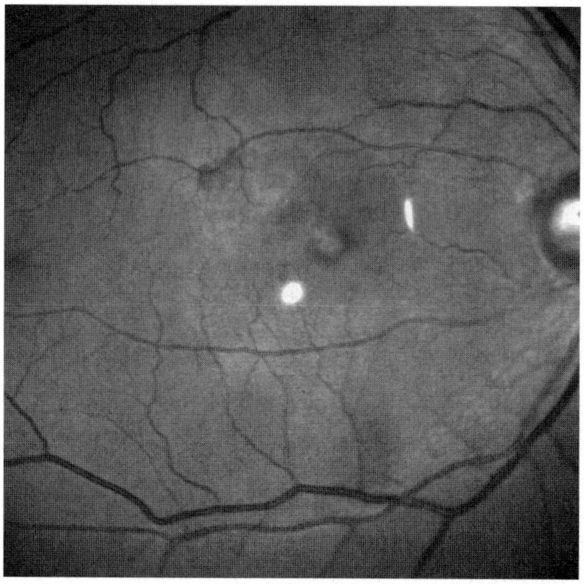

(a)

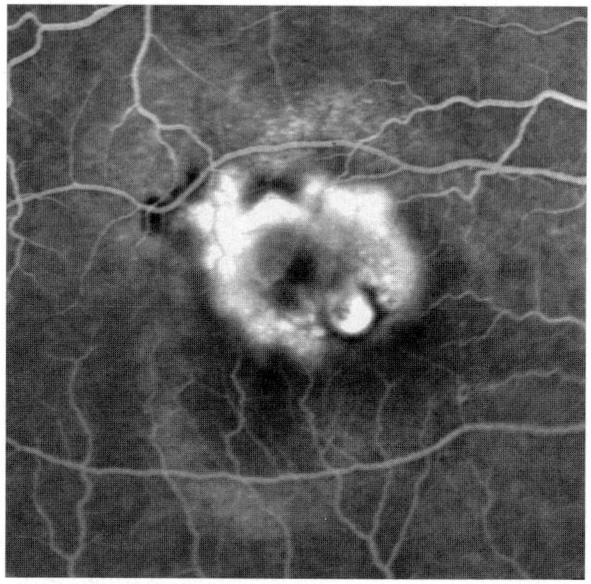

(b)

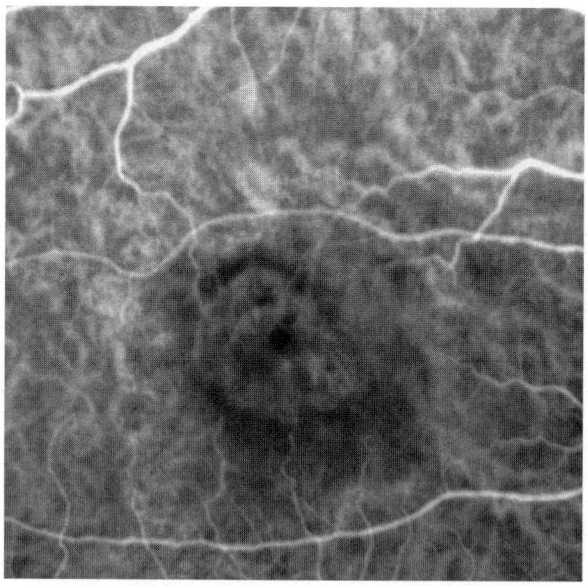

(c)

*Figure 3.*

*International Ophthalmology* **23:** 413–423, 2001.
*J.R. Sampaolesi (ed.), Laser Scanning: Update 1*, 231–241.



# Glaucomatous optic nerve head changes with scanning laser ophthalmoscopy

J. Marengo, R.A. Ucha, M. Martinez-Cartier & J.R. Sampaolesi
*Instituto de la Visión, Buenos Aires, Argentina*

*Key words:* SLO-optic nerve head-glaucoma-peripapillary capillary


## Abstract

*Purpose:* To determine whether there are angiographic differences among normal, preperimetric and advanced glaucoma eyes using indocyanine green angiography with SLO. This method was chosen because of its sensibility to detect peripapillary capillary vessels. *Methods:* Scanning laser opthalmoscopy was preformed on normal eyes, preperimetric glaucomas and advanced glaucomas. *Material:* The authors used a confocal SLO (Heidelberg Retina Angiograph-HRA) *Conclusion:* Several changes may be seen on peripapillary capillary vessels at the different glaucomatous stages. *Discussion:* In normal subjects HRT shows preservation of the disc/cup area ratio; indocyanine green angiography shows normal prepapillary plexus pattern on the neuroretinal rim and cup. Subjects on glaucomatous preperimetric stage reveal a decrease in the disc/cup area ratio as a result of an increase of the cup area secondary to a reduction of the neuroretinal rim area. ICG at this hipertensive stage shows an increase in prepapillary plexus visualization, which may be a consequence of increased blood flow while autoregulation is still operative. Subjects with advanced glaucoma show prominent decrease in the disc/cup area ratio as well as marked capillary droupout in ICG angiography.


## Introduction

In recent years several theories on glaucoma pathophysiology have been postulated. Decreased vascular flow has been linked to optic nerve head damage. On this line extensive research has been conducted for the assessment of optic nerve head flow by different methodologies.

It has been shown that focal decrease in capillary blood flow is associated with a predictable visual field defect.

We used indocyanine green angiography to assess optic nerve head flow on three patient populations: (1) normal volunteers; (2) preperimetric glaucoma, and (3) advanced glaucoma.

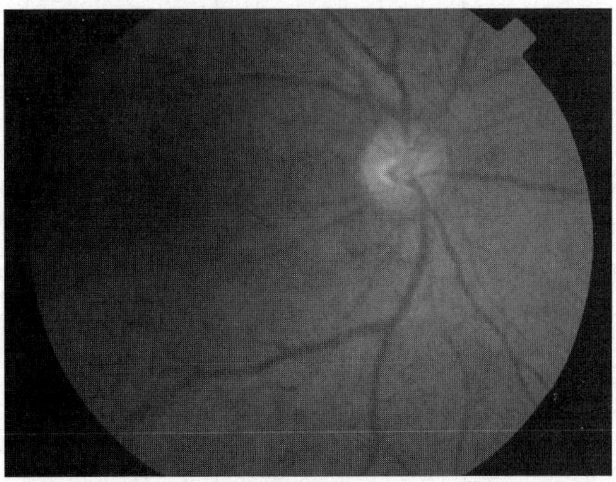

*Figure 1.*

## Patients and methods

A confocal scanning laser ophthalmoscope (Heidelberg Retina Angiograph) has been used with a laser source of 795 nm to excite the indocyanine green; a 810 nm filter was utilized as a barrier. Internal optics allow compensation between −12 to +12 diopters. The area of interest on the fundus can be scanned in

Interzeag OCTOPUS 1-2-3 V14.16
Seven-in-One

| | |
|---|---|
| Apellido: . . . . . . . . . . . . | Ojo / Pupila: Derecho (OD)/ |
| Nombre de pila: . . . . . . . . . | Fecha / Hora: 19.11.1999 / 19:49 |
| No de Ident. | Tiempo examen transcurrido: 2: 6 |
| Fecha nac: | Programa / Código G1 / 0 |
| Edad: | # de etapas / Fases: / 1 |
| Sexo masculino | Estrategia: TOP |
| Corr:esf/cil/eje: / / | Tamaño del estímulo: 3 |
| Agudeza: | No de estímulos / Repetic: 69 / 0 |
| PIO: | Trampas: pos 1/ 3, neg 0/ 4 |
| MDD corrección [dB]: | Código de diagnóst:. . . . . . . |

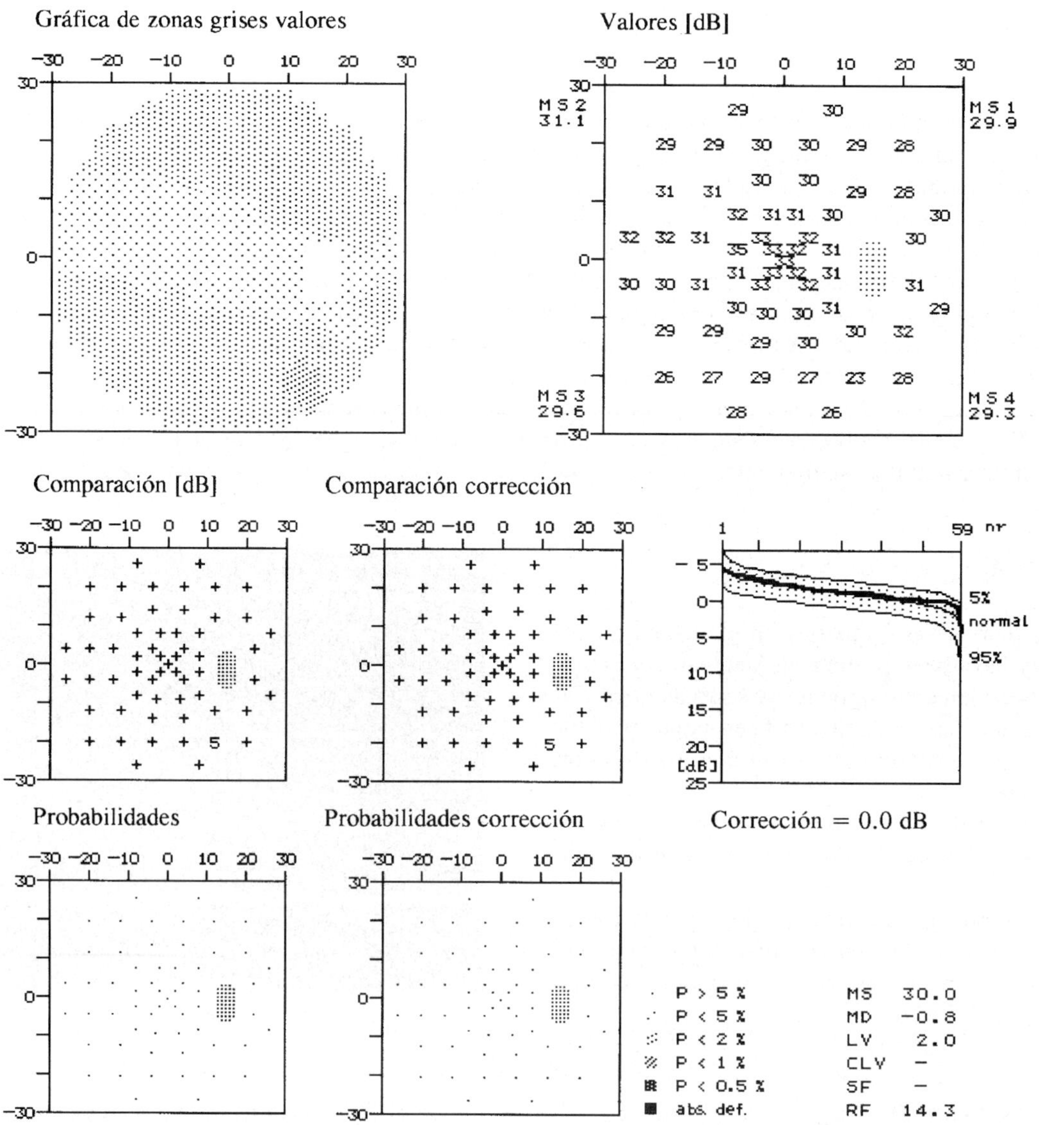


*Figure 2.*

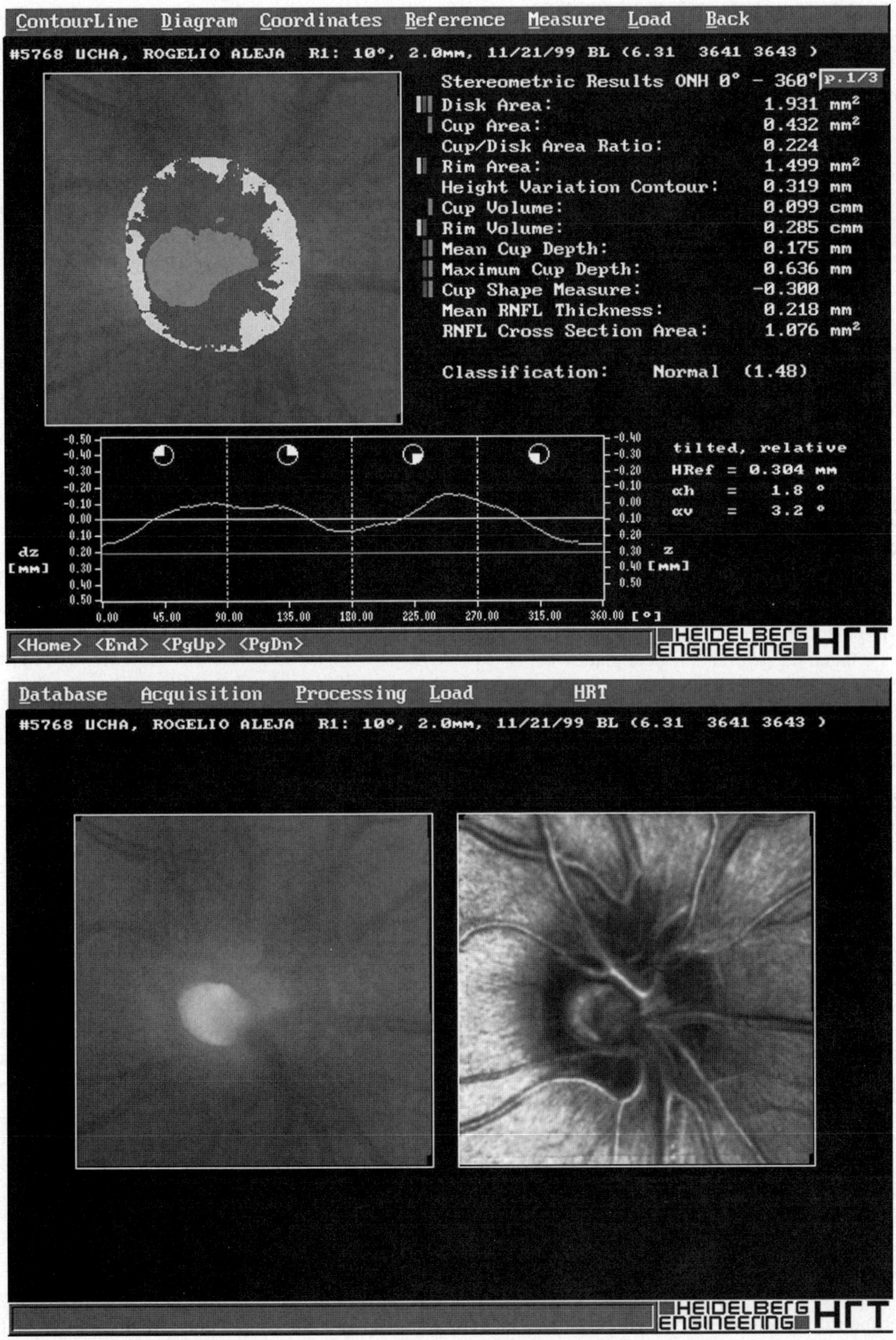
ContourLine Diagram Coordinates Reference Measure Load Back
#5768 UCHA, ROGELIO ALEJA R1: 10°, 2.0mm, 11/21/99 BL (6.31 3641 3643 )
Stereometric Results ONH 0° - 360° p.1/3
Disk Area: 1.931 mm²
Cup Area: 0.432 mm²
Cup/Disk Area Ratio: 0.224
Rim Area: 1.499 mm²
Height Variation Contour: 0.319 mm
Cup Volume: 0.099 cmm
Rim Volume: 0.285 cmm
Mean Cup Depth: 0.175 mm
Maximum Cup Depth: 0.636 mm
Cup Shape Measure: -0.300
Mean RNFL Thickness: 0.218 mm
RNFL Cross Section Area: 1.076 mm²
Classification: Normal (1.48)
tilted, relative
HRef = 0.304 mm
αh = 1.8 °
αv = 3.2 °
dz [mm]
z [mm]
0.00 45.00 90.00 135.00 180.00 225.00 270.00 315.00 360.00 [°]
<Home> <End> <PgUp> <PgDn>
HEIDELBERG ENGINEERING HRT
Database Acquisition Processing Load HRT
#5768 UCHA, ROGELIO ALEJA R1: 10°, 2.0mm, 11/21/99 BL (6.31 3641 3643 )
HEIDELBERG ENGINEERING HRT

*Figure 3.*

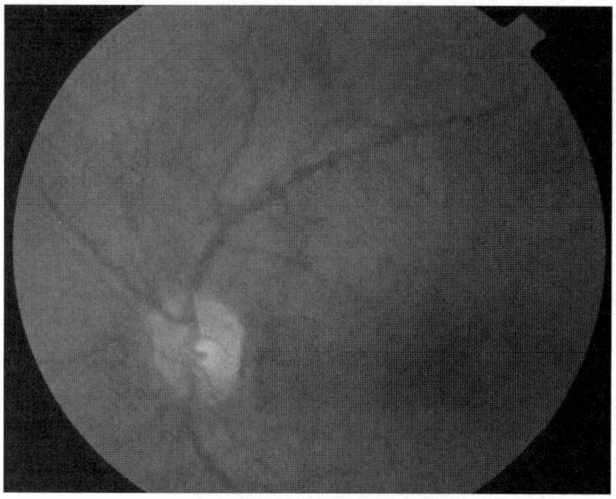

*Figure 4.*

a squared pattern of 10, 20 or 30 degrees. Images so obtained are stored on a hard disk for later processing. All optic nerve head images were taken using a 10 degree field.

Image acquisition was started as soon as contrast fluorescence was detected on angiography. Every optic nerve head was imaged on several confocal planes. A dose of 25 to 50 mg of indocyanine green was injected as a bolus (less than 5 seconds) diluted on 10 ml of normal saline.

Two glaucomatous patients in different stage of the disease were compared to a normal volunteer. All subjects underwent complete ocular examination, including best corrected visual acuity, slit lamp examination, IOP measurement, fundoscopy and gonioscopy. Also a diurnal IOP curve, automated perimetry (Octopus 500 perimeter, G1 program and Octopus 1-2-3 perimeter, G1X program), optic nerve head tomography with the HRT (ver.2.12) and fundus color photographs.

*Case 1:* Normal subject with open angle and deep anterior chamber. Average IOP was 15 mmHg OD and 14.5 mmHg OS. On funduscopy the optic nerve head shows pink, well-delineated neuroretinal rim; a cup-disk ratio of 0.2 (Figure 1). Automated perimetry was normal (MS 30.0; MD −0.8; LV 2.0) (Figure 2). HRT tomography revealed the following: disc area 1.931 mm2; cup area 0.432 $mm^2$; rim area 1.459 $mm^2$; cup volume 0.099 $mm^3$; rim volume 0.285 $mm^3$ (Figure 3).

*Case 2:* Preperimetric glaucoma; BCVA 20/20; deep anterior chamber; mean IOP 22.0 mmHg OD and 23.0 mmHg OS. On gonioscopy an open angle with wide ciliary body band and goniodysgenesis was observed. On funduscopy the optic nerve head showed pink, well-delineated neuroretinal rim; a cup-disk ratio of 0.2 (Figure 4). Automated perimetry was normal (MS 30.3; MD −1.0; LV 1.9) (Figure 5). HRT tomography revealed the following: disc area 2.199 $mm^2$; cup area **0.604 $mm^2$** (max. 0.60 $mm^2$); rim area 1.596 $mm^2$; cup volume **0.156 $mm^3$** (max. 0.12 $mm^3$); rim volume 0.375 $mm^3$; cup shape measure −0.333 (Figure 6).

*Case 3:* Advanced glaucoma; BCVA 20/200; deep anterior chamber; mean IOP 20.0 mmHg OD and 21.0 mmHg OS (with timolol 0.5% bid, dorzolamide 2% tid and latanoprost qd in both eyes). On gonioscopy an open angle with wide ciliary body band and scarce pigment in anterior trabecular meshwork was observed. On funduscopy the optic nerve head showed pale, well-delineated neuroretinal rim; a cup-disk ratio of 0.8 (Figure 7). Automated perimetry evidenced absolute scotoma (MS 25.4; MD 1.1; LV 14.1) (Figure 8). HRT tomography revealed the following: disc area 2.531 $mm^2$; cup area **1.909 $mm^2$** (max. 0.60 $mm^2$); rim area **0.622 $mm^2$** (min. 1.37 $mm^2$); cup volume **0.500 $mm^3$** (max. 0.12 $mm^3$); rim volume **0.062 $mm^3$** (min. 0.32 $mm^3$); cup shape measure **−0.083** (max. −0.15) (Figure 9).

## Results

On performing indocyanine green angiography in the normal patient, the optic nerve head capillary network can be observed uncompromised, and a normal hyperfluorescence of the optic nerve head can be seen (Figure 10). In preperimetric glaucoma, unlike what one might expect, an increased hyperfluorescence of the optic nerve head can be seen, due to an increase in the visualization of the papillary capillary vessels (Figure 11). In advanced glaucoma a decreased optic nerve head fluorescence was evident (vascular filling defect), due to a global decrease in the number of papillary capillary vessels (Figure 12).

## Discussion

Increased hyperfluorescence in preperimetric glaucoma compared to control patients might be the result of one of two possible causes. The first one might

Apellido: . . . . . . . . . . . .
Nombre de pila: . . . . . . . . .
No de Ident. .
Fecha nac:
Edad:
Sexo
Corr:esf/cil/eje: / /
Agudeza:
PIO:
MDD corrección [dB]:

Ojo / Pupila: Izquierdo (OS)/
Fecha / Hora: 18.11.1999 / 14:02
Tiempo examen transcurrido: 2: 7
Programa / Código G1 / 0
# de etapas / Fases: / 1
Estrategia: TOP
Tamaño del estímulo: 3
No de estímulos / Repetic: 69 / 0
Trampas: pos 0/ 3, neg 0/ 4
Código de diagnóst:. . . . . . . .

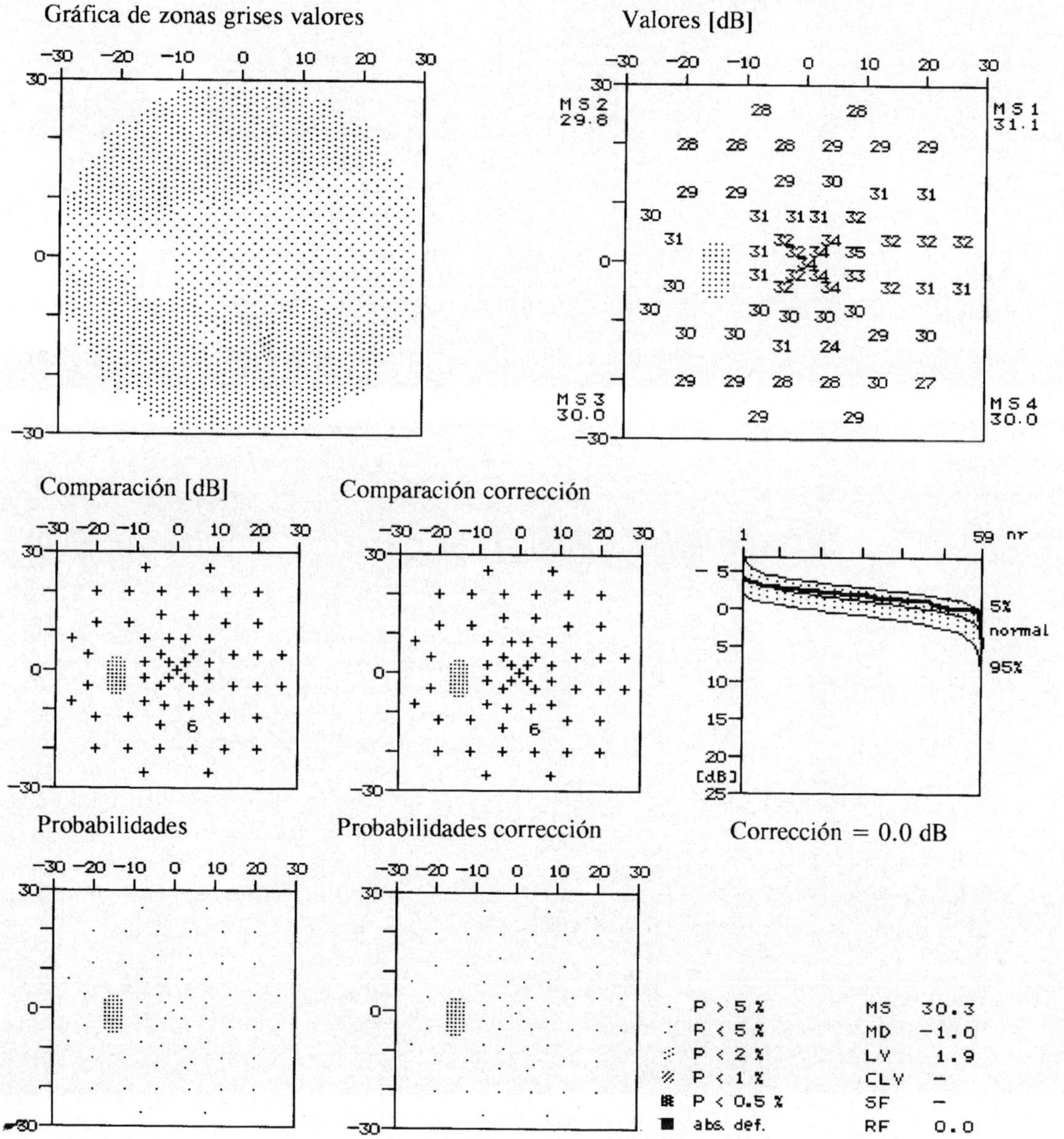


*Figure 5.*

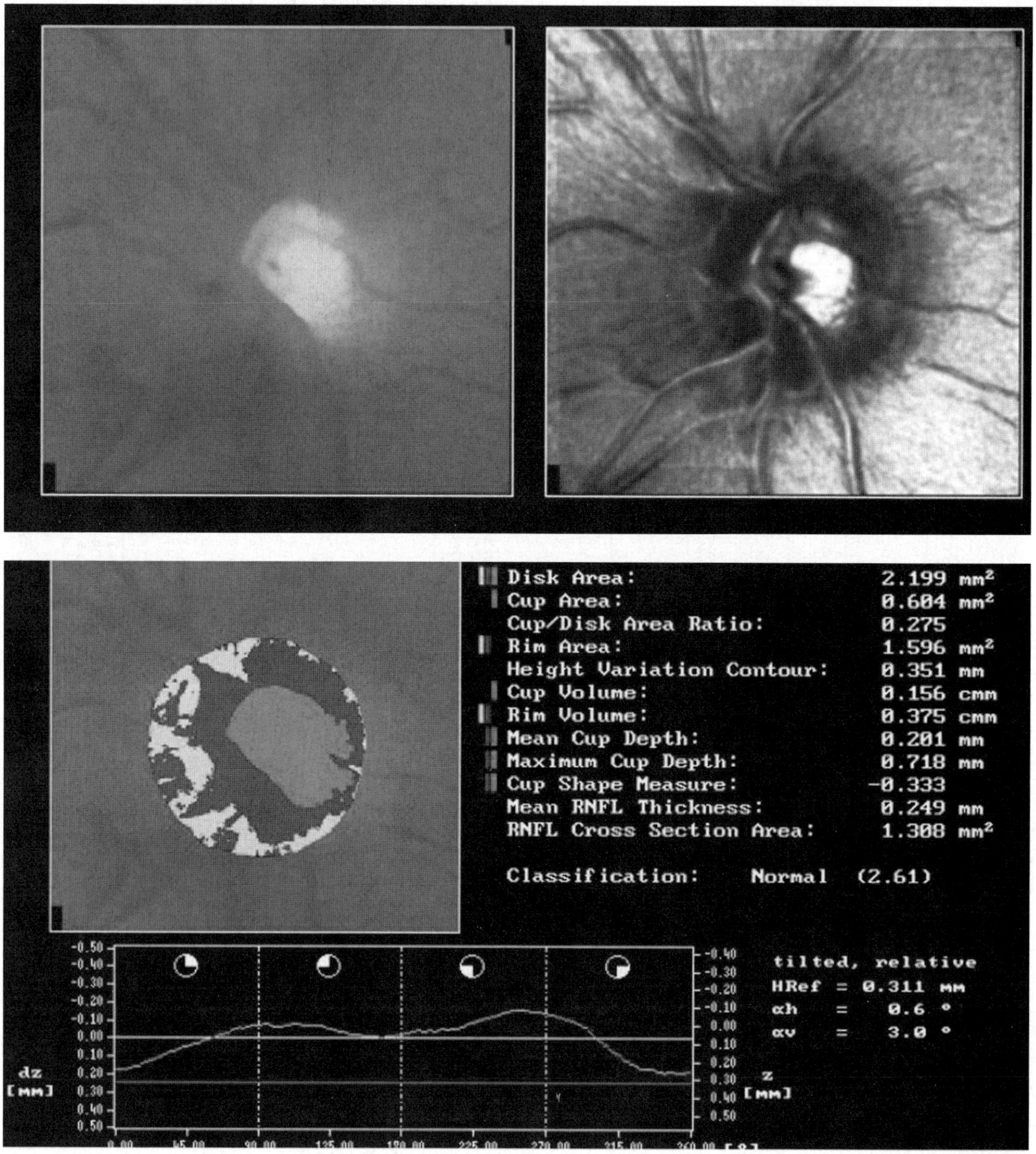
Disk Area: 2.199 mm²
Cup Area: 0.604 mm²
Cup/Disk Area Ratio: 0.275
Rim Area: 1.596 mm²
Height Variation Contour: 0.351 mm
Cup Volume: 0.156 cmm
Rim Volume: 0.375 cmm
Mean Cup Depth: 0.201 mm
Maximum Cup Depth: 0.718 mm
Cup Shape Measure: -0.333
Mean RNFL Thickness: 0.249 mm
RNFL Cross Section Area: 1.308 mm²
Classification: Normal (2.61)
tilted, relative
HRef = 0.311 mm
αh = 0.6 °
αv = 3.0 °
dz
[mm]
z
[mm]

*Figure 6.*

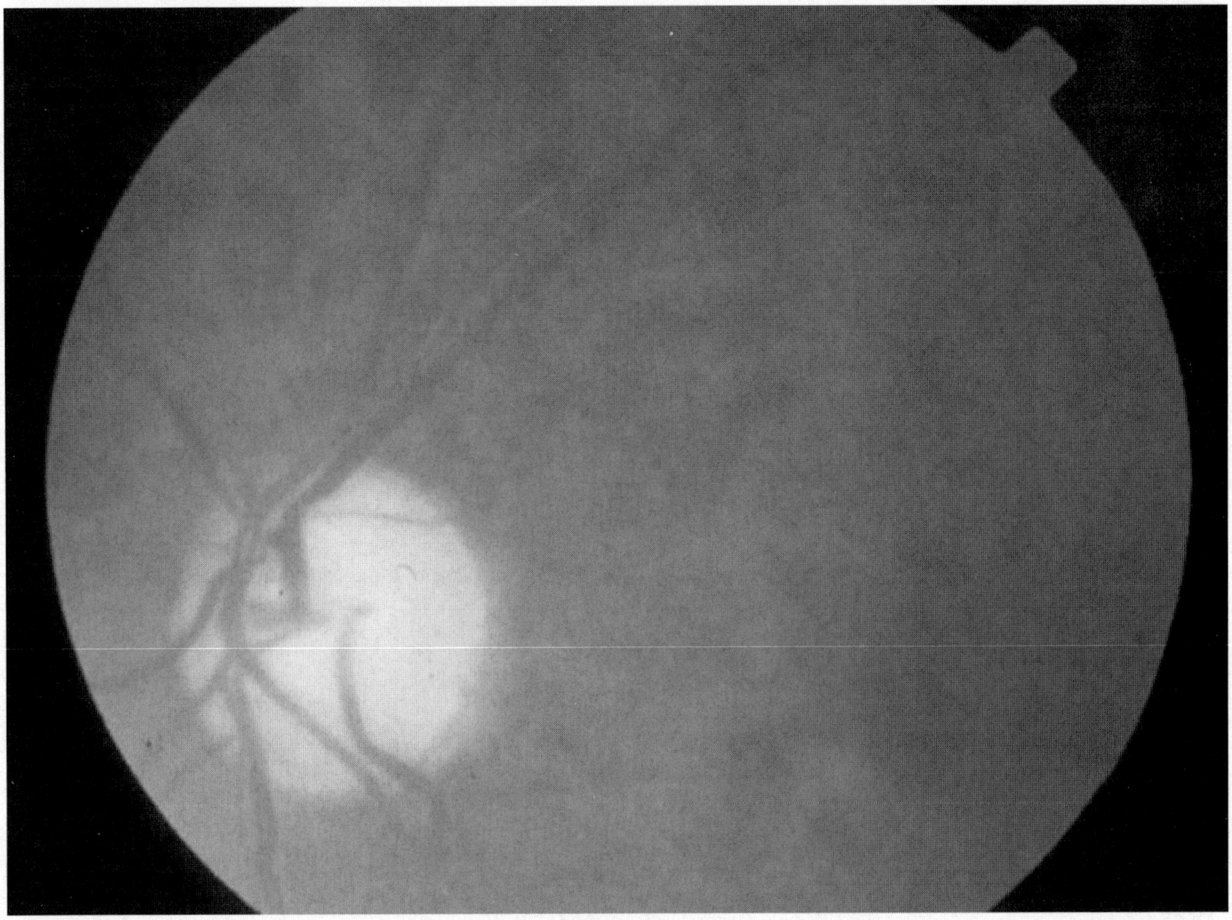

*Figure 7.*

be a decrease in the nerve fiber density, entailing an increase in the papillary capillary vessels visualization. The second one might be a hyperperfusion phenomenon, resulting from the autoregulatory mechanism brought into action by a sustained increase in IOP.

**Comments**

These findings are merely based on the observation of three cases. This paper is just a preliminary study, in which the number of patients does not represent a significative sample, in order to draw final conclusions. Therefore, we are presently enrolling new patients in order to be able to prove these hypotheses.

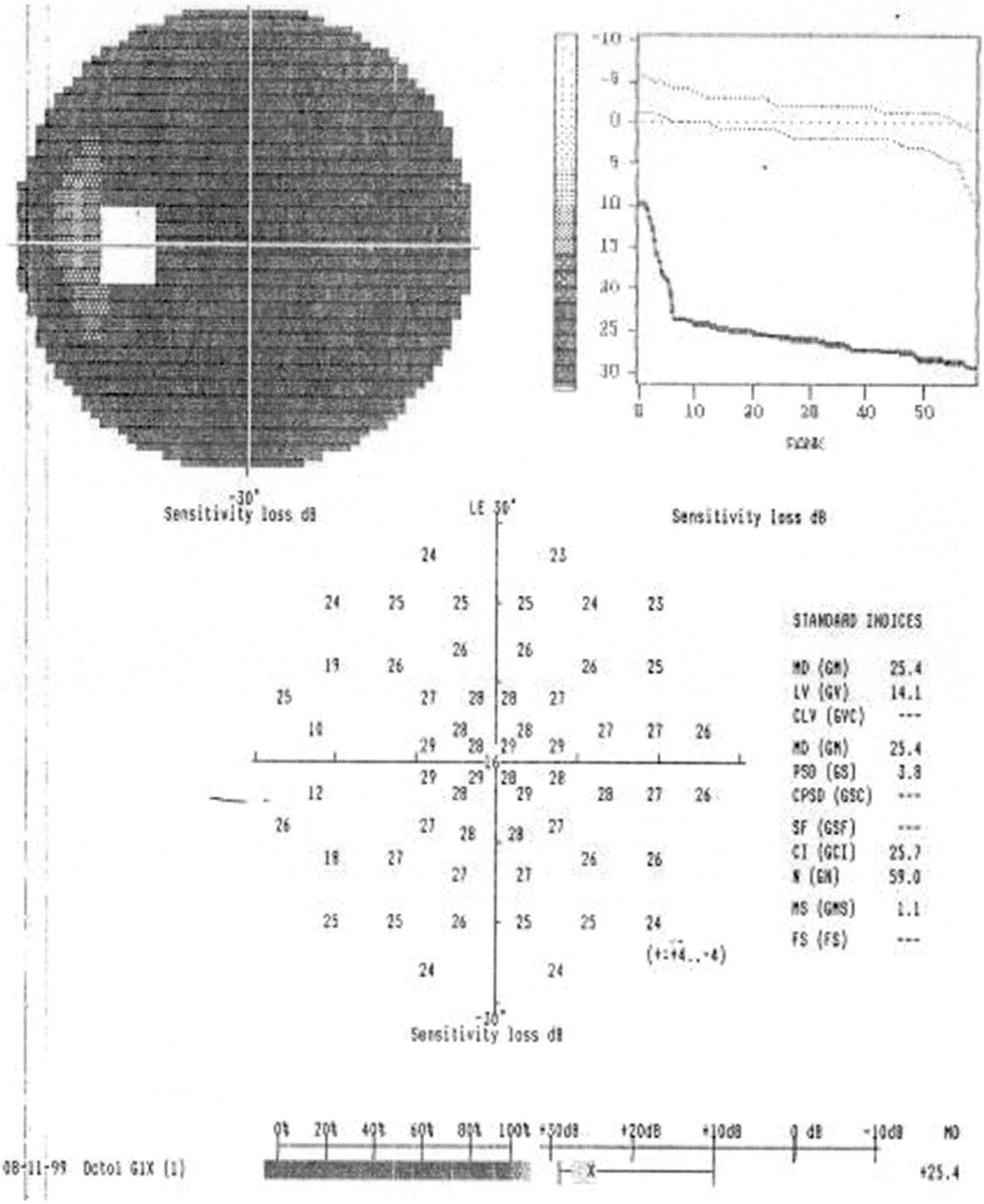
-30°
Sensitivity loss dB
LE 30°
Sensitivity loss dB
STANDARD INDICES
MD (GM) 25.4
LV (GV) 14.1
CLV (GVC) ---
MD (GM) 25.4
PSD (GS) 3.8
CPSD (GSC) ---
SF (GSF) ---
CI (GCI) 25.7
N (GN) 59.0
MS (GMS) 1.1
FS (FS) ---
-30°
Sensitivity loss dB
0% 20% 40% 60% 80% 100% +30dB +20dB +10dB 0 dB -10dB MD
LE 08-11-99 Octol G1X (1)
+25.4

*Figure 8.*

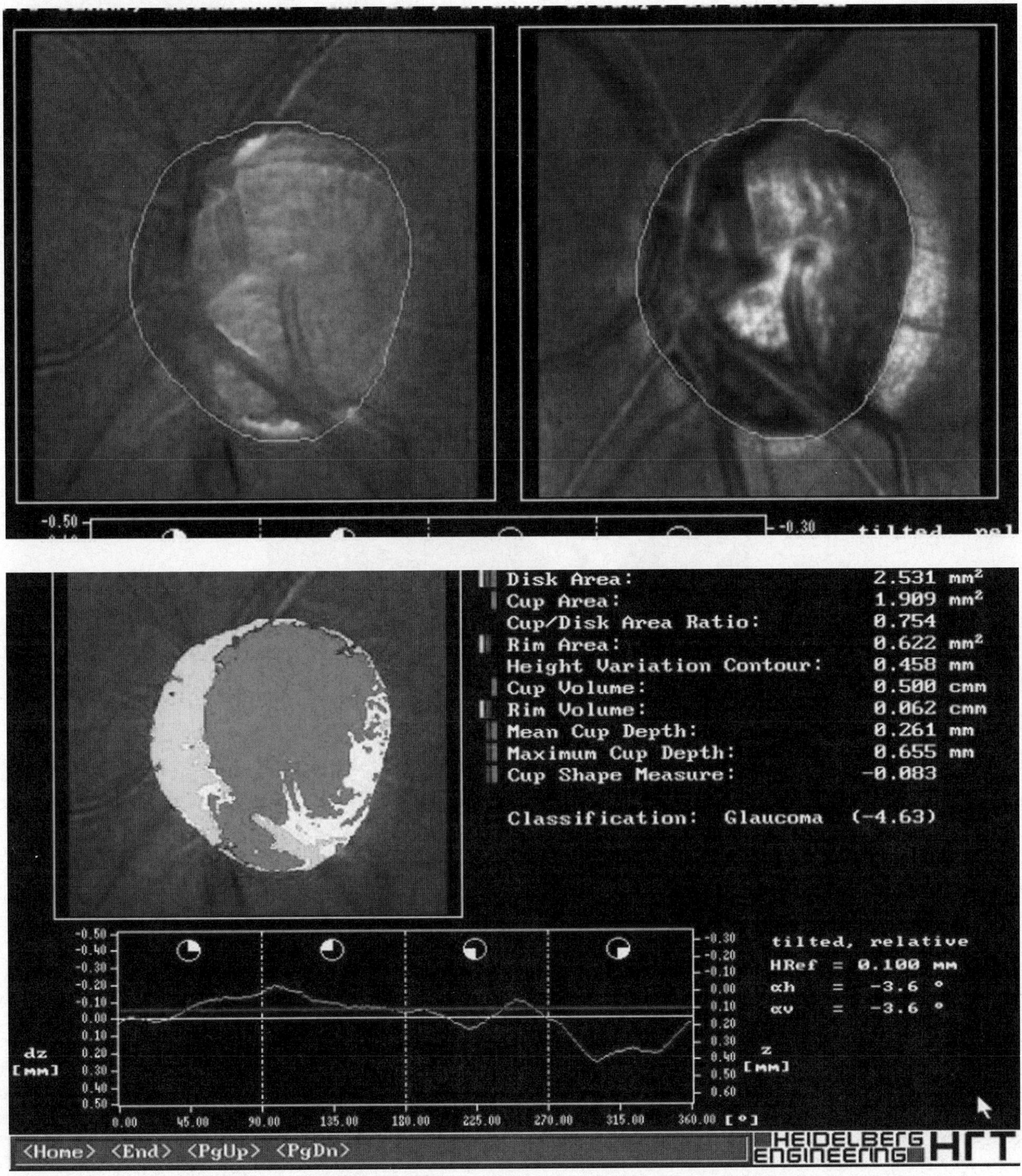
Disk Area: 2.531 mm²
Cup Area: 1.909 mm²
Cup/Disk Area Ratio: 0.754
Rim Area: 0.622 mm²
Height Variation Contour: 0.458 mm
Cup Volume: 0.500 cmm
Rim Volume: 0.062 cmm
Mean Cup Depth: 0.261 mm
Maximum Cup Depth: 0.655 mm
Cup Shape Measure: -0.083
Classification: Glaucoma (-4.63)
tilted, relative
HRef = 0.100 mm
αh = -3.6 °
αv = -3.6 °
dz [mm]
z [mm]
0.00 45.00 90.00 135.00 180.00 225.00 270.00 315.00 360.00 [°]
<Home> <End> <PgUp> <PgDn>
HEIDELBERG ENGINEERING HRT

*Figure 9.*

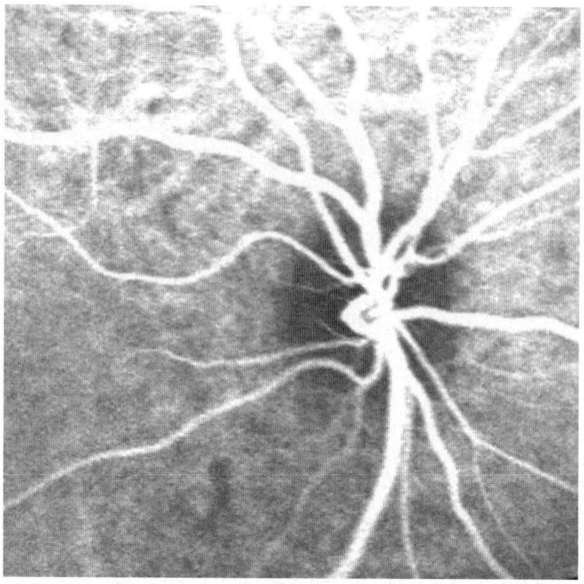

*Figure 10.*

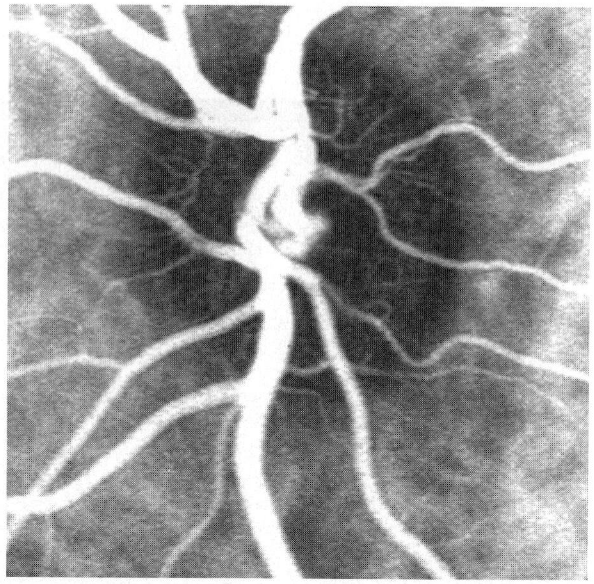

*Figure 11.*

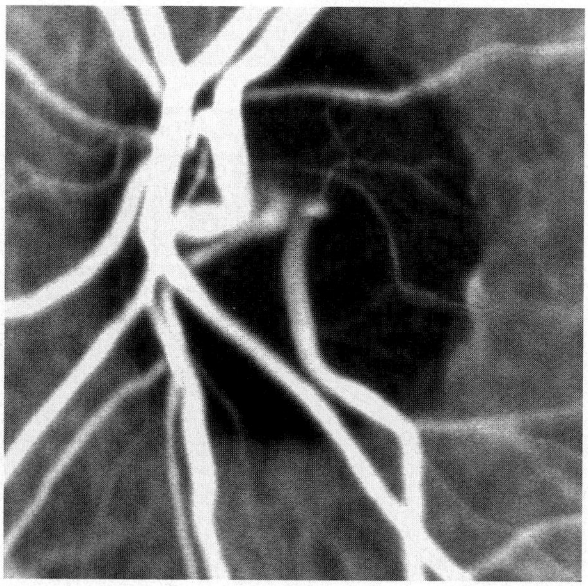

*Figure 12.*

## References


1. Guyer DR, Puliafito CA, Mones JM, Friedman E, Chang W, Verdooner SR. Digital indocyanine-green angiography in chorioretinal disorders. Ophthalmology 1992; 99: 287–291.
2. Scheider A, Schroedel C. High resolution indocyanine-green angiogrhaphy with scanning laser ophthalmoscope. Am J Ophthalmol 1989; 108: 458–459.
3. Hochheimer BF. Angiography of the retina with indocianine green. Arch Ophthalml 1971; 86: 564–565.
4. Hope-Ross M, Yannuzzi LA, Gragoudas ES, Guyer DR, Slakter JS, Sorenson JA, et al. Adverse eventes with indocianine green due. Ophthalmology 1994; 101: 534–541.
5. Webb RH, Hughes GW, Delori FC. Confocal scanning laser ophthalmoscope. Appl Opt 1980; 26: 1492–1499.

6. Weinreb RN, Bartsch DU, Freeman WR. Angiography of the glaucomatous optic nerve head. J Glaucoma 1994; 3: S55–S60.
7. Hayashi K, Nakase Y, Nishiyama A, Tokoro T, Yoshida T. Indocianine green fluorescence angiography, report 2: Studies on new interference filter. Acta Soc Ophthalmol Jpn 1982; 86: 1532–1539.
8. Bischoff PM, Flower W. Ten years experience with choroidal angiography using indocianine green dye: A new routine examination or an epilogue? Doc Ophthalmol 1985; 60: 235–291.
9. Yannuzzi LA, Slakter JS, Sorenson JA, Guyer DR, Orlok DA. Digital indocianine green videoangiography and choroidal neovascularization. Retina 1992; 12: 191–223.
10. Clark TM, Freeman WR, Goldbaum MH. Digital overlay of fluorescein angiogram and fundus images for treatment of subretinal neovascularization. Retina 1992; 12: 118–126.
11. Bartsch DU, Weinreb RN, Zinser G, Freeman WR. Confocal scanning infrared laser ophthalmoscopy for indocyanine green angiography. Am J Ophthalmol 1995; 120: 642–651.
12. Melamed S, Levkovitch-Verbin H, Krupsky S, Treister G. Confocal tomographic angiography of the optic nerve head in patients with glaucoma. Am J Ophthalmol 1998; 125: 447–456.
13. Cioffi GA, Robin AL, Eastman RD, Perell H F, Sarfarazi FA, Kelman SE. Confocal laser scanning ophthalmoscope. Reproductibility of optic nerve head opographic measurements with the confocal laser scanning ophthalmoscope. Ophthalmology 1993; 100: 57-62.
14. Yannuzzi S, Inoue Y, Yoshikawa K. Peripapillary fluorescein angiographic findings un primary open angle glaucoma. Br J Ophthalmol 1996; 86: 812–817.
15. Schwartz B. Circulatory defects of the optic disk and retina in ocular hypertension and high pressure open-angle glaucoma. S Of Ophthalmol 1994; 38: S23–S34.
16. Talusan ED, Schwartz B, Wilcox LI Jr. Fluorescein angiography of the optic disk. Arch Ophthalmol 1980; 98: 1579–1587.
17. Adam G, Schwartz B. Increase fluorescein filling defects in the wall of the optic disc cup in glaucoma. Arch Ophthalmol 1980; 98: 1590–1592.
18. Wolf S, Arend O, Sponsel WE, Schulte K, Cantor LB, Reim M. Retinal hemodynamics using scanning laser ophthalmoscopy an hemorheology in chronic open-angle glaucoma.

*International Ophthalmology* **23:** 425–428, 2001.
*J.R. Sampaolesi (ed.), Laser Scanning: Update 1*, 243–246.



# Scanning laser ophthalmoscopy for early diagnosis of vitreoretinal interfase syndrome

R. A. Ucha, E. J. Marengo & M. Martinez Cartier
*Instituto de la Visión, Buenos Aires, Argentina*

*Key words:* SLO-vitreoretinal interfase syndrome

## Abstract

*Purpose:* To describe the angiographic signs found using scanning laser ophthalmoscopy for the early diagnosis of vitreoretinal interface syndrome.This method is useful to visualize the inner retinal layers, being more sensitive than fundus biomicroscopy. *Material:* 61 patients with vitreoretinal interfase syndrome were evaluated. All of them had evidence of this disease using scanning laser ophthalmoscopy but four patients were referred without diagnosis of vitreoretinal interfase syndrome. These patients showed no biomicroscopic signs and diagnosis was made with SLO. *Methods:* Confocal scanning infrared laser ophthalmoscope (Heidelberg Retinal Angiograph assembled by Heidelberg Engineering). This SLO uses an infrared diode laser source of 795 nm. *Conclusion:* Patients included were referred with another diagnosis and with this method the correct diagnosis was made. In conclusion scanning laser ophthalmoscopy allows early diagnosis of this pathology for follow-up and treatment.

## Introduction

The vitreoretinal interfase syndrome represents traslucid fibrocellullar membranes located on the surface of the inner retinal layer. This membrane can develop after a local or complete posterior vitreous detachment, as well as after vitreoretinal surgery, uveitis, trauma, photocoagulation or cryotherapy, idiopathic, etc.

These membranes can shrink causing various degrees of inner retinal layer distortion, intraretinal edema, vascular tortuosity and/or vessel rectification, and degenerative processes of the underlying retina (Figure 1).

They are classified according to the retinal distortion they produce:

Grade 0: cellophane maculopathy
Grade 1: crinckled cellophane maculopathy
Grade 2: macular pucker

The diagnosis of this pathology is performed basically on the basis of funduscopy. But recently, since the appearance of SLO, this diagnosis can be established earlier, when the biomichroscopic findings are not yet present.

## Patients and methods

Over a total of 1987 angiographies of various pathologies, performed between 1998 and 1999, vitreoretinal interface syndrome was diagnosed in 61 eyes of 60 patients. Average age was 63 years.

A confocal scanning laser ophthalmoscope (Heidelberg Retina Angiograph) has been used with a laser source of 795 nm. Internal optics allow compensation between $-12$ to $+12$ diopters. The area of interest on the fundus can be scanned in a squared pattern of 10, 20 or 30 degrees. Images so obtained are stored on a hard disk for later processing. A first image was acquired with the infrared diode laser without contrast injection. Then, 5 cc of 10% solution of sodium fluorescein was injected IV in less than 5 seconds, and images were taken at choroid, arterial, early and late arterial-venous times, and also late time images, 30 minutes after injection.

All patients previously underwent biomicroscopic examination and colour fundus photography was obtained. Each patient's photograph and SLO obtained were separately evaluated by a trained ophthalmologist, in search for vitreoretinal interfase syndrome

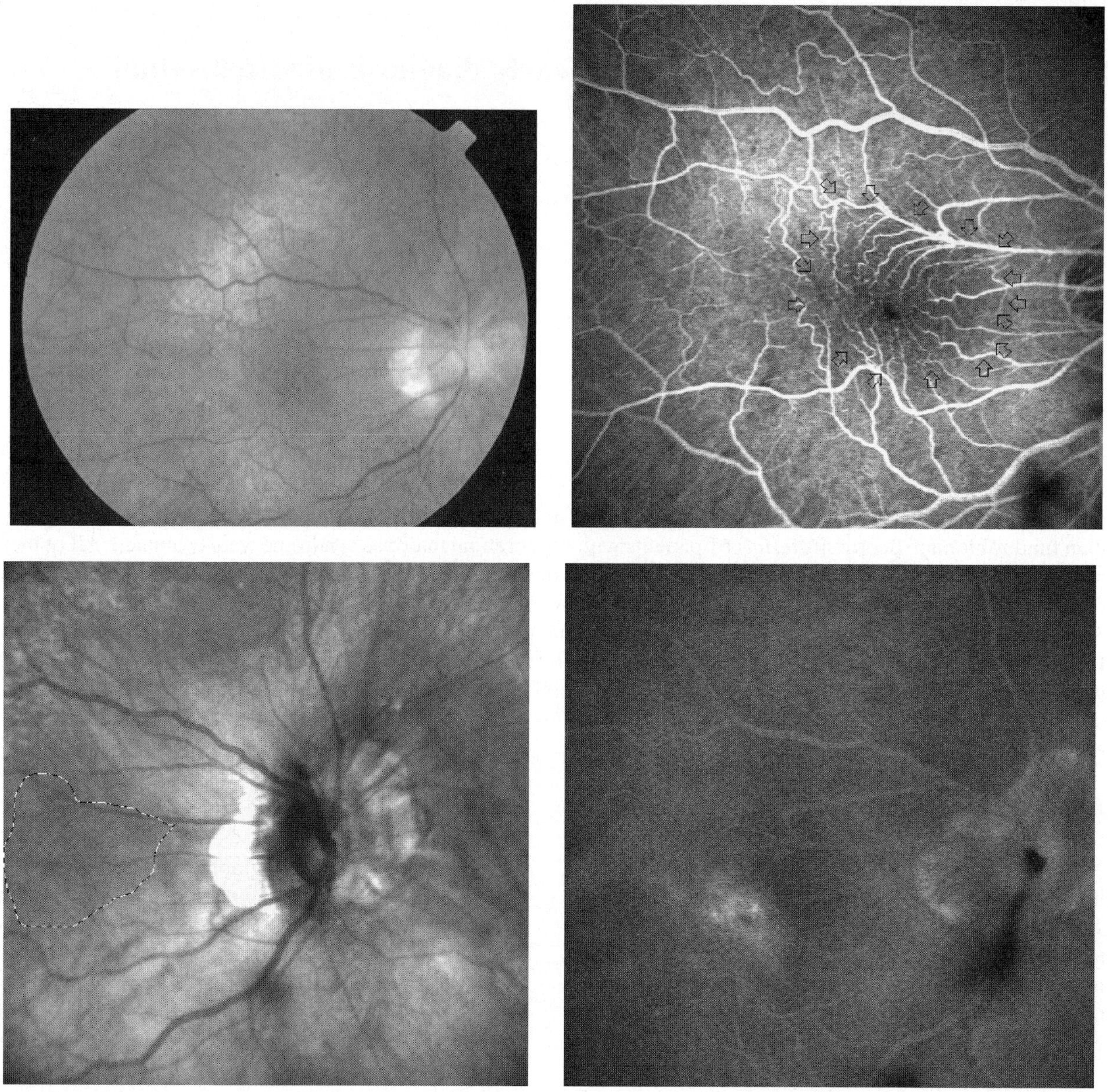

*Figure 1.*

signs, such as: cellophane reflex, macular edema, vascular tortuosity and/or rectification, retinal folds.

## Results

Over a total of 61 eyes with vitreoretinal interfase syndrome diagnosis, three were not diagnosed by funduscopy (Figure 2), while the presence of this syndrome became evident, even without contrast injection, when they were examined with the infrared diode laser device (Figure 3). A slight traction of macular vessels can be seen in the arterio-venous times on fluorescein angiography (Figure 4), as well as a filtration hyperfluorescence can be observed on late times (Figure 5).

## Discussion

There is a vast bibliography concerning early diagnosis of vitreoretinal interfase syndrome using SLO.

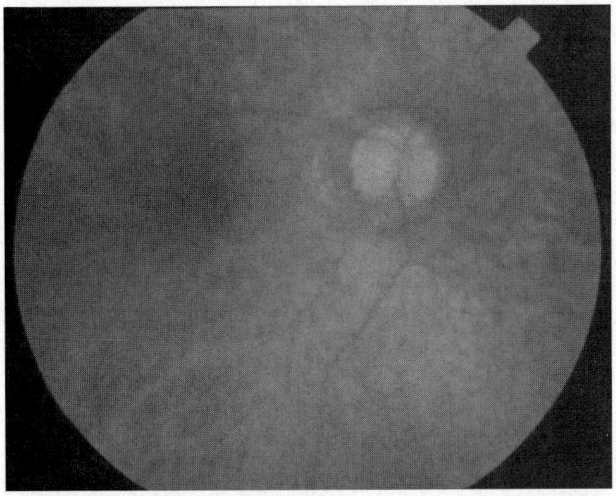

*Figure 2.*

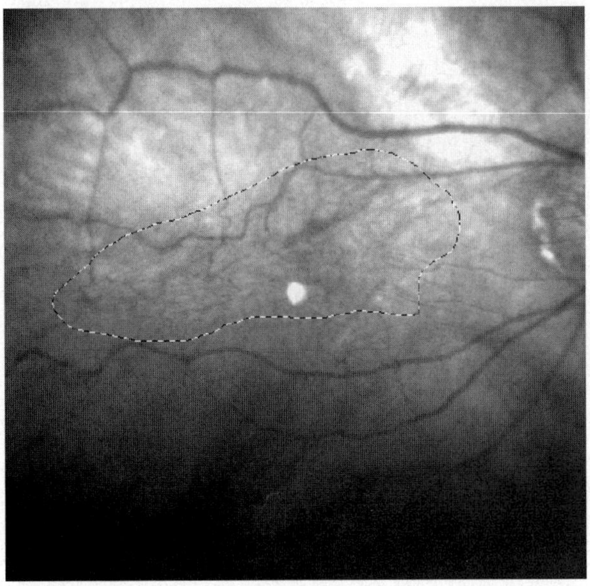

*Figure 3.*

We have reached the same conclusions, but we want to note that in our opinion and based on our experience, the images acquired with the scanning infrared diode laser ophthalmoscope without contrast injection might have a greater sensitivity in early stages of this entity.

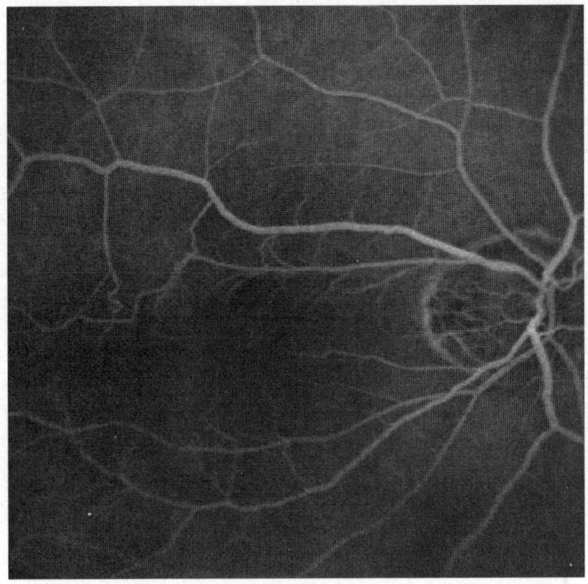

*Figure 4.*

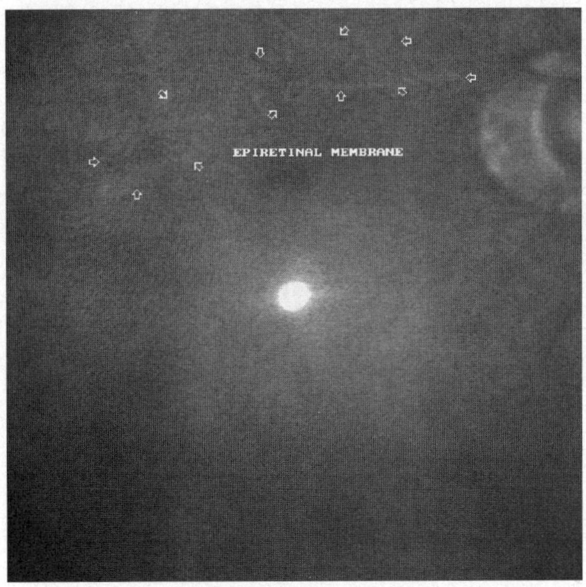


*Figure 5.*

## References


1. Kampic A, Kenyon KR, Michels RG. Epiretinal and vitreous membanes: comparative study of 56 cases. Arc. Ophthalmol 1981; 80: 1445–54..
2. Michels RG. A clinical and histopatology study of epiretinal membranes affecting the macula and removed by vitreous surgery. Tr Am ophthalmol Soc 1982; 80: 580–656.
3. Smiddy WE, Maguire AM, Green WR, et al. Idiopathic epiretinal membranes:ultrastructural characteristics and clinicopathologic correlation. Ophthalmol 1989; 96: 811–20.
4. Gass JDM, Stereoscopic Atlas of Macular Disease: diagnosis and treatment. St Louis, CV Mosby, 1987: 684–93.
5. Smiddy WE, Michels RG, Green WR. Morphology, pathology, and surgery of idiopathic vitreoretinal macular disorders. Retina 10: 288–296.
6. Webb RH, Hughes GW, Delori FC. Confocal scanning laser ophthalmoscope.Applied Optics 1987; 26: 1492–9.

7. Mainster MA, Timberlake GT, Webb RH, Hughes GW. Scanning laser ophthalmoscopy. Clinicsl aplication. Ophthalmology 1982; 89: 852–7.
8. Wolf S, Arend O, Toonen H, Eng D, Bertram B, Jung F, et al. Retinal capillary blood flow measurement with a scanning laser ophthalmoscope. Ophthalmology 1991; 98: 996–1000.
9. Ogura Y, Honda Y. Evaluation of idiopathic epiretinal membrane by a scanning laser ophthalmoscope. Br J Ophthalmol 1993; 77: 534–35.
10. Varano M, Billi B, Scassa C, et al. Scanning laser ophthalmoscopy in the early diagnosis of vitreoretinal interfase symdrome. Retina 1997; 17: 300–5.
11. Koenig F, Timberlake G, et al. Scanning laser ophthalmoscopy: Its value in macular diseases. J Fr Ophthalmol 1990; 13: 253–8.

# Corresponding contributors

Arenas, Archila Eduardo
University of Colombia, Transversal 21 #100-20, Cons 701, Bogota, Colombia
Phone: (57 1) 611 1711; Fax: (57 1) 215 7301; E-mail: earenas@cable.net.co

Argento, Carlos
Instituto de la Visión, Marcelo T. De Alvear 2261, 1122-Buenos Aires, Argentina
Phone: (54 11) 4822 1908; Fax: (54 11) 4823 5721; E-mail: vision@institutodelavision.com

Ashmann, Ross
Lions Eye Institute, Lasers & Bioengineering Group, 2 Verdun St. Nedlands, Perth WA 6009, Australia
Phone: (61 8) 9831 0756; Fax: (61 8) 9831 0700; E-mail: reinholz@eyellene.uwa.au

Darchuk, Vitali
Instituto de la Visión, Marcelo T. De Alvear 2261, 1122-Buenos Aires, Argentina
Phone: (54 11) 4822 1908; Fax: (54 11) 4823 5721; E-mail: vision@institutodelavision.com

Demailly, Philippe
Fondation Hôpital Saint Joseph, Inst. Glaucome, 185 rue Raymond-Losserand, 75674-Paris Cedex 14, France
Phone: (33 1) 44 12 34 20; Fax: (33 1) 44 12 32 85

Elsner, Ann E.
Schepens Eye Research Institute, 20 Stainford St., Boston, MA 02114-2500, USA
Phone: (617) 912 2595; Fax: (617) 912 0111; E-mail: elsner@vision.eri.harvard.edu

Hammer, Martin
University of Jena, Department of Ophthalmology, Bachstrasse 18, D-07740, Jena, Germany
Phone: (49 3641) 933 027; E-mail: martin.hammer@med.uni-jena.de

Marengo, Javier
Instituto de la Visión, Marcelo T. De Alvear 2261, 1122-Buenos Aires, Argentina
Phone: (54 11) 4822 1908; Fax: (54 11) 4823 5721; E-mail: vision@institutodelavision.com

Martinez, Haydee
Hospital de Clinicas, Av. Cordoba 1830, 1120-Buenos Aires, Argentina
Phone: (54 11) 4508 5766

Masters, Barry R.
Universitäts-Augenklinik, Inselspital, University of Bern, 3010 Bern, Switzerland
Phone: +41 31 632 8538; Fax: +41 31 381 7066

Meerhoff, Guillermo
Canelones 1228 Montevideo, Uruguay
Phone: (05982) 900 1037

Michelson, G.
Universität Erlangen-Nürnberg, Augenklinik, Schwabachanlage 6, D-91053 Erlangen, Germany
Phone: (49 9131) 85 33 001; Fax: (49 9131) 8536 401; E-mail: georg.michelson@augen.med.uni-erlangen.de

Mistlberger, Andrea
Department of Ophthalmology, County Hospital Salzburg, Müllner Hauptsdrasse 48, 5020 Salzburg, Austria
Phone: 0043-622-823 883; E-mail: amistlberger@aon.at

Mousalli, Maria Angelica
Instituto de la Visión, Marcelo T. De Alvear 2261, 1122-Buenos Aires, Argentina
Phone: (54 11) 4822 1908; Fax: (54 11) 4823 5721; E-mail: vision@institutodelavision.com

Mueller, Arthur
Augenklinik der Universität, Mathildenstr. 8, 80336 Munich, Germany
Phone: 49-0889-5160-31811; E-mail: arthur.mueller@ak-i.med.uni-muechen-de

Paris, Gianmarco
UTHSCSA/Ophthalmology, 7703 Flyd Curl Drive, San Antonio, TX, USA
Phone: (610) 567 8401; E-mail: Gianparis@hotmail.com

Riva, Charles E.
Institut de Recherches en Ophtalmologie, Caise Postale 4168, Av Grand Champsec 64, 1950 Sion, Switzerland
Phone: (41 27) 203 5971; Fax: (41 27) 203 5970; E-mail: charles.riva@iro.vsnet.ch

Sampaolesi, Juan R.
Parana 1239-1 Piso, 1018-Buenos Aires, Argentina
Phone: (54 11) 4811 7075; Fax: (54 11) 4814 2092; E-mail: imslo99arg@interar.com.ar and jsampaolesi@yahoo.com

Sampaolesi, Roberto
Parana 1239-1 Piso, 1018-Buenos Aires, Argentina
Phone: (54 11) 4811 7075; Fax: (54 11) 4814 2092; E-mail: imslo99arg@interar.com.ar

Schmid, Gregor
Institut de Recherches en Ophtalmologie, Caise Postale 4168, Av Grand Champsec 64, 1950 Sion, 4, Switzerland
Phone: (41 27) 203 5971; Fax: (41 27) 203 5970; E-mail: gregor.schmid@iro.vsnet.ch

Schweitzer, Dietrich Doz
University of Jena, Department of Experimental Ophthalmology, Bachstrasse 18, 07740-Jena, Germany
Phone: (49 3641) 933 027; E-mail: schweitzer@bach.med.uni-jena.de

Sponsel, William E.
Department of Ophthalmology, University of Texas Health Science Center, MS 6230, 7703 Floyd Curl Drive, San Antonio, TX 78229, USA
Phone: (210) 567 8401; Fax: (210) 567 8413; E-mail: sponsel@uthscsa.edu

Stanga, Paulo Eduardo
Moorfields Eye Hospital, Department Ophthalmology, Medical Retina, City Road, London ECIV 2PD, UK
Phone: (0171) 253 3411 Ext 2922; Fax: (0171) 251 9350; E-mail: p.stanga@ucl.ac.uk

Ucha, R. Alejandro
Instituto de la Visión, Marcelo T. De Alvear 2261, 1122-Buenos Aires, Argentina
Phone: (54 11) 4822 1908; Fax: (54 11) 4823 5721; E-mail: vision@institutodelavision.com

Vasco Posada, Jorge
Universidad de Antioquia, Calle 52, No. 40-156 Medellín, Colombia
Phone: (57 4) 239 7515; Fax: (57 4) 216 7738; E-mail: Mavasco@geo.net.co